Studies in the History of Knowledge
(Historische Wissensforschung)

edited by

Caroline Arni, Stephan Gregory, Bernhard Kleeberg,
Andreas Langenohl, Robert Suter † and Simon Teuscher

18

A New Organon

Science Studies
in Interwar Poland

Edited by
Friedrich Cain and Bernhard Kleeberg

Mohr Siebeck

Friedrich Cain, born 1985; historian and cultural scientist at the University of Vienna.

Bernhard Kleeberg, born 1971; Professor for the History of Science at the University of Erfurt.

This publication received financial support of the Centre of Excellence "Cultural Foundations of Social Integration" at the University of Konstanz.

ISBN 978-3-16-154315-9 / eISBN 978-3-16-154989-2
DOI 10.1628/978-3-16-154989-2

ISSN 2199-3645 / eISSN 2568-8383 (Historische Wissensforschung)

Die Deutsche Nationalbibliothek verzeichnet diese Publikation in der Deutschen Nationalbibliographie; detaillierte bibliographische Daten sind über *https://dnb.dnb.de* abrufbar.

© 2024 Mohr Siebeck Tübingen. www.mohrsiebeck.com

Das Werk einschließlich aller seiner Teile ist urheberrechtlich geschützt. Jede Verwertung außerhalb der engen Grenzen des Urheberrechtsgesetzes ist ohne Zustimmung des Verlags unzulässig und strafbar. Das gilt insbesondere für die Verbreitung, Vervielfältigung, Übersetzung und die Einspeicherung und Verarbeitung in elektronischen Systemen.

Das Buch wurde von Gulde Druck aus der Minion gesetzt, in Tübingen auf alterungsbeständiges Werkdruckpapier gedruckt und gebunden.

Der Umschlag wurde von Uli Gleis in Tübingen gestaltet. Umschlagabbildung: Henryk Berlewi: Kontrasty Mekanofakturowe (1924), Gouache, 83x109 cm. Image courtesy of The Henryk Berlewi Archive.

Printed in The Netherlands.

Contents

Friedrich Cain and Bernhard Kleeberg
Introduction – A New Organon: Science Studies in Poland   1

Part I: Historical Contexts   31

Jan Surman
Internationality in *Nauka Polska*: Infrastructure, Translation, Language . . .   33

Marta Bucholc
Strategies of Sincerity: Scientific Autobiographies in the Proceedings
of *Koło Naukoznawcze*   53

Friedrich Cain
A Grammar of Actual Thinking: Antoni B. Dobrowolski within and beyond
the Warsaw Science Studies Circle   71

Part II: Sources   99

Tul'si (Tuesday) Bhambry
Translator's Note   101

w/o author
Editorial Introduction 1923   103
Preface (1936)   105
Report on the activities of the Science Studies Circle (1929)   107
Second Report on the activities of the Science Studies Circle (1930)   111

Florian Znaniecki
The Subject Matter and Tasks of the Science of Knowledge (1925)   113

Maria Ossowska and Stanisław Ossowski
The Science of Science (1935)   171

Stefan Błachowski
The Problem of Scientific Creativity (1928)   183

Franciszek Bujak
The Man of Action and the Student (1929)   229

Paweł Rybicki
Science and the Forms of Social Life: Issues at the Intersection of Sociology
and Theory of Science (1929) 239

Tadeusz Kotarbiński
On the Skills of a Researcher (1929) 251

Bogdan Suchodolski
Investigation and Teaching (1936) 259

Antoni B. Dobrowolski
The Urgent Need for Mental Education in Poland: The Necessity for
a Fundamental Teaching Reform in Middle Schools and New Institutions
for Scientific Research in Relation to this Reform (1918) 289

Czesław Białobrzeski as 'C. B.'
An Autobiographical Sketch and Remarks on Scientific Creativity (1927) . . 303

Antoni B. Dobrowolski
An Archive of Materials for Research on Creativity (1927) 317

Antoni B. Dobrowolski as 'A. B. D.'
My "Scientific Biography" (1928) 319

Émile Borel
Contribution aux Documents sur la Psychologie de l'Invention
dans le Domaine de la Science (1936) 325

August Krogh
Visual Thinking: An Autobiographical Note (1938) 333

Sessions of the Science Studies Circle (1928–1938) 341

Part III: Commentaries 347

Jan Piskurewicz and Leszek Zasztowt
Science Studies in Poland before the Second World War: Institutional Frames   349

Andreas Langenohl
A look back on *Koło naukoznawcze* after the Science Wars 361

Paweł Kawalec
The Original Conception of the Science of Science and Innovation Studies . 373

Editorial Remarks and Sources 383
Biograms – A New Organon 387
Index 391

Introduction
A New Organon: Science Studies in Poland

Friedrich Cain and Bernhard Kleeberg

On 24 August 1965, at the Opening Session of the 11th *International Congress of the History of Science* in Warsaw, the crystallographers and historians of science John D. Bernal and Alan L. Mackay lined out ways towards a new discipline: a *Science of Science*. However, before they would set out into the deep history of the field, they did not only attribute the coinage of the term, but also the core concept of a Science of Science to two Polish authors who had published a pivotal text thirty years earlier.[1] Indeed, in 1935, Maria Ossowska and Stanisław Ossowski had offered a distinct vision in Warsaw.[2] Their idea of science studies lived to be republished several times in international journals – Bernal and Mackay themselves mentioned a 1964 edition in *Minerva* – and made its way into the self-narratives of the big science studies projects of the Cold War. Today, it again serves as tacit background of projects around the globe.[3] Yet, these references usually miss the original context of 1935, i.e., the extremely vivid science reflection in the intellectual landscape of the Polish interwar state.

This is where the present volume ties in, trying to open a view on this often-neglected landscape to complement studies in the history of science, or, more specifically, in the historical epistemology of science studies. Obviously, basing the shaping of society or the modern (nation) state on scholarly planning and scientific utopias is an old idea in Western traditions. Expertocratic regimes have been outlined over and over – from Aristotle and Plato to Bacon, from Condorcet to Saint-Simon and Spencer – and they became particularly elaborate in the 20th century.[4] All their political agendas presupposed a particular understanding of what scientific knowledge is, and how it can be applied. They were intricately linked to specific epistemologies that re-

---

[1] John D. Bernal and Alan L. Mackay: "Towards a Science of Science," *Organon*, vol. 3 (1966), 9–17: 9.

[2] ▶ **IN THIS VOL.** Maria Ossowska and Stanisław Ossowski: "The Science of Science" (1935), i.e. "Nauka o nauce," *Nauka Polska*, vol. 20 (1935), 1–12.

[3] See below, section III, footnote 88.

[4] For an overview see Kerstin Brückweh, Dirk Schumann, Richard F. Wetzell and Benjamin Ziemann (eds.): *Engineering Society. The Role of the Human and Social Sciences in Modern Societies, 1880–1980*, Basingstoke 2012; Thomas Etzemüller (ed.), *Die Ordnung der Moderne. Social Engineering im 20. Jahrhundert*, Bielefeld 2009; Theodore M. Porter, *Trust in Numbers. The Pursuit of Objectivity in Science and Public Life*, Princeton 1995; id., Dorothy Ross (eds.), *The Modern Social Sciences*, Cambridge e. a. 2003, esp. part IV.

flected the sciences in particular and knowledge in general. The plural is crucial, for over time a great number of approaches have emerged, drawn from fields as diverse as philosophy, psychology, sociology, and history, but also based on considerations on the proper organization of the scientific endeavor.

Besides the standard heroes in the history of these thoughts, there were others in the more Eastern areas of Europe, receiving far less attention, though sometimes highly original and sharing the same agenda of reflecting on the production, spreading and use of knowledge. Not only academic philosophy, but national projects of the late 19th century in particular praised science as harbinger of wealth and freedom. Emphasis was often added where nations were not in control of the territories they claimed. Research and learning became a mandatory way to form a nation's future through education, and as such they came to form an object of study themselves.[5] As the First World War sealed the decline of the continental empires in Europe, many initiatives took the geopolitical changes as a point of departure into a new, bright future, of which science and its mastering formed a crucial part.

This volume presents the telling example of an initiative from 1920s and 1930s Warsaw. In the Polish capital, the alleged top scientists of the nation assembled to discuss the foundations, the funding and future of their trade, thus creating a distinct version of science studies. And distinct it was: Apart from several congresses held on the organization of science between 1918 and 1939, the so-called *Koło Naukoznawcze* – the Science Studies Circle – organized regular meetings for over a decade. Established in 1928, its proceedings were published in two journals – the Polish *Nauka Polska* and the *Organon*, which collected original articles in French and English along with translations from Polish. However, for a long time, the layout of a *Science of Science* (Pol. nauka o nauce) by Maria Ossowska and Stanisław Ossowski was treated as a sole peak of the initiative.

The volume opens this intriguing chapter of science studies from 1920s and 1930s Poland to a wider audience. There are good reasons to do so: Firstly, science studies from interwar Warsaw as presented here stand out as an interdisciplinary project with a highly integrative agenda. Instead of concentrating on one angle, the programmatic aim was to integrate philosophy, psychology, sociology, and history in one new discipline that would not only study science, but also give directives for its proper organization on all levels and thus lay the grounds for a knowledge-based society. Although the project of a science of science did not survive the German oc-

---

[5] Cf. Martin Kohlrausch, Katrin Steffen and Stefan Wiederkehr (eds.): *Expert Cultures in Central Eastern Europe. The Internationalization of Knowledge and the Transformation of Nation States Since World War I*, Warsaw 2010, and with emphasis on the historical narratives and national economies ch. 8 in Balács Trencsényi, Maciej Janowski, Mónika Baár, Maria Falina and Michal Kopeček: *A History of Modern Political Thought in East Central Europe, Vol. I: Negotiating modernity in the 'Long Nineteenth Century'*, Oxford 2016, esp. 318–319. For the Polish case see Jan Surman: "The Contexts of Polish Positivisms, 1840s–1900s," in: *The Worlds of Positivism. A Global Intellectual History, 1770–1930*, ed. by Johannes Feichtinger, Franz L. Fillafer and Jan Surman, Cham 2018, 239–272.

cupation of Poland and the following socialist rule, Ossowska and Ossowski have been credited for the term and their original design. As mentioned above, the original article has been reprinted several times, and often referred to as a founding stone of science studies. Its actual context, however, fell into oblivion.

Secondly, as much as the materials from interwar Warsaw add to a general understanding of the genealogies of science studies, they also help to re-focus the history of systematic enquiry into the cultures of science. Not only in Poland, but also in the many new states founded in the former imperial spaces of Central, Eastern and Southeastern Europe, science figured as a great hope for technological, social, and political development with presidents, prime ministers, and ministers often being scholars themselves. The corresponding initiatives have only marginally been reflected in the history of (planned) scientification in the 20th century.[6] One reason may have been a language barrier, of which the Varsovian organizers were conscious themselves in 1936 when founding the *Organon* as a second, international revue to reach out to an international community.

Thirdly, the volume aims to draw attention to the highly diverse and productive epistemological landscape that Poland was during the 1920s and 1930s. On the one hand, it shows a certain strain of development of the Warsaw-Lwów School of philosophy and logic founded by Kazimierz Twardowski.[7] On the other hand, and quite different from the logicians Florian Znaniecki, the rising star of international sociology and founder of the first Sociological Institute in Poland, also contributed to Nauka Polska several times, at least in the early years. Interwar Poland was also the place where Ludwik Fleck developed his comparative epistemology, which did not make the impact on contemporary science studies that it has made since the 1970s. However, it was widely reviewed at the time, in Nauka Polska as well.[8]

Finally, this volume makes available new source material for studies in historical and political epistemology that aim at the moral economies[9] of the Warsaw initiative

---

[6] For a first overview see *The Past and Present of Political Epistemologies of (Eastern) Europe*, Special Section in *Historyka. Studia Metodologiczne*, vol. 49 (2019), ed. by Friedrich Cain, Bernhard Kleeberg and Jan Surman.

[7] Twardowski was former student of Franz Brentano and teacher to logician Jan Łukasiewicz, who in turn was not only an active figure in the Science Studies Circle, but also a teacher to young Alfred Tarski. For an overview cf. Jan Woleński: *Logic and Philosophy in the Lvov-Warsaw School*, Dordrecht 1989.

[8] Jan Dembowski: "[Review] Ludwik Fleck: Entstehung und Entwicklung einer wissenschaftlichen Tatsache, O obserwacji naukowej i postrzeganiu w ogóle, Zagadnienie teorii poznawania," *Nauka Polska*, vol. 24 (1939), 435–439. Dembowski reviewed Fleck's monograph *Entstehung und Entwicklung einer wissenschaftlichen Tatsache*, Basel 1935 and two articles: "O obserwacji naukowej i postrzeganiu w ogóle," *Przegląd Filozoficzny*, vol. 38 (1935), 58–76 and "Zagadnienie teorii poznawania," *Przegląd Filozoficzny*, vol. 39 (1936), 3–37. For context cf. Michał Kokowski: "The Science of Science (naukoznawstwo) in Poland: Defending and Removing the Past in the Cold War," in: *Science Studies during the Cold War and Beyond*, ed. by Elena Aronova and Simone Turchetti, New York 2016, 149–176: 151, and recently Sylwia Werner: *Lemberger Moderne. Studien zur Entstehung einer Wissenskultur*, Paderborn 2023.

[9] Cf. Lorraine Daston: "The Moral Economy of Science," *Osiris* (2nd Series), vol. 10 (1995), 2–24.

in particular, and similar concepts of scientific self-reflection in general. It assembles a selection of 17 texts (part II) that were originally published in the journals Nauka Polska and Organon. Some context to this material is given in three research articles that initiate a close reading in part I. The third part contains aries comments that assess the contemporary value of the original texts. The remainder of this introduction shall give an overview of the general historical context of the edited texts and introduce the source material briefly before finally laying out some paths of reception.

1. A brief history of Science Studies in interwar Warsaw

In April 1920, the aula of Warsaw University housed the *1st Congress for the Organisation and Progress of Polish Science* (Pol. I. Zjazd, poświęcony sprawom organizacji i rozwoju nauki polskiej). For four days, more than 500 delegates from all parts of Poland discussed the fate of Polish scholarship. The congress was organised by the *Academic Council* (Pol. Dział Naukowe) of a private foundation from Warsaw, the *Kasa Mianowskiego* (Engl. Mianowski Fund).[10] As the organisers hoped, the congress would start sustainable discussions on the development, organisation, and needs of scholarship in Poland.[11]

The II. Polish Republic, to whose development the congress meant to contribute, had been founded only eighteen months before. On 11 November 1918, Marshal Józef Piłsudski had assumed power in Poland, declaring an end to 123 years of partition. The territories, which had been under Tsarist, Prussian/German and Austro-Hungarian control since the late 18th century, were united in one sovereign state. After more than a century of non-existence, the newly erect state faced severe challenges, as modern structures had to be built to integrate the peripheries of three imperia. Next to general infrastructure, administration, the legal and the health system, the military had to be organised, and units of measurement, education and scholarship needed standardisation on national level. After the end of foreign, often suppressive rule, the deprivations of Polish scholarly life had to be overcome, and unified Poland had to work out an original programme for the development of science, which would consider the conditions and requirements of the young state. As the organizing committee for the congress made clear in its announcement, genuine scientific production was regarded a basic pillar for the "healthy development of a national organism" – for practical reasons, but even more for its values. Not only should the congress unite

---

[10] The full name was *Kasa Pomocy dla osób pracujących na polu naukowym im. d.ra med. Józefa Mianowskiego* (Engl. Aid Fund for People Working in the Scientific Field named after Dr. med. Józef Mianowski), or in short *Kasa (imienia) Mianowskiego* (Engl. Fund (named after) Mianowski).

[11] "Odezwa w sprawie zwołania I zjazdu, poświęconego sprawom organizacji i rozwoju nauki polskiej," *Nauka Polska*, vol. 3 (1920), 225–226: 225.

the forces of the most competent scholars of the country, but also help to formulate a sound programme for controlled development.[12]

The congress was one of many peaks in contemporary science organisation. In fact, the Mianowski Fund had already engaged in science promotion for forty years. Founded in 1881, it had funded several generations of scholars from Warsaw and other parts of the Russian partition with scholarships, printing subsidies, and institutional assistance. In November 1916, when Warsaw was under German control during the First World War, the Fund made a programmatic turn. Establishing the Academic Council with the Fund's secretary general Stanisław Michalski (1863–1935) at its centre, the Fund declared an end to the defensive agenda of safeguarding the small range of opportunities under Russian control, and instead devoted itself to a programme of active funding and promotion.[13] As a report noted in 1920, the foundation should become a "private ministry of science" and start big scale action to stimulate scientific development.[14] The Fund continued to subsidise individual scholars going abroad, accommodated others in two country houses for writing retreats, awarded prizes, helped equipping laboratories, libraries, or institutes, paid printing subsidies, and provided infrastructural assistance to other societies.[15]

The Academic Council appealed to Polish scholars of all kinds, asking them to write inventories of their respective disciplines. Over the course of two years, more than 44 reports arrived from all over the future state and beyond. A large portion of the collected reports was published in the first two volumes of a yearbook series that became known as the Fund's principal journal: *Nauka Polska. Jej potrzeby, organizacja i rozwój* (Engl. Polish Science: Its Needs, Organisation, and Progress[16]). From 1918 onwards, 24 volumes were published until the outbreak of the Second World War, followed by one final 25th volume in 1947. The 3rd volume contained the proceedings of the 1920 congress[17] and a report on the achievements of the Academic Council since its foundation. The report proudly stated that the Kasa had developed from a Varsovian initiative into an institution of national scope with Nauka Polska as its central organ.[18] The journal compiled original research work, reports on the national and international development of scientific institutions, science funding, and congresses, as well as

---

[12] Ibid.

[13] "Sprawozdanie z działu naukowego biura Kasy Mianowskiego," *Nauka Polska*, vol. 3 (1920), 239–256: 240. For a detailed auto-history see "Wiadomość o działalności Kasy pomocy dla osób, pracujących na polu naukowym imienia doktora medycyny Józefa Mianowskiego (1881-1916)," *Nauka Polska*, vol. 1 (1918), 523–546.

[14] "Sprawozdanie z działu naukowego biura Kasy Mianowskiego," *Nauka Polska*, vol. 3 (1920), 239–256: 241.

[15] Cf. *Sprawozdanie 50 z Działalności Kasy im. Mianowskiego*, Warszawa 1931, 4–8.

[16] Vol. 24 (1939) translates the title as *Science and Letters in Poland. Their Needs, Organization and Progress*.

[17] See for information on delegates, programme etc.: "Sprawozdanie z działu naukowego biura Kasy Mianowskiego," *Nauka Polska*, vol. 3 (1920), 239–256: 245–246.

[18] "Sprawozdanie z działu naukowego biura Kasy Mianowskiego," *Nauka Polska*, vol. 3 (1920), 239–256: 241, 243.

book reviews and a growing bibliography of relevant international literature.[19] Nauka Polska usually sold out quickly. The first two volumes, with runs of 1.500 copies each, were unavailable shortly after hitting the bookstores.[20] The consecutive volumes were also used in exchange relations with national and foreign libraries, institutions, and societies.[21]

From 1936 onwards, the Kasa published a second journal, which contained English and French texts exclusively. Before, starting with vol. 5 in 1925, Nauka Polska had been equipped with French or English title pages, short translations of contents and sometimes résumés.[22] More than just supplementing its Polish sister journal, the Organon was meant to strengthen international outreach.[23] Initial plans can be traced to letters from late 1932, when first drafts for an "international journal dedicated to the science of science" – or *naukoznawstwo* in Polish – circulated. As the project had no name yet, the editors were looking for a universal title, preferably something "in Latin."[24] The final title, of course, evokes Aristotle's work on logics as a tool of science, but, more likely, it refers to Bacon's *Novum Organon Scientiarum* of 1620, the second book of his *Instauratio magna*. Though the editors did not, like Bacon, necessarily demand that "the entire work of the understanding be commenced

---

[19] The more than 30 pages of the complete bibliography of articles from the first twenty volumes gives an overview about the topics and areas covered. Cf. addendum to *Nauka Polska*, vol. 20 (1935).

[20] Vol. 1 was published in August 1919 (1500 copies) and sold out in April 1920, vol. 2 sold out in November 1920 after being published in April that year. Cf. "Sprawozdanie z działu naukowego biura Kasy Mianowskiego," *Nauka Polska*, vol. 3 (1920), 239–256: 242.

[21] For example in 1938, the Fund cooperated with 39 regional scientific journals from Poland, receiving 250 volumes. Polish institutions and libraries sent 29 publications and 8 bibliographical overviews in all together 116 volumes. Moreover, 33 international institutions funding or organizing science and 14 editorial offices of scientific journals cooperated, sending 344 publications from abroad. Cf. *Sprawozdanie 57 z Działalności Kasy im. Mianowskiego*, Warszawa 1938, 32–33. See also: *Sprawozdanie 55 z Działalności Kasy im. Mianowskiego*, Warszawa 1936, 37.

[22] Vol. 17 (1933) introduces the journal in English: "Science and Letters in Poland. Their Needs, Organization and Progress: This periodical publication is devoted to the studies on Science, Letters and Learning in Poland, especially on the social substratum and psychological conditions regulating the rise and development of Science, Letters and Learning (Sociology of Learning, Psychology of scientific creation). Besides there are included in 'Nauka Polska' articles, describing the present state of Learning (Science and Letters) in Poland and abroad, its Organization, Development and Needs; the Chronicle of Polish and foreign Scientific Life and notes on the history of the Organization of Learning are added in every volume. – Editor: S. MICHALSKI." Vol. 18 introduces *La Science Polonaise. Ses besoins, son organization et ses progress* in French: "'La Science Polonaise' est une publication consacrée à l'étude de la science dans ses rapports avec l'ensemble de la culture humaine, et notamment à l'étude des origines sociales de la science et de ses conditions psychologiques (psychologie de la création scientifique). En outre, la 'Science Polonaise' publie des articles traitant de l'état actuel de la science (corps et institutions scientifiques) en Pologne et à l'étranger, de son organisation, de son développement et de ses besoins, une chronique de la vie scientifique en Pologne et à l'étranger et l'histoire de son organisation en Pologne. – Rédacteur en chef: S. MICHALSKI."

[23] For an overview of the contents cf. "Tables des matières des numeros 1, 1936 – 3, 1939 & 1, 1964 –37(40), 2008," *Organon*, vol. 37 vel. 40, (2008), 269–294: 269–270.

[24] Bibliothèque Polonaise à Paris: Materials of Zygmunt Lubicz-Zaleski, akc. 3874, 7: letter from Nauka Polska Editorial Office (St. Michalski), 23. Nov. 1932, no pagination.

afresh [...],"[25] they followed in his demand for a profound renewal of science, its tools, and its utilitarian objective. The programmatic idea of reflecting science and scholarship was paramount in Nauka Polska since 1918.[26] Other than the Polish journal, Organon had some starting problems. Though Zygmunt Lubicz-Zaleski (1882–1967), a literary scholar, translator, publicist, and de facto ambassador of the Mianowski Fund in Paris, reported on the great efforts he made at a Polish book shop and among French colleagues, only a handful copies were sold by early 1938, one of which to Moscow.[27]

As the unnamed authors of the introductions to the first volumes predicted, the shape and structure of Nauka Polska would change significantly over the years. Indeed, the editors refocused after the inventory of Polish scholarship in the first years. From the 4th volume onwards, a new approach whould foster the development of research and scholarly life. Science, as the introduction stated, should further be described and analysed as a *social* phenomenon, for a well-balanced science policy could only be developed upon precise knowledge of how institutions, schools, and (scientific) creativity (Pol. twórczość) interacted. A growing interest in the sociology of science since the 1920s included not only studying social interaction and dynamics of creativity, but also taking a broader look at the reputation of scholars and their disciplines within national society. As stated in one editorial, society's goodwill was indispensable for sound scientific development. Not only did society need high educational standards, but also understand science as a general value. The message was clear: neither should science funding be counted as luxury, nor should plain utilitarian approaches be favoured. Thus, the Mianowski Fund's responsibilities were further widened. It would have to promote science publicly, especially via Nauka Polska.[28]

Following the call for sociological studies of science, two eminent projects evolved within the Fund and its journals. The first was an archive. Already in 1918, geologist and pedagogue Antoni Bolesław Dobrowolski had insisted on "the urgent need of mental education."[29] In the first volume of Nauka Polska he developed a research

---

[25] See Francis Bacon: *The New Organon; or, True Directions for the Interpretation of Nature*, ed. by Lisa Jardine and Michael Silverthorne, Cambridge 2000, Author's Preface, 28: "... that the entire work of the mind be started over again; and from the very start the mind should not be left to itself, but be constantly controlled; and the business done [...] by machines." For Bacon's scientific utilitarianism and expertocratic thought see Robert K. Faulkner: *Bacon and the Project of Progress*, Maryland/London 1993.

[26] Cf. ▶ "Preface" (1936), 105, i.e. "Preface," *Organon*, vol. 1 (1936), V–VI.

[27] Bibliothèque Polonaise à Paris: Materials of Zygmunt Lubicz-Zaleski, akc. 3863, II, 1: letter from Organon Editorial Office (St. Michalski), 25. March 1938, no pagination. See also akc. 3863, II, 2: letter to St. Michalski, 17. Nov. 1937, no pagination, and letter to St. Michalski, 31. March 1938, no pagination.

[28] ▶ "Editorial Introduction" (1923), 103–104, i.e. "Wstęp redakcyjny," *Nauka Polska*, vol. 4 (1923), VII–IX.

[29] ▶ Antoni B. Dobrowolski: "The Urgent Need for Mental Education in Poland" 1918, i.e. "O pilnej potrzebie wychowania umysłowego w Polsce: o konieczności zasadniczej reformy nauczania w szkołach średnich oraz stworzenia w związku z ową reformą nowych placówek pracy naukowej," *Nauka Polska*, vol. 1 (1918), 489–502.

programme to examine the smallest units of thought, which, eventually, should be at the core of a new pedagogical method of education. Instead of teaching just the bare facts, Dobrowolski aimed to teach proper ways to think as he phrased it with reference to John Dewey.[30] Dobrowolski, who was connected to Nauka Polska from the very start, heavily leaned on the journal to propagate his project. He envisioned it as a rich collection of source material. Works from experimental psychology, accounts of past intellectual achievements, research reports and law suits, but also observations of children in school should help studying "intellectual strategies in different circumstances".[31] The most important sources, however, were to be assembled in an archive of "biographies of ideas".[32] Dobrowolski dreamt of honest accounts of the evolution of creativity in the arts and sciences, including all the dead ends, mistakes and coincidences that might occur in the creative process. His 1918 call culminates in an appeal to all creative workers to make biographies of their ideas available in Nauka Polska. A large collection would serve large systematic studies, and individual students, too, who could learn from role models. While Dobrowolski's interest went far beyond the 'great man', is aspect mostly attracted the journal editors. The call was republished several times in Nauka Polska, and supplemented with guidelines[33], while the first volume of Organon contained an English version.[34] Invitations to contribute were sent out and correspondence spread all over Europe. By 1939, about ten responses were published in both journals. Several of them are included in this volume.[35]

---

[30] Cf. John Dewey, *How We Think*, Boston 1910.

[31] ▶ Dobrowolski: "The Urgent Need for Mental Education in Poland" (footnote 30), 296.

[32] Ibid., 299.

[33] In November 1932, Stanisław Michalski, secretary general of the Fund and head of the Academic Council as well as of both journals, sent the following guidelines to a colleague in Paris: "The purpose of psychological autobiographies is to give material for the psychology of scientific creativity. Concerning the frame, there should be a psychological autobiography of scientific life against the background of personal life; in particular, it is about:

1/ to bring out the genesis of scientific creativity (origins of the first scientific ideas, the stimuli that triggered them), to illustrate its psychological course and to characterise the external (sociological) and internal (psychological) conditions of its development[.]

2/ [A]n important illustration of the psychology of creativity would be to give a biography of at least one thought or scientific idea which one considers to be the most original, presented in a manner reflecting strictly the actual state of affairs (facts, dates of origin, development and results achieved), refraining from drawing conclusions or giving interpretations." Bibliothèque Polonaise à Paris: Materials of Zygmunt Lubicz-Zaleski, akc. 3874, 7: letter from Nauka Polska Editorial Office (St. Michalski), 18. Nov. 1932, no pagination.

[34] Cf. ibid., editorial introduction.

[35] Cf. Stanisław Ossowski ["Student"]: "Kartka z życia studenta," *Nauka Polska*, vol. 4 (1923), 63–75, ▶ Czesław Białobrzeski ["C.B."]: "An autobiographical sketch and remarks on scientific creativity" (1927), i.e. "Szkic autobiograficzny i uwagi o twórczości naukowej," *Nauka Polska*, vol. 6 (1927), 49–76, Franciszek Bujak ["F.B."]: "Drogi mojego rozwoju umysłowego," *Nauka Polska*, vol. 6 (1927), 77–136, ▶ Antoni B. Dobrowolski: "My 'Scientific Biography'" 1928, i.e. "Mój 'życiorys naukowy,'" *Nauka Polska*, vol. 9 (1928), 68–216, anonymous ["X.Y."]: "Życiorys II-gi," *Nauka Polska*, vol. 9 (1928), 217–245, anonymous ["J.Z."]: "Wspomnienia o drogach do pracy naukowej," *Nauka*

By the end of the 1920s, the Academic Council initiated another central project. From June 1928 until 1939, the *Science Studies Circle* (Pol. *Koło Naukoznawcze*) would meet on a regular basis to discuss topics connected to the life of science: the psychology of scientific creativity, the sociology and organisation of science.[36] As the first report tells, the discussions should "[respond] to the perceived need for theoretical foundations for the activities undertaken by institutions that deal with the administration and management of research, in particular the Mianowski Fund."[37] All interested scholars and officials from private or governmental institutions connected to science organisation were invited. Meetings would start with a presentation followed by discussions. Protocols were regularly published in Nauka Polska, and most talks found their way into the journal. Though the topics were very heterogeneous, they did not lose attraction. The turnout was large, most sessions were attended by 20 people and more (cf. ▶ Sessions of the Science Studies Circle 1928–1938).[38]

Both the archive and the Science Studies Circle soon became central elements of the new research area *naukoznawstwo*. The name translates well as *science of science*, yet the compound noun is more complex. Its first part 'nauka' means science and scholarship, whereas the latter part '-znawstwo' alludes to the Polish equivalent of the English suffix '-logy' or the Greek 'logia', meaning the 'study of.' German equivalents would be the antique '-kunde' or the more contemporary '-forschung'. However, if referring to their project in English, interwar Warsaw scholars usually referred to a 'science of science'. The Polish adjective 'naukoznawcze' must have been self-explanatory by the late 1920s when the Circle was founded.[39]

Another project, which did not turn out successful, was the foundation of an institute dedicated to the science of science. Although not much material on these plans

---

Polska, vol. 9 (1928), 246–259, anonymous: "Życiorys I," *Nauka Polska*, vol. 15 (1932), 241–259, anonymous: "Życiorys II," *Nauka Polska*, vol. 15 (1932), 260–272, anonymous ["J.O."]: "Szkic autobiograficzny," *Nauka Polska*, vol. 16 (1932), 39–58, ▶ Emile Borel: "Contribution" (1936), i.e. "Contribution (Documents sur la Psychologie de l'Invention Dans Le Domaine De La Science)," *Organon*, vol. 1 (1936), 33–42), ▶ August Krogh: "Visual Thinking" 1938, i.e. "Visual Thinking," *Organon*, vol. 2 (1938), 86–94). Krogh's text was translated into Polish and published as "Myślenie wzrokowe," *Nauka Polska* vol. 24 (1939), 35–42.

[36] *Sprawozdanie 47 z Działalności Kasy im. Mianowskiego*, Warszawa 1928, 17.

[37] ▶ "Report on the activities of the *Science Studies Circle*" (1929), 107, i.e. "Sprawozdanie z działalności Koła Naukoznawczego," *Nauka Polska*, vol. 11 (1929), 353–355.

[38] Some smaller sessions dedicated to ongoing tasks of the Academic Council were not officially counted (*Sprawozdanie 47 z Działalności Kasy im. Mianowskiego*, Warszawa 1928, 17). See, however, for the mentioning of Ossowski's presentation in late 1929: ▶ "Second Report on the activities of the Science Studies Circle" (1930), 111–112, i.e. "Sprawozdanie drugie z działalności Koła Naukoznawczego," *Nauka Polska*, vol. 13 (1930), 166–169.

[39] In a contemporary translation of a text by Stanisław Ossowski, the *Koło Naukoznawcze* is referred to as "Debating Circle for Studies in Science" ("The Humanities and Social Ideology," *Baltic and Scandinavian Countries*, vol. 4/1 (1938), 68–76: 68. For the original see id. "Nauki humanistyczne a ideologia społeczna," *Nauka Polska*, vol. 22 (1937), 1–24. In the full bibliography of the first twenty volumes of Nauka Polska, Koło Naukoznawcze is translated into French as "Cercle d'Études sur la Science". Cf. the addendum to *Nauka Polska*, vol. 20 (1935), 4.

has been preserved, the Fund's annual report for the year 1928 states that the Mianowski Fund had been in contact with several governmental institutions to work out plans for such an institution.[40] One idea that was waged was to set up an institute in the Swiss town of Rapperswil. Close to Zürich, a Polish Museum had existed since 1870, where many Polish scholars stopped by when travelling west. Plans for an institute were made for several years, yet finally had to be dismissed due to a lack of funding during the global economic crisis of the late 1920s.[41]

Nevertheless, the activities organised under the auspices of the Fund's Academic Council did grow significantly. In 1923, a library for science organisation was created, whose bibliographical index reached a number of 31.500 cards in 1938. In that year, the library and archive compiled 650 titles from Polish and selected international journals, as well as bibliographical reports for relevant works in order to have them reviewed in Nauka Polska and Organon.[42] The annual chronicles in Nauka Polska registered activities in science organisation (congresses, workshops, founding and reactivation of societies or institutions, legislation concerning science etc.) both in and beyond Poland. Many texts were translated from foreign languages and revised by professional linguists, and often women whose names were not disclosed.[43] Moreover, the Academic Council conducted several surveys on science related issues.[44] After its foundation in 1916, it supported the multi-volume book series *Handbook to Autodidacts* (Pol. Poradnik dla Samouków), whose tradition led back to the times of Tsarist control. Between 1898 and 1932 more than two dozen volumes were published in three series. While the first should introduce general education in two stages, the later, much expanded series were conceived to provide more specialised, 'scientific' knowledge.[45]

---

[40] For a short note on the exchange with the Ministry of Religious Beliefs and Public Education and the Ministry for Foreign Affairs, as well as the Polish Commission of the International Committee on Intellectual Cooperation see *Sprawozdanie 49 z Działalności Kasy im. Mianowskiego*, Warszawa 1930, 17.

[41] Jan Piskurewicz: *W służbie nauki i oświaty. Stanisław Michalski (1865-1949)*, Warszawa 1993, 103. Piskurewicz extensively quotes Wanda Osińska's eulogy of Maria Ossowska ("Maria Ossowska (1896-1974)," *Kwartalnik Historii Nauki i Techniki*, vol. 20/2 (1975), 329-332: 332), who was supposedly chosen to organise the institute together with her husband. For their perspective see Maria's letter to Stanisław from 15 May 1928 in: *Intymny portret uczonych. Korespondencja Marii i Stanisława Ossowskich*, ed. by Elżbieta Neyman, Warszawa 2002, 271-272 and Stanisław Ossowski's diary entries from that time (*Dzienniki*, vol. 1: 1905-1939, ed. by Róża Sułek, Warszawa 2019, 174, 181, 183, 185).

[42] *Sprawozdanie 57 z Działalności Kasy im. Mianowskiego*, Warszawa 1938, 31-33.

[43] An office for French translations was mentioned in *Sprawozdanie 42 z Działalności Kasy im. Mianowskiego*, Warszawa 1923, 6-7. For a short mentioning of poet and graduate of English studies Felicja Kruszewska see *Sprawozdanie 58 z Działalności Kasy im. Mianowskiego*, Warszawa 1949, 30. A very central role had Janina Małkowska, who worked for the Academic Council from 1922 (Piskurewicz: *W służbie nauki i oświaty* (footnote 41), 80) and who preserved Stanisław Michalski's materials upon his death.

[44] For a survey on the organization of scientific congresses and 66 international responses cf. "Organisation des congrès scientifiques," *Organon*, vol. 2 (1938), 133-236, for an evaluation cf. Paweł Rybicki: "Ankieta Organonu," *Nauka Polska*, vol. 24, 397-410.

[45] Cf. Piskurewicz: *W służbie nauki i oświaty* (footnote 41), ch. 2.

Considering the broad range of its activities, and the great amounts of money that it moved, the Fund did fulfil the legacy of a "private ministry of education" that it had set for itself in 1920. Starting in 1924, it represented Poland at the *International Committee on Intellectual Cooperation* together with the *Polish Academy of Arts and Sciences* in Cracow (Pol. Polska Akademia Umiejętności).[46] However, the fund did not compete with the official *Ministry of Religious Beliefs and Public Education* (Pol. Ministerstwo Wyznań Religijnych i Oświecenia Publicznego, MWRiOP) as both were in fact tightly knit together. According to the annual reports, the Fund cooperated with legislative and executive bodies on several occasions.[47] In October 1936, Wojciech Świętosławski, professor of physical chemistry at the Warsaw University of Technology, who served as Minister of Religious Denominations and Public Enlightenment from 1935 to 1939, gave a talk at the Sciences Studies Circle.[48] Stanisław Michalski, the Fund's secretary general and head of the Academic Council, simultaneously served as the director of the Ministry's *Department of Science and Higher Schools*.[49] Being the fixed star of the milieu, Michalski became even more powerful in 1928, when he was appointed director to the new *Fund of National Culture* (Pol. Fundusz Kultury Narodowej). The Fund was founded under the auspices of Marshal Piłsudski, who had seized power in May 1926.[50]

Resulting from a private initiative in 1881, the Kasa Mianowskiego relied on subsequent private funding from Polish society. In 1925, Michalski asked the well-known novelist Stefan Żeromski to write a text in support of the Fund. Żeromski, who was hailed as the "conscience of Polish literature" at the time, did not disappoint and delivered an essay that was widely distributed and often quoted in subsequent publications of the institution.[51] Over the years, many grand and small donations were made,

---

[46] Cf. *Sprawozdanie 43 z Działalności Kasy im. Mianowskiego*, Warszawa 1924, 32–33 or *Sprawozdanie 45 z Działalności Kasy im. Mianowskiego*, Warszawa 1926, 31–42.

[47] Cf. *Sprawozdanie 43 z Działalności Kasy im. Mianowskiego*, Warszawa 1924, 7.

[48] Cf. ► Sessions of the Science Studies Circle 1928–1938.

[49] For more on Michalski see ► Jan Piskurewicz and Leszek Zasztowt: "Science Studies in Poland before the Second World War" and Piskurewicz: *W służbie nauki i oświaty* (footnote 41).

[50] For an account of the founding procedure cf. "Notatka o założeniu Funduszy Kultury Narodowej," in: *Fundusz Kultury Narodowej (1928-1937). Zarys działalności*, Warszawa 1937, 57–60.

[51] Stefan Żeromski: *Sprawa Kasy im. Mianowskiego*, Warszawa 1925. The most famous, often quoted passage (*Sprawozdanie 57 z Działalności Kasy im. Mianowskiego*, Warszawa 1938, front end) reads as follows: "It came into being against all odds and stood the test of time. It looked at the ruins of the conquerors' reign and awaited the end of captivity.

Should it not find support among the free Polish nation, now that all obstacles that stood against its creation and activity have been overthrown and reduced to ashes? Meritorious, dignified, noble fount of knowledge, should thou indeed dry out in an independent Poland that was freed from the irons by a beautiful twist of fate? […]

Hard science has to become a regulator and manager to our modern life that is so rich, exuberant, unrestrained. She only can help, help us working with the vivid substance, which is yet looted by the invaders, ruffled by poor cultivation, mistreated with blatant stupidity divided over the smallest issue or sitting idle. She only can help to start building a new inner content to systematically, a new breadth and luxuriance of mind, which would be conscious, purely national, our own, and which would be based on the most excellent and undoubted experience and distinguished models. Only

and for some time in the early 20th century the Fund had even received money from a Baku oil well that belonged to one of the patrons. Money also came from selling publications, which often sold out quickly.

The Fund remained a central player in Polish science organisation until the Second World War, which abruptly limited the scope of its activities. As a consequence of the annihilation of Warsaw upon the uprising in 1944, the bibliographical index, large parts of the library, and the bookkeeping department were irretrievably destroyed. The Academic Councils's office and collections had suffered before. During the attacks on Warsaw in September 1939, the printing department was severely damaged, causing the loss of the freshly printed 25th volume of Nauka Polska and the 3rd of Organon.[52] However, the staff managed to save other brand-new publications, as for example the Polish translation of Marie Skłodowska-Curie's *Radioactivité*. These publications were secretly sold under German occupation, and the income used to help scholars in need. Furthermore, organisational work was accomplished and allegedly the Science Studies Circle met as well.[53]

After the war, the Fund faced hard times. Stanisław Michalski, who had been detained in Auschwitz and was already in his eighties, tried to re-establish the structures corresponding from a sanatory bed in Cracow. At the same time, some of the remaining veterans from the interwar period did their best to transpose the old structures into the new political realities of emerging People's Poland. In 1947, an actual 25th volume of Nauka Polska was finally published, which was an attempt to determine the state of affairs after the war in order to start anew.[54] It turned out to be the Fund's last major publication. In 1951, the Kasa was incorporated into the *War-*

---

with the help of science can we preserve the tremendous fortunes that our nation once already called its own, and only with her help can succeed to discover more.

In a free Polish state science must not be obtained by begging, copied, smuggled, it has to be Polish. Science – this crystal-clear mountain lake, which reflects the infinity of the heavens, has to be ours, our own, coming out of our homeland.

We do have an institution that serving exactly this goal. It was formed in the mists of captivity by honourable fathers, who handed it to us, who live under a boundless sun, in golden freedom. And today this very institution – does not have enough resources.

Let us safe! Let us support! Let us form anew the Mianowski Fund!"

[52] *Sprawozdanie 58 z Działalności Kasy im. Mianowskiego*, Warszawa 1949, 26, 31. For some of the efforts during the war the Fund's 1949 annual reports give detailed numbers, for example concerning expenses for printing, scholarships etc. For a partly reconstruction of the contents of the lost Organon volume cf. "Tables des matières" (footnote 23), 270.

[53] Cf. *Sprawozdanie 58* (footnote 52), 25–31. The Polish translation *Promieniotwórczość* was officially published in 1939, yet it was secretly set and printed during the first year of the occupation. For other works cf. ibid. 28–29. The Physicist Ludwik Wertenstein, who had translated Curie's text, secretly received money from the Fund. Cf. Friedrich Cain: *Wissen im Untergrund. Praxis und Politik klandestiner Forschung im besetzten Polen (1939–1945)*, Tübingen 2021, 423–424.

[54] As before 1939, the volume assembled original research articles, chronicles on the fate of certain disciplines during the war and reports on new initiatives to continue interwar *naukoznawstwo*. A section on "Secret work in times of occupation" presented a general report on institutions of higher learning as well as the results of a survey on academic organization conducted during the war. Cf. J. Z.: "Nauka i szkolnictwo wyższe", *Nauka Polska*, vol. 25 (1947), 196–213, "Memoriał w

saw *Society of Arts and Sciences* (Pol. Towarzystwo Naukowe Warszawskie), which one year later joined the *Polish Academy of Arts and Sciences* in Cracow, establishing the new *Polish Academy of Sciences* (Pol. Polska Akademia Nauk, PAN) in Warsaw. Forty years later, in October 1991, the Fund was re-established under the name of Kasa im. Józefa Mianowskiego – Fundacja Popierania Nauki (Engl. Józef Mianowski Fund – Foundation for the Promotion of Science), which took over publishing Nauka Polska again the following year.[55]

Although the Fund could not thrive in Socialist Poland, *naukoznawstwo* did not phase out. In the new socialist state, which filled the central positions of science organization with its own officials, many central interwar figures remained in the game and certainly took on important roles. Michał Kokowski has recently given an overview over a nested history, in which he described the situation of science studies in Socialist Poland as a cohabitation of "old" and "new" traditions.[56] Many protagonists of interwar science studies, such as Maria Ossowska, Stanisław Ossowski, Józef Chałasiński and Tadeusz Kotarbiński, were connected to the Academy *Institute of Philosophy and Sociology* (Pol. Instytut Filozofii i Socjologii) which was founded in 1956 during the Thaw period. In 1958, Kotarbiński became the founding chair of a *Research Centre for the General Problems of Labour Organization* at the Academy (Pol. Pracownia Ogólnych Problemów Organizacji Pracy), whose members established research in what they called *praxeologia*, a theory of efficient or purposeful human action.[57] In 1956, a department for the History of Science and Technology was established and chaired by Bogdan Suchodolski. In 1964, the still existing *Committee of the Science of Science* (Pol. Komitet Naukoznawstwa PAN) was established, which ran as *Commission* for the first three years.[58] That same year, a *Centre for Research on the Science of Science* (Pol. Pracownia Badań Naukoznawczych) was founded, which in 1965 collaborated with Kotarbiński's unit to form a *Seminar of the Science of Science* (Pol. Konwersatorium Naukoznawcze PAN) to revive the Science Studies Circle tradition.[59] In 1973, the big praxeological working unit, which in the meantime had grown into a department, was dissolved, while two years later a *Re-*

---

sprawie wyższych uczelni," *Nauka Polska*, vol. 25 (1947), 213–219, and "Ankieta w sprawie badań i organizacji szkół akademickich," *Nauka Polska*, vol. 25 (1947), 219–235.

[55] Cf. Piotr Hübner, Jacek Soszyński, Jan Piskurewicz and Leszek Zasztowt: *A History of the Józef Mianowski Fund*, Warszawa 2013, 116–118, 133–134.

[56] Cf. Kokowski: "The Science of Science (naukoznawstwo) in Poland" [2016] (footnote 8). See also id.: "The Science of Science (naukoznawstwo) in Poland. The Changing Theoretical Perspectives and Political Contexts – A Historical Sketch From the 1910s to 1993," *Organon*, vol. 47 (2015), 147–237.

[57] Kokowski: "The Science of Science (naukoznawstwo) in Poland" [2016] (footnote 8), 157. One of the central texts of this praxeology was Tadeusz Kotarbiński: *Traktat o dobrej robocie*, Łódź 1955. Cf. Wojciech Gasparski, Tadeusz Pszczołowski (eds.): *Praxiological Studies: Polish Contributions to the Science of Efficient Action*, Dordrecht, Boston, Lancaster, and Warsaw 1983.

[58] Kokowski: "The Science of Science (naukoznawstwo) in Poland" [2016] (footnote 8), 158 and id. "The Science of Science (naukoznawstwo) in Poland" [2015] (footnote 56), 171.

[59] Id.: "The Science of Science (naukoznawstwo) in Poland" [2016] (footnote 8), 158–159.

search Centre for History of Organization of Science at the Department for the History of Science and Technology (Pol. Pracownia Historii Organizacji Nauki Zakładu Historii Nauki i Techniki PAN) was established. Further restructuring brought forward the new *Department of Praxeology and the Science of Science* (Pol. Zakład Prakseologii i Naukoznawstwa) in 1980.[60] Today, several committees in Polish Academies follow in the footsteps of the Science Studies Circle.[61]

The different institutions established several journals. In 1953, the Academy restarted Nauka Polska as its own official journal. However, apart from the main title, it did not have much in common with the former journal. For example, it contained all protocols of Academy meetings[62]. Since 1956, the *Kwartalnik Historii Nauki i Techniki* (Engl. Quarterly Journal of the History of Science and Technology) was up and running, which still is published today at the *Ludwik and Aleksander Birkenmajer Institute for the History of Science* (Pol. Instytut Historii Nauki im. L. i A. Birkenmajerów). In 1965, the *Committee of the Science of Science* started a journal called *Zagadnienia Naukoznawstwa: Studia i Materiały* (Engl. Problems of the Science of Science: Materials and Studies). One year before, an *Information Review on the Science of Science* (Pol. Przegląd Informacji o Naukoznawstwie) had been launched. The *Organon* had already been reactivated in 1964 under the auspices of the Institute of the History of Science at the Polish Academy of Science.[63]

2. The structure of the book

The 25 volumes of Nauka Polska and two of Organon compile around 10.000 pages and a great wealth of topics. In order to provide some context for the selected source materials, this edition provides research articles and commentaries from historical, sociological and philosophical perspectives. The three parts of this edition are interrelated on many levels and can, but do not need to be read in consecutive order.

1) The texts in part I draw a background picture for the sources in part II. They point to aspects of structure and subjectivity, crucial for the situation and self-conception of interwar Polish science and science studies. Jan Surman outlines the infrastructural and linguistic situation in the academic realm of the new state. He points to the critical role of educational and academic institutions such as Nauka Polska as places to reflect the relation between the internationalisation of science and the ongoing attempts to build up national educational infrastructures. In respect to scientific

---

[60] Ibid., 164, 167.
[61] Cf. below, section III, footnote 77.
[62] The new title of Nauka Polska. was "Journal for Questions of the Progress of Science in Poland" (Pol. Czasopismo poświęcone zagadnieniom rozwoju nauki w Polsce). Kokowski: "The Science of Science (naukoznawstwo) in Poland" [2015] (footnote 56), 164.
[63] Ibid., 180. For the later development of these publications see ibid., 195–196. For an overview of the contents until 2008 cf. "Tables des matières" (footnote 23), 271–294.

productivity and specific Polish thought styles, he stresses the fundamental importance of the multilingualism and a related multiculturalism of Polish academia, which was rebuilt by scholars who had been educated and worked in other countries, when a Polish state did not exist. However, mastering several academic languages did not make translating scholarly work easy.

Marta Bucholc and Friedrich Cain focus on the way creativity was conceived as the fundamental subjective prerequisite to support Polish science. Basing her study on the material in part II, Bucholc highlights the authors' reflexion on the sociogenetic context of their writings, the role of scientific biographies and of creativity, but particularly of the genesis and development of scientific methods, theories, and concepts. Within the larger context of the Science Studies Circle, Cain lines out the background of the *Archive of Materials for research on creativity*, which Antoni B. Dobrowolski had established to collect "biographies of scientific ideas". Finally, he discusses the epistemology and pedagogy of Dobrowolski's studies in creativity and his delineations of a pragmatic "grammar of thought" to direct educational policies.

2) Part II contains the source material, i.e. original texts from the interwar period, published in Nauka Polska and Organon. As it is impossible to cover the whole variety of material[64], some areas were selected to capture the topical vastity of both journals along the following guidelines: *First*, this volume assembles programmatic texts from both journals. *Second*, and closely connected to this, three basic interests stated over and over in the journals are represented, namely the psychology of creative scientific production, the social embeddedness of science and scholarship, as well as questions concerning the organisation of a well-functioning (national) system of research and education. The *third* aim was to exhibit the great expenditure of resources for *naukoznawstwo*. In *fourth* place, next to representing the original strategy of internationalisation, access should also be granted to texts dwelling behind the Polish language barrier. Therefore, several texts are made available in English for the first time. *Fifth* and finally, the volume should also show the contemporary impact of *naukoznawstwo* and represent also some well-established, and still influential scholars, who committed themselves to the project.

In this introduction, the texts will be introduced only briefly. Detailed context informations are given in editorial introductions. The source section starts with three short introductory texts to illustrate the tone of Nauka Polska. The *Editorial Introduction* to the 4th volume reminisces the original character of an inventory and then marks the turn to a new programme of science studies in 1923. The second introductory text is the *Preface* to vol. 1 of Organon in 1936. It lays out the programme of the journal, and of the Warsaw initiative in general. Selections from the two initial reports on the Science Studies Circle continue to sketch the project's outline. The first

---

[64] Thanks to exemplary digitalisation projects, both journals are available online at *Śląska Biblioteka Cyfrowa* today: For Nauka Polska see https://www.sbc.org.pl/dlibra/publication/27486/edition/24558#structure, for Organon https://www.sbc.org.pl/dlibra/publication/68770/edition/64987#structure (acc. July 2023).

(11th vol., 1929) contains a brief introduction and several short abstracts from the first year's sessions. Lateron, the abstracts were supplemented by revised protocols of the discussions.

The second report (13th vol., 1930) introduces a preliminary stage of the most prominent text from both journals. Maria Ossowska's and Stanisław Ossowski's *The Science of Science* was first published in Polish in 1935 and opened vol. 1 of the Organon one year later. The authors proposed an interdisciplinary approach to the study of science resting on philosophy, psychology, sociology, and history. All four disciplinary backgrounds should engage in a fifth area of studies, which aimed at the application of the new discipline's insights. However, Ossowski had presented a first draft of the science of science in November 1929 at a time, when the idea of a distinct institute, presumably in Swiss Rapperswil, had not yet been buried. This session of the Science Study Circle did not go into the official count.

After some conceptual changes, the 1935 version of Ossowska's and Ossowski's project experienced a remarkable career, making its way into several international publications after being translated in 1936. It was cited as a forebearer to projects in international science studies around the globe. The 1964 reprint in *Minerva* was introduced as a 30-years old precursor of the "disciplined academic study of the social, political, administrative and economic aspects of scientific and scholarly research" that had spread in the US, Great Britain, France, Japan, Sweden and India "only recently."[65] In 1965, the article appeared not only in a volume on *Science and Society*, again with a note hinting on its remarkable actuality and the fact that it had been written offside the centers of science studies of its time.[66] That same year, Mackay and Bernal made their reference at the international congress on the history of science in Warsaw, framing it as a conceptual anticipation of Derek de Solla Price's work.[67] One year later, the economist and founding father of soviet scientometrics Gennadi M. Dobrov mentioned it as a central venture point for modern science studies.[68] In his introduction to Robert K. Merton's *The Sociology of Science*, Norman W. Storer pointed out the Ossowskian origin of the term "science of science".[69] A 1977 volume on the sociology of science in Europe edited by Merton and Jerry Gaston highlighted

---

[65] Maria Ossowska and Stanisław Ossowski: "The Science of Science," *Minerva*, vol. 3/1 (1964), 72–82: 72.

[66] "Prologue to the Present: Introduction," in: *Science and Society*, ed. by Norman Kaplan, Chicago 1965, 9–13: 10–11.

[67] Bernal, Mackay: "Towards a Science of Science" (footnote 1), 9. The authors referred to Derek de Solla Price: *Little Science, Big Science*, New York 1963.

[68] Gennadij M. Dobrov: *Nauka o nauke: Vvedenie v obščee naukoznanie*, Kiev 1966, 18.

[69] Norman W. Storer: "Introduction," in: Robert K. Merton: *The Sociology of Science. Theoretical and Empirical Investigations*, Chicago, London, 1973, xi–xxxi: xxx.

the historical relevance of the text, too.[70] In 1988, a general introduction into science studies from the German Democratic Republic still credited the Ossowskis.[71]

Preceding Ossowska and Ossowski by ten years, in 1925, sociologist Florian Znaniecki, who at the time was still travelling back and forth between Western Polish Poznań and Chicago, had already outlined a first sketch for a *Science of Knowledge* as he imagined it. Three years later, psychologist Stefan Błachowski, who was also based in Poznań, published a lengthy article on aspects of scientific creativity that drew heavily on materials published in Nauka Polska before (see below). After these programmatic outlines, the volume presents several papers with more specific questions. These texts were usually published in Nauka Polska upon presentation at the Science Study Circle. In December 1928, Franciszek Bujak from Lwów, who was an expert in economic, political and social history of Poland, made an attempt to define the activist and the researcher and their social functions. The text was published in vol. 11 of Nauka Polska and later made its way into vol. 1 of Organon. In February 1929, Lwów sociologist Paweł Rybicki gave a paper on *Science and the Forms of Social Life*, in which he narrowed down *Issues at the Intersection of Sociology and Theory of Science*. The text was published right after Bujak's in vol. 11 of Nauka Polska, which opened with the Tadeusz Kotarbiński's text *On the Skills of a Researcher*. Again, it was based on a talk that the Warsaw philosopher had given at the Science Study Circle in April 1929. Bogdan Suchodolski, who made not only the most appearances, but also gave the first talk in the recorded history of the Circle widened the discussion with a presentation on *Investigation and Teaching* in November 1935, which was published one year after.

The six remaining texts in part III of the volume belong to the project of the *Archive of Materials for Research on Creativity* initiated by Antoni B. Dobrowolski in 1918. The first text was published in vol. 1 of Nauka Polska, and an English translation of the passage, in which the author explains his idea of "biographies of ideas" could be found in Organon in 1936. The project found great support among the editors, and so the Archive opened in vol. 6 of Nauka Polska (1927) with two autobiographical contributions by physicist Czesław Białobrzeski and Franciszek Bujak, who only signed with their initials instead of full names. Here, Białobrzeski's *Autobiographical Sketch and Remarks on Scientific Creativity* is reprinted along with a short note on the archive by Dobrowolski. One year later, Dobrowolski published his own *Scientific Curriculum Vitae*, from which an episode on the history of his idea for the Archive is presented in this volume.[72] Two further biographies follow to show the broad range of responses.

---

[70] Tadeusz Krauze, Zdzisław Kowalewski and Adam Podgórecki: "The Sociology of Science in Poland," in: *The Sociology of Science in Europe*, ed. by Robert K. Merton and Jerry Gaston, Carbondale e.a. 1967, 193–223.

[71] Günter Kröber (ed.): *Grundlagen der Wissenschaftsforschung*, Berlin 1988, 54–55.

[72] ▶ Dobrowolski: "The Urgent Need for Mental Education in Poland" (footnote 29), ▶ Białobrzeski: "An autobiographical sketch and remarks on scientific creativity" (footnote 35) and ▶ Dobrowolski: "My 'Scientific Biography'" (footnote 35).

French mathematician Emile Borel contributed to the project in vol. 1 of Organon, and Danish zoophysiologist and nobel laureate August Krogh in vol. 2. As these texts show, the authors partly struggled to meet Dobrowolski's requirements. Instead of writing about the histories of their ideas, they wrote about themselves in the tradition of biographies of 'great men'. Ironically, Dobrowolski himself did not live up to his own demands, when writing his report in 1928. As the project's patron, Dobrowolski did never publish an original analysis of the material in Nauka Polska, though he did analyse it in his pedagogical works, which were only published posthumously.[73] Thus, Błachowski's article seems to be the only contemporary analysis of the material.

With this volume, most of these texts are being made available in English for the first time. Tul'si (Tuesday) Bhambry has translated the biggest part with great empathy. In her commentary she describes some of the hardships of translating scientific texts from a long-passed world with specific vocabularies. An English translation of Znaniecki's text was published in a 1982 edition of *Polish Contributions to the Science of Science*. Concerning the earlier translations of the texts by the Ossowskis, Bujak, and Suchodolski the names of the translators could not be established. The same holds true for the authorship and sometimes translation of the introductory texts in the first part of the source section.[74]

3) The final part of the volume assembles three reflections on interwar Science Studies from today's perspective. Leszek Zasztowt and Jan Piskurewicz reassess the history of this initiative within an institutional longue durée. They do this from a historical point of view, but also as important representatives of the re-established Kasa Mianowskiego. Andreas Langenohl takes a similar look, yet from the perspective of sociology, and especially a sociology of knowledge. Finally, Paweł Kawalec comments on the history of naukoznawstwo from a philosophical perspective, and as a member of the aforementioned *Committee of the Science of Science* at the Polish Academy of Science. He lines out opportunities for the Science of Science in the 21st century.

3. Paths of reception: innovation, auto-genealogy, history and contemporaneity

The phenomenon of science studies in Warsaw was by no means restricted to interwar times or, for that matter, to the Second Polish Republic between 1918 and 1939. In fact, it can be well described as a ground to flourish, as a context in which a precise understanding of science was granted formerly unseen attention. However, this is a question of perspective. This final part of the introduction lines out more contexts

---

[73] Antoni B. Dobrowolski: *Ustrój oświatowy* (= Pisma pedagogiczne, vol. I), Wrocław 1958, id.: *Nowa dydaktyka* (= Pisma pedagogiczne, t. II), Wrocław 1960, and id.: *Moralność umysłowa (materiały)* (= Pisma pedagogiczne, t. III), Wrocław 1964.

[74] See above, p. 10, esp. footnote 43.

and histories, for which the Warsaw initiative – or the initiatives grouped under one roof in 1920s and 30s Warsaw – bore and still bears relevance.

While the programmatic cohabitation of old and new approaches existed until the 1980s[75], historiographical interest rose during the 1980s, and first pivoted during the transformation period, when the Mianowski Fund was reactivated in 1991. Thus, the (institutional) history of the Fund, its activities and central members is rather well documented in Polish publications.[76] More recently, interwar science studies have become relevant as a hinge for new attempts at a systematic study and promotion of research with both Polish Academies sponsoring units taking up their leads.[77] These committees, research units and further initiatives engage in systematic studies, offer expertise for science organization and policies, and seek to train future science managers.[78] They thus continue interwar and postwar heritage.[79]

---

[75] Cf. Kokowski: "The Science of Science (naukoznawstwo) in Poland" [2016] (footnote 8), 169.

[76] Apart from ▶ Piskurewicz and Zasztowt: "Science Studies in Poland before the Second World War" cf. for the history of the Fund Hübner, Soszyński, Piskurewicz, Zasztowt: *A History of the Józef Mianowski Fund* (footnote 55), Leszek Zasztowt (ed.): *Kasa Mianowskiego 1881–2011*, Warszawa 2011, and Piotr Hübner, Jan Piskurewicz and Leszek Zasztowt: *Kasa im. Józefa Mianowskiego. Fundacja Popierania Nauki 1881–1991*, Warszawa 1992. For the patron see Leszek Zasztowt: "Józef Mianowski. *Biografia konserwatysty*, Warszawa 2021. For the central figure Michalski see Jan Piskurewicz: *W służbie nauki i oświaty* (footnote 41), id.: "Stanisław Michalski w dziejach nauki polskiej pierwszej połowy XX wieku," *Kwartalnik Historii Nauki i Techniki* vol. 35/1 (1990), 55–92. For conceptual histories of science studies see Stefan Zamecki: *Problematyka naukoznawcza na łamach periodyku "Nauka Polska. Jej Potrzeby, Organizacja i Rozwój" – Studium historyczno-metodologiczne. Lata 1918–1947*, Warszawa 2016, id.: *O niektórych potrzebach nauki polskiej omawianych na łamach periodyku "Nauka Polska. Jej Potrzeby, Organizacja i Rozwój" – Lata 1918–1947, Aneks*, Warszawa 2017 or Grażyna Wrona: "Nauka Polska. Jej Potrzeby, Organizacja i Rozwój, 1918–1939," *Rocznik Historii Prasy Polskiej*, vol. 7/2 (2004), 19–47. Earlier works are Jan Piskurewicz: "Refleksja naukoznawcza w środowisku uczonych II Rzeczypospolitej 1989," *Zagadnienia Naukoznawstwa*, vol. 99–100 (1989), 601–612 and Piotr Hübner: "Instytucje i programy naukoznawcze w Polsce w latach 1945–1949," *Zagadnienia Naukoznawstwa. Studia i Materiały*, vol. 19/4 (1983), 446–454. For the Postwar era cf. Joanna Schiller-Walicka, Leszek Zasztowt (eds.): *Instytut Historii Nauki Polskiej Akademii Nauk w latach 1953–2003. Księga jubileuszowa z okazji pięćdziesięciolecia działalności*, Warszawa 2004.

[77] The Polish Academy of Science in Warsaw houses a *Committee* of the *Science of Science* (Pol. Komitet Naukoznawstwa), a *Committee for the History of Science and Technology* (Pol. Komitet Historii Nauki I Techniki) as well as a Research Center for Science Studies (Pol. Pracownia Naukoznawstwa). The Polish Academy of Learning in Cracow opened a *Commission for the History of Science* (Pol. Komisji Historii Nauki) in 1998.

[78] For another initiative cf. the project *Management of research and development at educational units* (Pol. Zarządzanie systemem B+R w instytucjach naukowych), a cooperation between the Catholic Univ. of Lublin and Lublin Business School starting in 2008. Among several conference volumes, methodological works on science of science education and a dictionary (Józef Herbut, Paweł Kawalec (eds.): *Słownik terminów naukoznawczych. Teoretyczne podstawy naukoznawstwa*, Lublin 2009), the project organized a 240h postgraduate course. Cf. the two-volume textbook on the basics of science studies: Paweł Kawalec, Piotr Lipski, Rafał Wodzisz (eds.): *Podstawy naukoznawstwa, t. 1: Skrypt dla studentów studiów literackich*, Lublin 2011 and eid. (eds.): *Podstawy naukoznawstwa, t. 2: Skrypt dla studentów studiów magisterskich*, Lublin 2011.

[79] Cf. ▶ Paweł Kawalec: "The Original Conception of the Science of Science and Innovation Studies". Cf. also the recent Piotr Hübner: *Encyklopedia polskiej nauki akademickiej* (3 vol.s), Toruń 2023.

Earlier historiographical research provides the general context for this volume on interwar Poland, for example on the structures of organizing science and education in the Second Republic.[80] Recent work on the development of specific disciplines, and especially from the humanities and the social sciences, provide more information on protagonists and their cultural embedding.[81] The latter aspect is also sustained in literature on specific figures, for example the *expert* or *modernity*[82], but also on broader cultural conflicts.[83] More and more biographical studies help to draw a detailed picture of interwar Poland's intellectual landscapes.[84] Recently, a broader view

---

[80] For science organization in general cf. Bogdan Suchodolski (ed.): *Historia Nauki Polskiej V: 1918-1951*, Wrocław e.a. 1992, Bohdan Jaczewski: *Organizacja i finansowanie nauki polskiej w dwudziestoleciu międzywojennym*, Wrocław, Warszawa 1971, id.: *Polityka naukowa Państwa Polskiego w latach 1918-1939*, Wrocław, Warszawa 1978. For the history of education in Poland cf. Danuta Krześniak-Firlej, Stanisław Majewski and Waldemar Firlej: *Organizacja szkolnictwa w II Rzeczypospolitej 1918-1939*, Kielce 2014, Maciej Fic, Lech Krzyżanowski and Miłosz Skrzypek (eds.): *Dwa dwudziestolecia Rzeczypospolitej. Oświata, religia, kultura i społeczeństwo. Próba bilansu*, Katowice 2010, Mirosław S. Szymański: *Pädagogische Reformbewegungen in Polen 1918-1939. Ursprünge - Verläufe - Nachwirkungen*, Köln e.a. 2002, Krzysztof Baranowski: *Alternatywna Edukacja w II Rzeczypospolitej. Wolna Wszechnica Polska*, Warszawa 2001, Joanna Sadowska: *Ku szkole na miarę Drugiej Rzeczypospolitej. Geneza, założenia i realizacja reformy Jędrzejewiczowskiej*, Białystok 2001, Władysława Szulakiewicz: *Historia oświaty i wychowania w Polsce 1918-1939. Studium historiograficzne*, Toruń 2000, Danuta Drynda (ed.): *Inspiracje dla współczesnej edukacji w dydaktyce Drugiej Rzeczypospolitej*, Katowice 2000, Leonard Grochowski: *Studia z dziejów polskiej szkoły i polskiej pedagogiki lat międzywojennych w kontekście europejskim*, Warszawa 1996, Karol Poznański (ed.): *Oświata, szkolnictwo i wychowanie w latach II Rzeczypospolitej*, Lublin 1993, Danuta Drynda (ed.): *Studia z historii polskiej pedagogiki. Koncepcje pedagogiczne w Drugiej Rzeczypospolitej*, Katowice 1993, id.: *Pedagogika Drugiej Rzeczypospolitej. Warunki, orientacje, kontrowersje*, Katowice 1987, Franciszek Bereźnicki: *Innowacje pedagogiczne w Polsce (1918-1939)*, Szczecin 1984, Feliks W. Araszkiewicz: *Ideały wychowawcze Drugiej Rzeczypospolitej*, Warszawa 1978.

[81] Cf. for the social sciences Katherine Lebow, Małgorzata Mazurek and Joanna Wawrzyniak: "Making Modern Social Science: the Global Imagination in East Central and Southeastern Europe after Versailles," *Contemporary European History*, vol. 28/2 (2019), 137–142 or Friedrich Cain: "On Racism and Scholarship. Stanisław Ossowski Between Science Studies and Sociology," in: *Spotkania z Ossowskim*, ed. by Antoni Sułek, Warszawa 2020, 100–131. For ethnography see Olga Linkiewicz: "Applied Modern Science and the Self-Politicization of Racial Anthropology in Interwar Poland," *Ab Imperio*, vol. 2 (2016), 153–181, id.: "Scientific Ideals and Political Engagement: Polish Ethnology and the 'Ethnic Question' Between the Wars," *Acta Poloniae Historica*, vol. 114 (2016), 5–27 and id.: "Judasze wśród ludu. Żydzi galicyjscy w polskiej etnologii i tekstach etnograficznych z okresu międzywojnia," *Kwartalnik Historii Żydów*, vol. 273 (2020), 109–127, for literary studies Michał Mrugalski, Schamma Schahadat and Irina Wutsdorff (eds.): *Central and Eastern European Literary Theory and the West*, Berlin, Boston 2023.

[82] Cf. Kohlrausch, Steffen, Wiederkehr (eds.): *Expert Cultures in Central Eastern Europe* (footnote 5) and Friedrich Cain: "Moderne errichten. Über Experimente in der Stadt Warschau, ca. 1918–1927," in: *Laboratorien der Moderne. Orte und Räume des Wissens in Zentraleuropa*, ed. by Sylwia Werner and Bernd Stiegler, München 2016, 253–288.

[83] Cf. Paul Brykczynski: *Primed for Violence. Murder, Antisemitism, and Democratic Politics in Interwar Poland*, Madison 2016 and Dobrochna Kałwa: "Poland," in: *Women, Gender and the Extreme Right in Europe*, ed. by Kevin Passmore, Manchester 2003, 148–167.

[84] Cf. for example for Ludwik Fleck the special issue "Gestalt, Ritus, Kollektiv. Ludwik Fleck im Kontext der Ethnologie, Gestaltpsychologie und Soziologie seiner Zeit," *NTM*, vol. 22/1–2 (2014), ed. by Bernhard Kleeberg and Sylwia Werner as well as Werner: *Lemberger Moderne* (footnote 8). For

was taken on the epistemologies of science studies in Central, Eastern and Southeastern Europe.[85]

As much as the realia of the Mianowski Fund and its projects that are in question here may suggest a rather enclosed phenomenon, this introduction started from one of many connections of interwar Warsaw and greater historical narratives. In fact, Bernal and Mackay linked two main aspects in their 1965 presentation. Not only did they integrate the programme by Ossowska and Ossowski into their historical account of modern science studies, but they also reactivated it by comparison to de Solla Price's influential *Little Science, Big Science* published two years earlier.[86] Whether or not this was a courtesy to their hosts at the congress in Warsaw, their talk helps opening a panoramic view on the history of modern science and systematic knowledge. While Bernal and Mackay gesture at deep histories of "Graeco-Jewish-Christian-Roman traditions of Western Europe" and other, "Chinese, Islamic, Hindu, Japanese or Russian experiences"[87], their perspective may also be stretched right into our nowadays realities.

In 2018, a group of fourteen authors from a wide range of disciplinary backgrounds published a paper in *Science* introducing an "emerging field" that would break down disciplinary boundaries between scientometrics, sociology of science, and innovation studies: a *science of science*, or *SciSci* in short. According to the authors, SciSci would gather insight on the "fundamental drivers of science" and eventually discover the universal dynamics of all sciences.[88] It could "enhance the prospects of science as a whole to more effectively address societal problems"[89] and "make science flourish for society". Therefore, science needed to be studied as an ecosystem of research and publications, of "communicators, teachers, and detail-oriented experts."[90] SciSci would boost individual career paths and the performance of organisations, enhance funding schemes and could help maneuvering the reproducibility crisis.[91] SciSci is presented as *the* future science here, leading the way towards mind-machine partnerships, appreciating and incentivizing intellectual exchange and curiosity, and sub-

---

Stanisław Ossowski see the recent volume Antoni Sułek (ed.): *Spotkania z Ossowskim*, Warszawa 2020. Further Katrin Steffen: *Blut und Metall. Die transnationalen Wissensräume von Ludwik Hirszfeld und Jan Czochralski im 20. Jahrhundert*, Göttingen 2021.

[85] Cf. the guest section "The Past and Present of Political Epistemologies of (Eastern) Europe," in *Historyka. Studia Metodologiczne* (footnote 6).

[86] Cf. above, footnote 67.

[87] Ibid., 12.

[88] Santo Fortunato, Carl T. Bergstrom, Katy Börner, James A. Evans, Dirk Helbing, Staša Milojević, Alexander M. Petersen, Filippo Radicchi, Roberta Sinatra, Brian Uzzi, Alessandro Vespignani, Ludo Waltman, Dashun Wang and Albert-László Barabási: "Science of science," in: *Science*, vol. 359/6379 (2018), 1–3. Cf. in the same issue the one page review summary.

[89] Ibid., 1.

[90] Ibid., 6.

[91] Ibid., 1, 5.

stantially furthering "our understanding of human imagination by revealing the total pipeline of creative activity."[92]

The authors in *Science* name data availability and "collaborations among natural, computational, and social scientists" as two key factors in the rise of SciSci, which can now "capture the unfolding of science, its institutions, and its workforce."[93] Interestingly, science here is devoid of any historical dimension. Of the 106 citations, only 11 were published before 1990, and 80 were not older than a decade.[94] Even though the paper does not employ a historical perspective, this bibliometric finger exercise reflects both on the explicit and implicit contexts of emergence that it gives. It is, of course, de Solla Price's 1963 work which is introduced as an early study on the "exponential growth in the volume of scientific literature"[95]. Other earlier work is quoted for references to bibliometry and citation indexes (Eugene Garfield, de Solla Price) and classical, apparently non-overcome sociological theorems, such as Robert K. Merton's "Matthew effect" or "the ongoing tension between productive tradition and risky innovation" as described by Thomas S. Kuhn and Pierre Bourdieu in the 1970s.[96]

This narrative presents SciSci as a historical effect of growth and institutionalization. While big science and its undeniable relevance for politics and policies on the one hand and its equally indisputable inclination to produce tremendous administrative hinterlands on the other satiates the epistemological needs of SciSci (as comprised of scientometrics, sociology of science, and innovation studies), it inextricably links the new project to the Cold War era. Recent historiography seems to agree on a further sharp increase in scientization of social systems since the middle of the 20th century, at least from the viewpoint of regimes of modernity from the Northern hemisphere and its colonizing grab for the globe and outer space.[97]

---

[92] Ibid., 5.

[93] Ibid., 1.

[94] Ibid., 6–7. One dates back to the 1950s, three to the 1960s, six to the 1970s and one to the 1980s. Three quoted items were published in the 1990s, and 12 in the 2000s.

[95] Ibid., 1.

[96] Ibid., 2. Cf. Robert K. Merton: "The Matthew effect in science," *Science*, vol. 159 (1968), 56–63, Thomas S. Kuhn: *The Essential Tension: Selected Studies in Scientific Tradition and Change*, Chicago 1977 and Pierre Bourdieu: "The specificity of the scientific field and the social conditions of the progress of reasons," *Social Science Information*, vol. 14/6 (1975), 19–47.

[97] Cf. for a general overview Brückweh et al. (eds.), *Engineering Society* (footnote 4), and Roberto Sala "Verwissenschaftlichung des Sozialen – Politisierung der Wissenschaft? Zum Verhältnis von Wissenschaft und Politik in der Geschichtsschreibung des 19. und 20. Jahrhunderts," *Berichte zur Wissenschaftsgeschichte / History of Science and Humanities*, vol. 40/4 (2017), 333–349. For an introduction into the debate on science colonialism see Londa Schiebinger: "Forum Introduction: The European Colonial Science Complex," *Isis*, vol. 96 (2005), 52–55, and further Laurelyn Whitt: *Science, Colonialism, and Indigenous Peoples. The Cultural Politics of Law and Knowledge*, Cambridge 2009 or Leonardo Viniegra-Velázquez: "Colonialism, science, and health," *Boletín Médico del Hospital Infantil de México*, vol. 77/4 (2020): 166–177. For the colonization of outer space see Albert K. Lai: *The Cold War, the Space Race, and the Law of Outer Space: Space for Peace*, Milton Park 2021 and Alexander C.T. Geppert, Daniel Brandau and Tilmann Siebeneichner (eds.): *Militarizing Outer Space. Astroculture, Dystopia and the Cold War*, London 2021.

Apart from the thus disclosed historical starting point, SciSci shares a second characteristic of Cold War social sciences. Despite all ideological conflicts of the era, and the (mostly institutional) development connected to the great divide between the blocks, the social sciences and especially those dealing with science itself show an example of the permeability of the infamous Iron Curtain or the vast common ground shared on all sides. The ideal relation between science and the state was a joint focal point in both Western and Eastern discourses, which often enough overlapped. Intense debates about rationality, scientific attitude and expert knowledge, about new technologies and applied sciences were happening practically everywhere.[98] These debates were part of a growing institutionalization of scientific self-reflection in form of new disciplines like science studies, philosophy of science, and history of science.[99]

This is the historical background of de Solla Price's multidisciplinary, scientometrics based approach to science policymaking. In March 1965, only a couple of months before Mackay and Bernal referred to him in Warsaw, de Solla Price gave the 1st Annual Lecture of the *Science of Science Foundation*, in which he pointed out a disproportion between growing expertise in science policy and a lacking scientific basis of science policy.[100] Science and technology, "once the condiments of our civilization," had to "be reckoned as the very meat and potatoes of our economy."[101] The growth of science and human knowledge needed statistical backing, specifically in the ongoing transition of the industrial to a knowledge society – it was to "do for science policy

---

[98] From the vast body of literature cf. Eglė Rindzevičiūtė: *The Will to Predict. Orchestrating the Future through Science*, Ithaca, London 2023, Mark Solovey, Christian Dayé (eds.): *Cold War Social Science. Transnational Entanglements*, Cham 2021, Elena Aronova: "Recent Trends in the Historiography of Science in the Cold War," *Historical Studies in the Natural Sciences*, vol. 47/4 (2017), 568–577, Eglė Rindzevičiūtė: *The Power of Systems. How Policy Sciences Opened Up the Cold War World*, Ithaca, London 2016, Elena Aronova, Simone Turchetti (eds.), *Science Studies during the Cold War and Beyond. Paradigms Defected*, New York 2016, Naomi Oreskes, John Krige (eds.): *Science and Technology in the Global Cold War*, Cambridge (MA), 2014, Jamie Cohen-Cole: *The Open Mind: Cold War Politics and the Sciences of Human Nature*, Chicago 2014, Orit Halpern: *Beautiful Data. A History of Vision and Reason since 1945*, Durham, London 2014, Paul Erickson, Judy L. Klein, Lorraine Daston, Rebecca Lemov, Thomas Sturm and Michael Gordin: *How Reason Almost Lost Its Mind: The Strange Career of Cold War Rationality*, Chicago 2013. Specifically on the relation of science and state see Fa-ti Fan: "Science, State, and Citizens: Notes from Another Shore," *Osiris*, vol. 27 (2012), 227–249 and the issue "Science and National Identity" of *Osiris*, vol. 24 (2009), ed. by Carol E. Harrison and Ann Johnson. See also James C. Scott: *Seeing Like a State. How Certain Schemes to Improve the Human Condition Have Failed*, New Haven and London 1998.

[99] Cf. the special issue "The Fate of Disciplines" edited by James Chandler and Arnold I. Davidson in *Critical Inquiry*, vol. 35/4 (2009), 729–1102, esp. Lorraine Daston: "Science Studies and the History of Science" (798–813) and Mario Biagioli: "Postdisciplinary Liaisons: Science Studies and the Humanities" (816–833). See also Theodore M. Porter: "How Science Became Technical," *Isis*, vol. 100/2 (2009), 292–309.

[100] The lecture was delivered at the Royal Institution, London, 25 March 1965. Derek J. de Solla Price: "The Scientific Foundations of Science Policy," *Nature*, vol. 233 (1965), 233–238: 233. Cf. Susan E. Cozzens: "Derek Price and the Paradigm of Science Policy," *Science, Technology & Human Values*, vol. 13/3–4 (1988), 361–372: 363.

[101] de Solla Price: "The Scientific Foundations of Science Policy" (footnote 100), 237.

what economics had done for business."¹⁰² Later that year in Warsaw, Bernal and Mackay seconded this position in making an argument that the ongoing changes in the scientific production of knowledge were as fundamental as those of the scientific revolution and thus required to be reflected in a similar way as Francis Bacon had done, calling for a new logic of scientific discovery.¹⁰³

At the same time, similar conceptions were cast in the Soviet Union.¹⁰⁴ In 1966, cybernetics expert, economist and member of the executive committee of the Academy of the Sciences of the Ukrainian SSR, Gennadij M. Dobrov, published his soon to be classic *Nauka o nauke* (Engl. Science of Science), which was quickly translated into several languages. The epigraph in Dobrov's first chapter quotes the famous aphorism commonly attributed to Francis Bacon's *Meditationes Sacrae* (1597) and accentuates it with an exclamation mark: "Knowledge is power!"¹⁰⁵ In the 1960s, about 350 years later, the author goes on, Bacon's view had been confirmed. Mankind gazed into space and explored the microscopic secrets of life as a consequence of the rampant scientific-technological progress. Scientific development sustained economic growth and cybernetical analysis opened new horizons for the human mind. Science had become an all-encompassing endeavor: "the colossal power of the modern apparatus of science is now directed to studying literally all aspects of reality." However, all the power knowledge had or could give needed control: "[S]cience itself is an object worth of scientific investigation."¹⁰⁶ Dobrov identified philosophy and history of science as the two disciplines that had developed an advanced understanding of the scientific process over centuries. He went on committing them to a joint analysis of "both the scientific process as a whole and scientific activity as a professional activity", mainly by quantitative and structural research methods.¹⁰⁷

While clearly adapting the contemporary idiom of state socialist discourse, much like his Western colleagues, Dobrov's initial paragraphs are universal. After a fair share of references to Marx, Engels and Lenin and the Communist Party of the Soviet Union, his history of science studies leads back to Warsaw twice, first to Maria Ossowska, Stanisław Ossowski and Tadeusz Kotarbiński, and second to Mackay's and Bernal's lecture at the 1965 Congress of the History of Science, which Dobrov

---

¹⁰² Ibid.

¹⁰³ Bernal, Mackay: "Towards a Science of Science" (footnote 1), 10–11; cf. Bacon's call for an "art to invent new arts" in book 9 of his *Advancement of Learning*, and his demand of a new logic of scientific discovery in his *Novum Organon Scientiarium*, book 1, 11.

¹⁰⁴ Cf. Elena Aronova: "The politics and contexts of Soviet science studies (Naukovedenie): Soviet philosophy of science at the crossroads," *Studies in East European Thought*, vol. 63/3 (2011), 175–202 and id.: "Big Science and 'Big Science Studies' in the United States and the Soviet Union during the Cold War," in: *Science and Technology in the Global Cold War*, ed. by Naomi Oreskes and John Krige, Cambridge (MA), 2014, 393–429.

¹⁰⁵ Cf. Dobrov: *Nauka o nauke* (footnote 68), 5. Dobrov's book was translated into Czech (1968), German, Polish, Serbian (1969), and Hungarian (1973).

¹⁰⁶ Ibid.

¹⁰⁷ Ibid., 6.

attended himself.[108] While Dobrov praised the Ossowskis and Kotarbiński for coining the term, he especially highlighted an other contribution of Bernal. Accordingly, Bernal had given the first thorough and comprehensive definition of the problems a science of science would have to take on in his 1939 *The Social Function of Science*.[109] Next to the technocratic programmatic, it was Bernal's further political warnings that received praise. Dobrov highlighted the farsightedness concerning the confidentiality and militarization of science, which, due to deficient organization and information exchange, could threaten the existence of science itself.[110]

Such mutual awareness had a history. In 1964, Stevan Dedijer, Yugoslav physicist and pioneer of business intelligence emphasized the Baconian inspiration and the practical importance of a Science of Science, which in the footsteps of Ossowska and Ossowski and others would not only help organizing research in leading nations, but also be relevant for "underdeveloped countries".[111] Charles Percy Snow's 1959 Rede Lecture took a slightly different route. The *Two Cultures* was not only a profound critique of the British educational system, but also an appraisal of its American and especially Soviet counterparts.[112] From this perspective, the Sputnik crisis of 1957 and following Space Race seemed to be just a matter of time. Twenty years earlier, it had been Bernal to stress that the Soviet Union had first understood the important

---

[108] Kokowski: "The Science of Science (naukoznawstwo) in Poland" [2015] (footnote 56), 181–182. At one panel, both Dobrov and de Solla Price gave presentations. Cf. Gennadij M. Dobrov: "Tendentsii razvitiya organizatsii nauki," *Organon*, vol. 2 (1965), 227–242 and Derek J. de Solla Price: "Regular Patterns in the Organization of Science," *Organon*, vol. 2 (1965), 243–248. However, De Solla Price and Dobrov had already met two years earlier at a symposium on general problems in the history of science and technology in Jabłonna (near Warsaw). For an overview see "Le symposium consacré aux problèmes généraux d'histoire de la science et de la technique Jabłonna (près de Varsovie) de 17 à 21 septembre 1963," *Organon*, vol. 1 (1964), 5–7, for de Solla Price's initial contribution Derek J. de Solla Price: "The History of Science as Training and Research for Administration and Political Decision-Making," *Organon*, vol. 1 (1964), 21–24. The volume contains all other contributions by de Solla Price, Dobrov and the other participants.

[109] Cf. Dobrov: *Nauka o nauke* (footnote 68), 18–19. The author refers to John D. Bernal: *The Social Function of Science*, London: Routledge 1939.

[110] Ibid.

[111] Stevan Dedijer: "The Science of Science: A Programme and a Plea," *Minerva*, vol. 4/4 (1966), 489–504. Dedijer referred both to the first English edition of Ossowska's and Ossowski's programme and to the 1964 reprint in *Minerva*. Cf. ▸ Ossowska, Ossowski: "The Science of Science" (1935), editorial introduction. For further definitions of the Science of Science, he refers to *Zagadnienia Naukoznawstwa*, vol. 1/1 (1965), as well as he mentioned: (1) Stephen Toulmin: "Towards a Natural History of Science," *New Scientist*, vol. 20/364 (1963), 315–316, (2) id.: "Science Policy as a Focus of Academic Study," *Teknisk-Vetenskaplig Forskning*, vol. 35/5 (1964), 155–161, (3) Derek J. de Solla Price: "The Scientific Foundations of Science Policy," *Nature*, vol. 206/4981 (1965), 233–238, (4) id.: "The Science of Science," *Bulletin of the Atomic Scientists*, vol. 21/8 (October 1965), 2–8, and (5) Adam L. Mackay and John D. Bernal: "Towards a Science of Science," *The Technologist*, vol. 2/4 (1966), 319–328. Finally, he referred to an unpublished text by F. Nekola and B. Vobornik ("Nauka o nauke, programa islledovanija roli nauki v sovremenom obshestve" [sic!] Engl. "Science of science, a research programme on the role of science in contemporary society"), likely a Russian translation of J. Nekola, Ladislav Tondl, B. Voborník: *Věda o vědě*, Praha 1964.

[112] Charles Percy Snow: *The Two Cultures and the Scientific Revolution*, Cambridge 1959.

connection of technological innovation and (social) planning to overcome "the incredible muddle and confusion of present-day application of science". Bernal highlighted that the "relatively vast sums of money spent" for the development of science organization and scientific education had not been unrecognized in England, the USA, France, Japan and other countries.[113] However, Bernal continued, the underlaying groundwork of dialectic materialism was only fully realized in Western Europe in 1931. At the *International Congress for the History of Science and Technology* in London, the soviet delegation "showed what a wealth of new ideas and points of view for understanding the history, the social function, and the working of science could be and were being produced by the application to science by Marxist theory."[114] Of all soviet delegates, it was especially Boris Hessen with his talk on the social and economic roots of Newtons *Principia* that made an immediate impact.[115]

Having met the soviet delegates in London, Bernal kept contact with chef de mission Nikolai I. Bukharin, who since 1932 headed the Institute for the History of Science and Technology at the Academy of Sciences in Moscow. Between 1931 and 1936 Bukharin was also editor in chief of *SoReNa*, a popular journal that covered politics and theory of science, for which Bernal wrote an article on X-ray crystallography.[116] During that time, in Nauka Polska, the Soviet Union surely was an issue for reports on science policy and educational matters. However, apart from some reviews, soviet science did not seem to make a greater impression, not even as a negative example. In Nauka Polska, Hessen's text was not reviewed, at least not in the volumes preserved.[117]

---

[113] John D. Bernal, *The Social Function of Science*, London: Routledge 1946 [1939], 392–393.

[114] Ibid.

[115] Cf. Boris Hessen: "The Social and the Economic Roots of Newton's 'Principia,'" in: *Science at the Crossroads. Papers presented to the International Congress of the History of Science and Technology 1931 by the Delegates of the U.S.S.R.*, ed. by Nikolai I. Bukharin, London 1931. For the impact of Hessen's talk cf. the editors' introduction "Classical Marxist Historiography of Science: The Hessen-Grossmann-Thesis," in: *The Social and the Economic Roots of Newton's 'Principia': Texts by Boris Hessen and Henryk Grossmann*, ed. by Gideon Freudenthal and Peter McLaughlin, New York 2009, 1–40: 26–34. Next to Bernal, also de Solla Price credited Hessen (Derek De Solla Price: "The Science of Science," in: *The Science of Science. Society in the Technological Age*, ed. by Maurice Goldsmith and Alan L. Mackay, London 1964, 195–208: 203.) and Robert K. Merton published his reaction some years after (Robert K. Merton: "Science, Technology and Society in Seventeenth Century England," *Osiris*, vol. 4 (1938), 360–632).

[116] SoReNa is an acronym for *Socialisticheskaja rekonstrukcija i nauka* (Engl. Socialist Reconstruction and Science). Bernal's article is mentioned in Andrew Brown and Alan L. Mackay: "J. D. Bernal and the replication of the genetic material – hindsight for foresight," *Journal of Biosciences*, vol. 30/4 (2005), 407–409: 409. For a descriptive overview on SoReNa cf. Sergej P. Strekopytov: "Zhurnal 'socialisticheskaja rekonstrukcija i nauka' ('SORENA') kak istochnik po istorii organizacii nauki v sisteme VSNH – Narkomtjazhproma SSSR (1931–1936 gg.), in: *Vspomogatel'nye istoricheskie discipliny*, ed. by Valerij A. Shishkin, Leningrad 1991, 73–87. For the institute cf. Aronova: "The politics and contexts of Soviet science studies (Naukovedenie)" (footnote 104), 181–183.

[117] However, vol. 23 contains a review of a preliminary version of Merton's approach (Robert Merton: "Some Economic Factors in Seventeenth Century English Science," *Scientia*, vol. 62/305-9, 142–52.): J. D. "Recenzja: Robert K. Merton: *Some Economic Factors in Seventeenth Century English Science*," *Nauka Polska*, vol. 23 (1938), 341.

Western, non-communist scholars seemed to be more crucial for the projects in Warsaw.[118] Henryk Grossman (1881–1950), a Polish born economist, statistician and historian, who developed an argument akin to Hessen's in 1935[119], was not discussed either. However, he had been a discussant in the panel "Poland as a topic in the humanities" at the *Congress for the Organisation and Progress of Polish Science* in April 1920.[120] Grossman then worked at the *Central Statistical Office* (Pol. Główny Urząd Statystyczny) and later taught at the Free Polish University in Warsaw. Facing continuous harassment and several arrests for his communist activities, he left Poland in 1925 and settled at the Institute for Social Research in Frankfurt.[121]

The list of projects for systematic collection and organization of science could be extended with many more examples. The body of Nova Organa Scientiarum – attempts to unite knowledge production and universify the languages of its production – has grown ever since Francis Bacon adapted the Aristotelian term. In interwar Warsaw, the capital of a newly founded national state, a specific mix of (intellectual) traditions and interests formed its very own project. The Mianowski Fund formed a specific forum, on which people with very different backgrounds assembled: offshoots of Polish Romanticism and Positivism with the sincere Lvov–Warsaw School of (Analytic) Philosophy, conservatives, liberals and left-leaning scholars, some of them nomenclatura to be in the People's Republic after 1945.

The Fund, together with its journals and working groups, may serve as a historical looking glass into the overwhelmingly dynamic world of science reflection. Starting from the idea of an inventory at the 1920 Congress, initially, the journal Nauka Polska took a step back from the ordering principle of encyclopedic projects such as the

---

[118] The comprehensive, systematic index of the first 20 volumes of Nauka Polska (cf. footnote 19, 22–26) lists only three texts on "Soviet Russia" in the section *Scientific Organisation and Life Abroad*, but more than ten each for England/Great Britain, Germany, Italy and the USA. All together 24 countries were covered. In the four following volumes published until 1939, only one article on the Soviet Union was added (Józef Reutt: "Akademia Nauk ZSSR o potrzebach i organizacji nauki sowieckiej," *Nauka Polska*, vol. 22 (1937), 257–260), but several for each of the other countries. An analysis of the ongoing bibliographical section on international science studies literature (Pol. Bibliografia Naukoznawcza) might produce further context. For the strong inclination to Western, and especially German literature cf. ▶ Ossowska, Ossowski: "The Science of Science" (1935). For the role of Russian as a language cf. ▶ Jan Surman: "Internationality in *Nauka Polska*: Infrastructure, Translation, Language", 49, 50.

[119] Henryk Grossmann: "Die gesellschaftlichen Grundlagen der mechanistischen Philosophie und die Manufaktur," *Zeitschrift für Sozialforschung*, vol. 4/2 (1935), 161–231 and id.: "The Social Foundations of the Mechanistic Philosophy and Manufacture," in: *The Social and the Economic Roots of Newton's 'Principia': Texts by Boris Hessen and Henryk Grossmann*, ed. by Gideon Freudenthal and Peter McLaughlin, New York 2009, 103–156. Supposedly, Grossmann learned about Hessen's text only in 1938. Cf. Freudenthal and McLaughlin: "Classical Marxist Historiography of Science" (footnote 115), 1.

[120] Józef Ujejski: "Polska jako przedmiot nauk humanistycznych," *Nauka Polska*, vol. 3 (1920), 153–174: 173.

[121] Rick Kuhn: "Henryk Grossman: A Biographical Sketch," in: *The Social and the Economic Roots of Newton's 'Principia': Texts by Boris Hessen and Henryk Grossmann*, ed. by Gideon Freudenthal and Peter McLaughlin, New York 2009, 239–252: 243–245.

Handbook to Autodidacts series (see above) in Poland or the *International Encyclopedia of Unified Science* of Otto Neurath and others.[122] Before (re-)ordering knowledge, the first three volumes of Nauka Polska were meant to assess the state of the art of science and learning as the subtitle stated: the "Needs, Organisation, and Progress" of science in Poland. The Science Studies Circle was installed only several years later to systematize efforts. Only then comparisons to the Brussels *Mundaneum* of Paul Otlet and Henri La Fontaine, who had been serious about the practical organization of research for some time, would fit, as they would to the *International Committee on Intellectual Cooperation* of the League of Nations, in which Marie Skłodowska-Curie was serving.[123]

When looking back from 2018 SciSci through the conceptions of de Solla Price, Dobrov and Bernal into the era before the Second World War, it seems that this very influential strain of science studies over time dismissed some aspects that were dear to the sources of their own genealogical stories. Dobrov's account of the historical and statistical primacy leaves out the sociological, cultural and psychological dimensions of science that Florian Znaniecki and Ossowska and Ossowski had advocated for in the 1920s and 1930s, and which seem to be emblematic for the Science Studies Circle. However, it is this observation that helps establishing the breadth of the history of science studies, which spans across the whole 20th century at least. Apart from quantitative and structural analysis, and questions of application, it covers much broader contexts than Dobrov could dream of in 1966, as shows alone the development of Soviet *naukovedenie* or *Wissenschaftswissenschaft* in the GDR.[124] In Poland, discussions about the theoretical closure, institutional outline and applicability of naukoznawstwo, which was never a homogenous undertaking, but rather a field, reached well into the 1980s.[125]

At this point, other, new, and quite different initiatives had taken off elsewhere. The broad field of STS as we know it today had started to form, prominent feminist and postcolonial critiques of science gained ground, and historical epistemology emerged from the classic history of science. For over half a century now, these different strands have diversified and converged at times, while constantly engaging in more or less fierce boundary work against their subjects and each other.[126] Among

---

[122] Otto Neurath, Rudolf Carnap and Charles W. Morris (eds.): *International Encyclopedia of Unified Science (vol. I/1–5)*, Chicago 1938.

[123] For the relations of Mundaneum and the League of Nations Committee cf. Eva Hemmungs Wirtén: *Making Marie Curie. Intellectual Property and Celebrits Culture in an Age of Information*, Chicago, London 2015, ch. 4.

[124] Cf. for the Soviet Union Aronova: "The politics and contexts of Soviet science studies (Naukovedenie)" (footnote 104), for the GDR Friedrich Cain: "Authority Claims. Situating Socialist Science Studies in the GDR," *Berichte zur Wissenschaftsgeschichte / History of Science and Humanities*, vol. 44/4 (2021), 352–372.

[125] Cf. Kokowski: "The Science of Science (naukoznawstwo) in Poland" [2016] (footnote 8), 164–167. See also Bohdan Walentynowicz (ed.): *Polish Contributions to the Science of Science*, Boston, 1982 and Gasparski, Pszczołowski (eds.): *Praxiological Studies* (footnote 57).

[126] See for example the instructive analysis of Karl R. Popper, Robert K. Merton and Thomas S.

others, questions of boundary work are also of interest for recent research focusing on the political epistemologies in which science – and science studies – develop.[127]

***

Editing this volume has taken some time. The editors like to thank everyone, who embarked on the journey. Tul'si (Tuesday) Bhambry has more than fulfilled the expectations with her brilliant translation and also helped with copy editing. The guests at a workshop in Konstanz helped creating a productive atmosphere delivering papers and discussing, some of them contributed as authors to this volume. Jonas Brüderlin, Max Gathemann and Simon Leimeister took care of the workshop infrastructure in Konstanz and were great support in the early editing process, as were Jonas Brüderlin, Klara Valentina Fritz, Lorenz Hartung, Paul Stoll, and especially Carmen Wójcik in the final process. We thank Christopher Kasparek, who translated the text by Florian Znaniecki and supported the reprint, as well as Znaniecki's grandchildren, Theodora Menasco and Stefan Lopata. Leszek Zasztowt was very generous with source material and photographs from the Archive of the Polish Academy of Science. He also supported the project without hesitation as the director to the Mianowski Fund. All this would not have been possible without the financial support of the Centre of Excellence *Cultural Foundations of Social Integration* at the *University of Konstanz*, making possible the workshop, translations, and printing the book. Finally, Stephanie Warnke-De Nobili, Kendra Mäschke, Susanne Mang and Tobias Stäbler helped at various stages in the process of finishing the book for printing at the publishing house.

Kuhn in Thomas F. Gieryn, "Boundaries of Science," in: *Science and the Quest for Reality*, ed. by Alfred I. Tauber, Houndmills et al. 1997, 293–332: 294–305.

[127] Cf. the special issue "Scientific Authority and the Politics of Science and History in Central, Eastern, and Southeastern Europe" of *Berichte zur Wissenschaftsgeschichte / History of Science and Humanities*, vol. 44/4 (2021) ed. by Friedrich Cain, Dietlind Hüchtker, Bernhard Kleeberg, Karin Reichenbach and Jan Surman.

Part I:
Historical Contexts

Internationality in *Nauka Polska*: Infrastructure, Translation, Language

Jan Surman

The end of the First World War brought about serious changes for the scientific landscape of Europe. The breakdown of the Empires meant a reconfiguration of state boundaries, crossing through the colourful Central European linguistic canvas, and a rewriting of the fine threads linking them. Rewriting in two senses, both as future-oriented reorientation from the previous imperial linkages as well as historicization of previous contacts in nationalist wording, concentrating on imperial dependence. Changes in the scholarly landscape were deep, influencing not only institutions, but also international cooperation and finally also research programs, especially those which had previously relied on the political or personal infrastructure of the empire.[1] The late nineteenth century had already seen the strengthening of an internationalist agenda that included the imperially governed Polish community. This agenda now received another push, often joined with visions of science as a panacea for nationalism and a means of preventing new wars.[2] The Polish government as well as Polish academic institutions seized the opportunities independence offered. The beginning of the new statehood intensified efforts to have Poland represented at international meetings of all sorts and in the newly established associations. Władysław Natanson, the representative of the Polish Academy of Sciences and Arts in Cracow (henceforth PAU), attended the founding meeting of the *Conseil international de recherches*, and the presence of Marie Curie-Skłodowska and Oskar Halecki made Polish interests loudly heard at the League of Nation's *International Committee on Intellectual Cooperation* (ICIC).[3] From 1922 onwards, the *Józef Mianowski Fund for the Promotion*

---

[1] See, e.g. Deborah Coen on the example of Habsburg-Slovenian seismologist Albin Belar: Deborah R. Coen: *The Earthquake Observers: Disaster Science from Lisbon to Richter*, Chicago 2013, 159–160.

[2] Warden Boyd Rayward (ed): *Information beyond borders. International Cultural and Intellectual Exchange in the Belle Époque*, Farnham, Surrey 2014. Already at this time this view was strongly supported by philanthropic organizations, see Helke Rausch and John Krige (eds): *American Foundations and the Coproduction of World Order in the 20th Century*, Göttingen 2012.

[3] Małgorzata Willaume: "Udział Polski w pracach Union Académique Internationale (1919–1939)," *Annales Universitatis Mariae Curie-Skłodowska* (Sectio F: Historia), vol. 46/47 (1991/1992), 369–386: 370; Jan Piskurewicz: *Między nauką a polityką. Maria Skłodowska Curie w laboratorium i w Lidze Narodów*, Lublin 2007, 85–222. Halecki was regarding himself not only as a representative of the Polish state, but of Central Europe altogether: Andrzej M. Brzeziński: "Oskar Halecki – rzecznik krajów Europy Środkowej i Wschodniej w Komisji Międzynarodowej Współpracy Intelektual-

of Science (Pol. Kasa im. Józefa Mianowskiego – Fundacja Popierania Nauki) represented ICIC's liaison in Poland, and in 1923, when the National Commission of ICIC was established, it was closely affiliated with the Warsaw-based Mianowski Fund.[4] Responsibility for international relations was in the hands of the government. The Mianowski Fund, which had been established to grant scholarships and publication support, turned into an intellectual think-tank organizing and publishing discussions on science, and became an important voice on all scholarly topics. Thus an analysis of its publishing organ, *Nauka Polska*, gives important insights into how Polish academics perceived internationality and what sorts of projects they associated with it, both on a concrete institutional level and on a theoretical or intellectual level.

But internationalism and internationalization can signify many different things. They can imply a transnational endeavour, at least officially pursuing communication without (political) borders, as imagined by the savants of the Republic of Letters.[5] They can also suggest Olympic Internationalism,[6] where nations compete in bloodless battles for the honour of being the most cultured. In the interwar period, according to Geert Somsen, another turn took place: While science reflected the greatness of a nation in the Olympic model, in this time it began more and more to be presented as 'an *antidote* to nationalist sentiments'.[7] (Of course, fascist states presented an exception to this rule). Paul Otlet's *Mundaneum* or Henri Léon Follin's *République universelle* (later *supranationale*) are well-known examples of this transnationalist project which in comparison to ICIC for instance relied more on the supranationality of ideas than on the cooperation of their national exponents.[8]

But how does Central Europe, the paradigmatic realm of nationalism, fit into this picture, apart from the fact that a *Mundaneum* branch called *Poloneum* was planned

---

nej Ligi Narodów (1922–1925)," *Studia z Dziejów Rosji i Europy Środkowo-Wschodniej*, vol. 48 (2013), 141–156.

[4] Andrzej M. Brzeziński: "Polska w systemie międzynarodowej współpracy intelektualnej Ligi Narodów (1922–1939)," in: *Dzieje najnowsze*, vol. 34/2 (2002), 3–22: 4. Ironically, PAU, which was best linked internationally during the Habsburg period, was allowed to represent Polish science in international organizations only 1921, and although still important, lost the function of primary representative of Polish science abroad, see Danuta Rederowa: "Formy Współpracy Polskiej Akademii Umiejętności z zagranicą (1872–1952)," *Studia i Materiały do Dziejów Nauki Polskiej* (Seria A: Historia Nauk Społecznych), vol. 10 (1966), 77–171: 83.

[5] On republic of letters see Hans Bots, Françoise Waquet: *La République des* Lettres, Paris, Bruxelles 1997, and Lorraine Daston: "The Ideal And Reality of The Republic if Letters in the Enlightenment," *Science in Context*, vol. 4 (1991), 367–386.

[6] Geert Somsen: "A History of Universalism: Conceptions of the Internationality of Science, 1750–1950," *Minerva*, vol. 46 (2008), 361–379. The Olympic metaphor comes from Paul Forman: "Scientific Internationalism and the Weimar Physicists: The Ideology and Its Manipulation in Germany after World War I," *Isis*, vol. 64 (1973), 151–180.

[7] Somsen: "History of Universalism" (footnote 6), 370.

[8] See Jolles Gillen et al. (eds): *Paul Otlet, fondateur du Mondaneum (1868–1944): Architecte du savoir, Artisan de paix*, Brussels 2010, on Follin see Stéphanie Manfroid and Jacques Gillen: "Les papiers personnels de Paul Otlet," in: ibid., 51–73: 66–73.

but eventually not built in Warsaw?[9] Surprisingly, the relation between nationalism and internationalism has not yet seen deeper inquiry, though international relations have always been noted as a vital component of the scientific landscape of Central Europe, be it Poland or Czechoslovakia. Internationalism may in fact never have been absent (for long) – it was merely nationalized, which is to say that the benefits of translingual contacts were seen as necessary for national development, especially as science grew in the nineteenth century to be a major source of national pride.[10] When the 'Polish' economic boycott of Prussia in 1908 was expanded to include culture, scholarship was explicitly excluded, because the resulting scientific isolation was considered detrimental.[11] Stanisław Madeyski-Poray (1841–1910), professor of civil law in Cracow and from 1893 to 1895 Habsburg Minister of Education and Religion, wrote on this occasion:

> This is precisely the essence of the nation, that after attaining higher civilization through the use of the native tongue, it finds means to develop fully the spiritual forces that lie in its distinct national character. Given the continuous development of civilization, a nation can maintain its position among the nations if it contributes, together with them and with results in its own spirit, to civilization's general development.[12]

Somsen writes that in the 'West' socialist internationalism only developed in the interwar period. But in the 'East' it arose earlier, simply because in imperial times scholars were able to address particularist aims by appealing to universalist science. This appeal to universal value of science grew strong in the non-imperial languages in nineteenth century as a response to claim of especially German that only science produced and published in "Kultursprachen" (mostly German and French) was of value. But Madeyski-Poray's phrasing also echoes the model of *Res publica litteraria*, which had fostered communication among the intellectuals of the Age of Enlightenment and was definitely outdated in Madeyski-Poray's time. The ways in which internationalism was understood by different people at different times deserves careful

---

[9] Leonard Grochowski: "Udział polski w międzynarodowym ruchu pedagogicznym w okresie międzywojennym," *Rozprawy z dziejów oświaty*, vol. 27 (1984), 177–211: 199. More on the conception of regional branches of *Mundaneum* in Nader Vossoughian: "The Language of the World Museum: Otto Neurath, Paul Otlet, Le Corbusier," *Transnational Associations*, vol. 1–2 (2003), 82–93.

[10] Julian Dybiec: *Nie tylko szablą: Nauka i kultura polska w walce o utrzymanie tożsamości narodowej 1795–1918*, Kraków 2004; Jan Surman: "Science and its publics. Internationality and National Languages in Central Europe," in: *The Nationalization of Scientific Knowledge in the Habsburg Empire, 1848–1918*, ed. by Mitchell G. Ash und Jan Surman, Basingstoke 2012, 30–56. See also, on the politically prominent issue of German-Polish relations, Julian Dybiec: "Prześladowca i nauczyciel. Niemcy w nauce i kulturze polskiej 1795–1918," in: *Literatura, kulturoznawstwo, uniwersytet. Księga ofiarowana Franciszkowi Ziejce w 65. Rocznicę urodzin*, ed. By Bogusław Dopart, Jacek Popiel and Marian Stala, Cracow 2005, 455–468.

[11] See the discussions in *Przegląd Powszechny*, vol. 100 (1908), 1–24, with several scholars commenting on the issue; most speakers agreed that the boycott should not include scientific publications, instruments, participation in congresses and organizations, or studies at German universities.

[12] Ibid., 24.

analysis. Besides my present study, further research should examine the historical semantics and trace the use and discursive embedding of this term.

In this article I will illustrate this nationalization-internationalization discussion on the examples from articles printed in *Nauka Polska* – from early postwar discussions on the organization of science in the newly established state up to the 1930s, when the topic of internationalization largely disappeared from the pages of the journal. I want to concentrate on how scholars imagined international contacts, how they placed Polish science in the European and global arena, and finally on the theoretical models they applied while describing international relations. I give special attention to the question of translation, which in my view paints a neat picture of the complexity of the inter/nationality topic, since it addresses the dialectics of openness and closeness of Polish science. This article does not intend to look deeper into the prehistory or archaeology of science theorists and sociologists who were active at the time, such as Ludwik Fleck, Maria and Stanisław Ossowski or Florian Znaniecki. But the debates referenced here do add another facet to discussions on the situatedness of Polish *naukoznawstwo* (science studies), for instance on the motives behind Fleck's stressing of openness in interlingual communication.[13]

One issue has to be kept in mind: Poland, more precisely the Second Polish Republic, was a pluricultural state with peoples of several nationalities coexisting, communicating and even freely exchanging national 'masks,' or assumed identities.[14] Stefan Baley – also known as Stepan Maksym Volodymyrovyč Balej – who I will discuss later in this article, was one of many who identified with more than one linguistic group.[15] With an ethnic minority population of around 30 % – according to official censuses[16] – the Second Polish Republic was a multicultural amalgam, and Piłsudski's government endorsed this diversity, at least officially.[17] In *Nauka Polska*, inter-

---

[13] Cf. Deborah Coen: "Rise, Grubenhund: On Provincializing Kuhn," *Modern Intellectual History*, vol. 9 (2012), 109–126, who links Fleck's openness with the specific situation of the Habsburg Empire.

[14] Polish interwar writer Witold Gombrowicz sees masks as social, cultural, or intellectual positions one puts on while confronting others. In comparison to other social metaphors about his process, like Erving Goffman's Theatre, Gombrowicz's notion of the mask did not imply an authentic self beneath, which is why he is often associated with postmodernism. In this sense it is reminiscent of Brubaker & Coopers's idea of identification. See Witold Gombrowicz, *Ferdydurke*, transl. by Danuta Borchardt, New Haven 2000; Rogers Brubaker, Frederick Cooper: "Beyond 'Identity'," *Theory and Society*, vol. 29 (2000), 1–47.

[15] M. M. Vernikov: „Zhittja i naukova dijal'nist' akademika Stepana Baleja," in: *Akademik Stepan Balej. Zibrannja prac*, vol. 5, L'viv, Odessa 2002, 19–79.

[16] Henryk Chałupczak and Tomasz Browarek: *Mniejszości narodowe w Polsce: 1918–1995*, Lublin 1998, 21–22.

[17] As argued by Patrice M. Dabrowski, even if paradigmatic 'nationalizing states' as Poland, there were possibilities for multiculturalism (Patrice M. Dabrowski: "Multiculturalism, Polish style: glimpses from the interwar period," in: *Understanding multiculturalism. Central Europe and the Habsburg experience*, ed. by Johannes Feichtinger and Gary B. Cohen, New York 2014, 85–100), cf. also Anna Landau-Czajka: *Syn będzie Lech... Asymilacja Żydów w Polsce międzywojennej*, Warszawa 2006.

national relations were also discussed frequently, though internal cultural plurality was not addressed at all. This appears surprising, as some commentators openly celebrated diversity as productive but at the same time did not recognize the importance of contacts between scholars identifying with different nationalities within one state. But one must acknowledge the complexities behind the fact that they overlooked internal cultural variety with all its problems and merits. These were in the first place the long-lasting cultural hierarchies and internal colonialism, as well as the more current question of acculturation. But in fact they pointed more toward the problematic status of plurilinguality and pluriculturality at the time, and toward the issue of accepting communicational variety and equality of languages. Polish scholars who commented on science abroad openly endorsed such ideals. Yet it seems that where internal plurality in Poland was concerned, the discussion was overshadowed by political and cultural tensions, even if these conflicts did not openly challenge the idea of science's universalism. These intriguing double standards in the discussion had already marred Polish-Ruthenian relations a half-century earlier, when Polish scholars claimed the importance of plurilinguality when addressing their German or French counterparts, but stressed the importance of having one leading literary language for scholarly publications while referring to Ruthenian scholars.[18]

From debates on dependence to intentional internationalization: Nauka Polska on its way to Organon

There were several moments when internationality was debated, beginning with the aftermath of the First World War. Already in 1916, the Mianowski Fund organized a survey on how to perfect science within the new political structures; it also organized a section on Polish and international sciences on the last day of the 1922 *First Congress on the Organization and Development of Polish Science* (Pol. I Zjazd poświęcony zagadnieniom organizacji i rozwoju nauki polskiej).[19] In the early years of *Nauka Polska*, post-imperial unification made visible the disagreements between scholars who had been socialized and worked in different empires.[20] Internationality was addressed from a utilitarian perspective and the spirit of universality hovered not only above the natural sciences, medicine or classical philology – disciplines commonly acknowledged as international endeavours – but also for example over philosophy, which Tadeusz Kotarbiński called 'a general, international terrain' where scholars'

---

[18] Jan Surman: "Symbolism, communication and cultural hierarchy. Galician discourses of language hegemony at the beginning of the second half of the nineteenth century," *Historyka. Studia Metodologiczne*, vol. 42 (2012), 151–174.

[19] On the survey see "Wstęp," *Nauka Polska*, vol. 1 (1918), I–XVI. Proceedings from the conference were reprinted in 3rd number of *Nauka Polska*.

[20] One of such topics was university reforms, see Urszula Perkowska: "La genèse et la caractéristique de la loi sur les écoles supérieures du 13 juillet 1920," *Zeszyty Naukowe Uniwersytetu Jagiellońskiego. Prace Historyczne*, vol. 79 (1985), 95–107.

work is based on 'a general, European tradition'.[21] One key concern were how to make international contacts fruitful for the development of Polish science. Another priority was to balance the inward, pro-autarchic struggle for national identity against the outward call for a universalist agenda and intercultural exchange. Change came with the journal's tenth anniversary in 1928 and the editors' declaration of success of the first decade of Polish science, which they attributed to the 'ever closer contact of Polish culture with international culture.'[22] From now on, internationalism becomes less and less prominent in debates and controversies; it comes to be a part of everyday scholarly practice, causing *Nauka Polska* to become a hub for international science of science. The discussion between Adam Krokiewicz and Hermann Noack about the latter's 1936 book *Symbol und Existenz der Wissenschaft*, where the German national-socialist philosopher answered to a Polish-language review defending himself against the accusation of nationalism, shows that this idea was realized with some success.[23]

But the early re-thinking of Polish science in international terms stood under the motto of unification and autonomization, with post-imperial traits like nativization and ethnicization on the agenda. Characteristic here was criticism of Polish science's dependence on imperial powers, their media and styles of thought and the subsequent claim that in order to meet the demands of the independent state, Polish science should be independent as well, if not even autarchic. The anthropologist Jan Czekanowski wrote about Polish science as a 'subsidiary to foreign science' especially German and Russian and compared it to a 'dependent province.'[24] Romuald Minkiewicz similarly stressed in terms reminiscent of dependency theory that Polish science should not be 'taking from foreigners, storing and using of foreign influences, but making a great Polish contribution to the fount of human knowledge.'[25] But criticism of previous conditions did not boil down to mere complaints – it also offered rhetorical support for political claims, like the one calling for the autonomy of scientists and scientific institutions, where previous politicization of scholarly en-

---

[21] Tadeusz Kotarbiński: "W sprawie potrzeb filozofji u nas," *Nauka Polska*, vol. 1 (1918), 443–451: 443. Many of his contemporaries would, however, not agree.

[22] "Coraz ściślejsza styczność kultury polskiej z kulturą międzynarodową". Redakcja: "Po dziesięciu latach," *Nauka Polska*, vol. 10 (1929), XXIII–XXVIII: XXIV.

[23] Adam Krokiewicz: "[Review:] Hermann Noack 'Symbol und Existenz der Wissenschaft. Untersuchungen zur Grundlegung einer philosophischen Wissenschaftslehre,' Halle/Saale: Max Niemeyer Verlag (1936)," *Nauka Polska*, vol. 23 (1938), 278–288; Hermann Noack: "[Letter to the editors]," *Nauka Polska*, vol. 24 (1939), 523–525, Adam Krokiewicz: [Letter to the editors and answer to Noack], ibid., 525–526. On Noack's political allegiance see Jens Thiel: "Von 'ärgerlichen Äußerlichkeiten' und 'innerlichem Unberührtsein'. Hermann Noack im 'Dritten Reich'," in: *Philosophie im Nationalsozialismus*, ed. by Hans-Jörg Sandkühler, Hamburg 2009, 253–269.

[24] Jan Czekanowski: "W sprawie potrzeb nauk antropologicznych w Polsce," *Nauka Polska*, vol. 1 (1918), 201–223: 202.

[25] Romuald Minkiewicz: "O polską twórczość naukową," *Nauka Polska*, vol. 1 (1918), 503–514: 503.

deavours under imperial conditions was mentioned as a warning against politicians' ideas of close ministerial supervision over academic institutions.[26]

However, it was not the dependence, but pluralism of international relations that was most strongly emphasized. This was by no means unsurprising, since 'Polish' scholars after the First Word War were socialized and had worked in many different places in the world, and many were now returning from France, Switzerland or Belgium.[27] Not all scholars returned to Poland, and of those who did, not all stayed permanently. Marie Curie-Skłodowska is the most prominent person who decided to remain abroad. The question of how to help returning scholars establish careers and working conditions comparable to the ones they had abroad was frequently discussed in the first years of interwar period.

Efforts at pluralization included descriptions of research structures and organizations in foreign countries. These were meant to allow to compare how science works in different political and social contexts. First the program of the Ministry for Religious Denominations and Public Enlightenment was launched while academic reforms were planned. There followed reports sent directly to *Nauka Polska*.[28] Not only scientific superpowers like France, Great Britain, Germany, Italy or the United States were scrutinized (the last report was written by Znaniecki himself), but also China, Belgium, Bulgaria, Greece, Hungary or Switzerland. In most cases these reports scrutinized both state and private science funding and described the most important institutions. Clearly, the question how to organize efficient research structure and how the scholarly productivity related to state supervision and international relations were the leading issues for the editors and that was what the authors were paying special attention to. Only in rare cases authors provided more depth analysis of history of respective sciences systems, when they helped to understand the current situation.[29] *Nauka Polska* also took a clear stance towards educational questions. Concerns about students' diminishing foreign language skills were voiced frequently, and several petitions to the government on this issue were discussed and supported in the journal. Early language learning (gymnasia and below) and foreign scholarships were seen as a special necessity. In 1920 Michał Siedlecki proposed that only those students should be admitted to doctoral programmes who were reasonably

---

[26] See, most clearly in Michał Siedlecki: "Nauka polska na terenie międzynarodowym," in: *Nauka Polska*, vol. 3 (1920), 188–196: 189.

[27] Dorota Mycielska: "Drogi życiowe profesorów przed objęciem katedr akademickich w niepodległej Polsce," in: *Inteligencja polska pod zaborami: studia*, ed. by Ryszarda Czepulis-Rastenis, Warszawa 1979, 243–290.

[28] Bohdan Jaczewski: *Polityka naukowa państwa polskiego w latach 1918–1939*, Wrocław 1978, 60–66.

[29] See, e.g. on Spain: Eugeniusz Frankowski: "Organizacja Nauki w Hiszpanii," *Nauka Polska*, vol. 14 (1931), 213–262, on the crisis of universities in 17th–18th centuries 221–226; on Italy, with special attention to entanglement of historiography, philosophy and educational policy Bohdan Kieszkowski: "Organizacja Nauki w Hiszpanii," in: *Nauka Polska*, vol. 19 (1934), 185–199, here 192–199.

well-versed in English, French and German literatures on their topics.[30] No hard data was presented on scholars' knowledge of languages. Only in a discussion of a 1930 survey of literary writers, printed in *Nauka Polska* in 1933, the reviewer mentioned a significant drop in foreign language skills and experience abroad from generation to generation.[31] One can easily imagine that similar processes occurred among scholars and scientists, even if substantial efforts were undertaken to prevent it.

The problem of internationalization was certainly not limited to an understanding of Poland as a mere recipient of ideas. Instead, many scholars foresaw Poland's active role as a forerunner of internationalist agendas.[32] In this initial period ideas for internationalist endeavours were already formulated: an international series in classical languages was proposed, as well a mathematical journal that should include translations of less known works "written in 'not-international' languages [...] particularly Polish works, which go wasted because unnoticed."[33] But the author of this statement, Zygmunt Janikowski, thought differently than his predecessors who had been publishing a general 'Polish' journal in foreign languages – the *Bulletins Internationaux*.[34] For him, the growth and differentiation of the mathematical sciences required specialized journals and these ought to be international. *Fundamenta Mathematicae*, established in Warsaw in 1920, is considered the first specialized mathematical journal (here for set theory and its applications). It was the outcome of his proposals.[35]

An article on the needs of Polish biology and embryology by Emil Godlewski junior shows certain tensions between international cooperation and national pride.[36] The issue at stake was the establishment of a Polish maritime zoological institute (one of the questions asked in the 1916 survey), supported by several scholars, including the previously mentioned Siedlecki.[37] Godlewski, however, opposed the project. In

---

[30] Siedlecki: "Nauka polska" (footnote 26), 193.

[31] Tadeusz Makowiecki: "[Review] 'Życie i Praca pisarza polskiego. Na podstawie ankiety Związku Zawodowego Literatów Polskich w Warszawie', Warszawa 1932," *Nauka Polska*, vol. 17 (1933), 358–362: 361.

[32] The idea of smaller nations taking up leading roles in internationalisation processes was quite popular among intellectuals in these nations, see Rebecka Lettevall, Geert Somsen and Sven Widmalm (eds): *Neutrality in Twentieth-Century Europe: Intersections of Science, Culture, and Politics after the First World War*, New York 2012.

[33] Zygmunt Janiszewski: "O potrzebach matematyki w Polsce," *Nauka Polska*, vol. 1 (1918), 11–18.

[34] This journal was published by the Polish Academy of Sciences and Arts in Cracow from 1889 in 'world languages,' which by the time were mostly French and German. It was later divided into one publication in the sciences and one in the humanities. For more details see Krystyna Stachowska: "Z działalności wydawniczej Polskiej Akademii Umiejętności. Starania o upowszechnienie za granicą polskiej myśli naukowej w latach 1873–1952," *Rocznik Biblioteki Polskiej Akademii Nauk w Krakowie*, vol. 19 (1973), 39–71.

[35] Małgorzata Przeniosło: "'Fundamenta Mathematicae' – pierwsze polskie czasopismo matematyczne o wąskiej specjalizacji (1920 – 1939)," *Nauka*, vol. 2 (2006), 167–184. According to Kazimierz Kuratowski Janikowski expressed this idea already 1912 (Ibid, p. 169).

[36] Emil Godlewski, mł.: "O potrzebach biologii i embrjiologii," *Nauka Polska*, vol. 1 (1918), 193–200.

[37] Michał Siedlecki: "Potrzeby nauki polskiej w zakresie zoologii," *Nauka Polska*, vol. 1 (1918),

his eyes the creation of a separate Polish maritime station would jeopardize the intense cooperation with the Zoological Station in Naples (now *Statione Zoologica Anton Dohrn*), a privately owned station that was commonly regarded as 'German' in Polish discourse. And this would bring more harm than good, since "the station in Naples always had so many workers, especially in the spring, that it became something of a continuing international congress. This permanent contact with natural scientists from all over the world made it possible to exchange scientific ideas and to take advantage of many new fields."[38] Since Poland had no financial means to establish such an institute on its own, and no access to a particularly attractive location, either, Godlewski considered it better to send scholars abroad.

In the same year, 1918, Antoni B. Dobrowolski underscored the importance of international comparisons for the "creative-thought science" (Pol. nauka o myśli twórczej)[39], reiterating the link between international-universalist science and pedagogical reforms in Poland.[40] Dobrowolski imagined that the Mianowski Fund should be the place where scholars could "send their reports of the actual order and circumstances of their thoughts in particular discoveries and inventions," which, if enough material for comparison were gathered, would form the basis for a "psychology of the mind,"[41] which had hitherto lacked scientificity and was limited to anecdotal relations and biographical reconstructions. Of course, this endeavour should be an international one, since only a broad perspective allows to identify the external conditions under which inventiveness flourishes.[42] These stories should be published in an international journal – clearly the one edited in Warsaw – and from 1928 onwards, *Nauka Polska* increasingly took on that function, even if academic biographies remained only one of many types of contributions.

The year 1928 marked not only the tenth anniversary of the journal and the declared success of Polish science. Beginning with this issue, *Nauka Polska* started to turn itself into an international centre for science of science. Sections on science of science abroad, such as 'Voices in the press about the role of science' became more frequent, and from 1928 onwards an international bibliography on philosophy, psychology and sociology of science compiled by Janina Małkowska was printed in every journal and published as a separate brochure. This change went hand in hand with

---

171–192: 185–186. At the same time Siedlecki clearly stated, that international contacts with other maritime institute should be intensified.

[38] Godlewski: "O potrzebach biologii" (footnote 36), 193–194.

[39] ▶ **IN THIS VOL.** Antoni B. Dobrowolski: "The Urgent Need for Mental Education in Poland" (1918), 299, i. e. "O pilnej potrzebie wychowania umysłowego w Polsce: o konieczności zasadniczej reformy nauczania w szkołach średnich oraz stworzenia w związku z ową reformą nowych placówek pracy naukowej," *Nauka Polska*, vol. 1 (1918), 489–502: 501.

[40] More details on this project in ▶ Friedrich Cain "A Grammar of Actual Thinking".

[41] ▶ Dobrowolski: "The Urgent Need for Mental Education in Poland" (footnote 39), 291, 299. Pol.: "zwierzenia z rzeczywistych kolei i okoliczności swych myśli w poszczególnych odkryciach i wynalazkach".

[42] Ibid., 299.

other events within Polish historiography of science. In 1929 philosopher and theologian Konstanty Michalski and Aleksander Birkenmajer, professor of the History of Exact Sciences at the Jagiellonian University, proposed that the *Union Académique Internationale* should publish a series on *Corpus Philosophorum Medii Aevi*, stressing its importance for the history of science and for the task of exploring the influence of Jewish and Arab scholars on European thought.[43] This project, addressing 'olympic' internationalism, signaled that Polish science of science, and here its particular branch history of science, had matured to be on a par with its contemporaries abroad.

Finally, in 1936, the Mianowski Fund and *Nauka Polska* decided to publish an international, predominantly anglo- and francophone journal called *Organon: International Review*. Its aim was not to reprint Polish articles in translation, but to form a core medium for the young international community of science of science scholars (*naukoznawcy*). The journal was to conform to the notion of science as an interdisciplinary international activity, and to offer reflections across disciplines and political boundaries.[44] This approach, formulated just when international cooperation in Central Europe grew increasingly complicated, was also mirrored in the composition of its authors: Emile Borel, August Krogh and Sébastien Charléty submitted their contributions for two volumes published before the outbreak of the War forced the journal into a hiatus that lasted until 1964.[45] An interesting detail in the first issue was a letter to the editor consisting of extracts from Antoni B. Dobrowolski's above-mentioned theoreticization of scientific biographies.[46] Only a short part of the Polish-language article was translated, stopping a paragraph short of the moment where in the original version Dobrowolski suggests that the Mianowski Fund ought to be the outrider of internationalization. Still, it clearly placed *Organon* in the orbit of Dobrowolski's ideas. Since Borel and Krogh sent their autobiographies to the first two volumes, the beginning of this cooperation was indeed rather impressive.[47]

In 1937 *Organon* initiated another project, which led to no ensuing discussion. An international and interdisciplinary survey on the importance of scientific conferenc-

---

[43] *Projet de l'Académie Polonaise des Sciences et des Lettres pour la publication d'un Corpus Philosophorum Medii Aevi*, Kraków 1929 and Archiwum Nauki PAN i PAU, Kraków, I–158, "Uzasadnienie wydania Corpus Philosophorum Medii Aevi" (typescript, undated). The project was originally proposed 1927 to the PAU, but was considered to extensive for one country.

[44] "Preface," *Organon*, vol. 1 (1937), V: V.

[45] From 1964 *Organon* was published by Institute of History of Science at the Polish Academy of Sciences in cooperation with the Division of History of Science and Technology of International Union of History and Philosophy of Science and published original works in French, English and Russian only (until 1973, when first article in German was printed). While the editors mentioned pre-Second World War *Organon* in their introduction to the first volume, the journal was not an official continuation of the *Kasa Mianowskiego* endeavour. ("Note de L'Éditeur," *Organon*, vol. 1, 1964, 3).

[46] ▶ Dobrowolski: "The Urgent Need for Mental Education in Poland" (footnote 39).

[47] ▶ Emile Borel: "Contribution" (1936), i.e. "Contribution (Documents sur la Psychologie de l'Invention Dans Le Domaine De La Science)," *Organon*, vol. 1 (1936), 33–42, ▶ August Krogh: "Visual Thinking" (1938), i.e. "Visual Thinking," *Organon*, vol. 2 (1938), 86–94.

es for scientific progress and on organizational questions concerning such meetings (like specialization or pre-circulation of papers) brought 66 responses printed in *Organon* the following year, of which 25 came from abroad. The survey was an outcome of internal Polish debates on the usefulness of scientific meetings – a debate that had begun in the early 1930s, thus bridging the national-international levels of reflections on science.[48] Paweł Rybicki analysed the responses in *Nauka Polska* (published thus only in Polish), highlighting the idea of a sociology of scientific congresses. This idea had been proposed by indologist Stefan Stasiak, which once more shows the importance of the theoretization of science as social endeavour among Polish academics.[49]

We should not forget that this internationalization project also met with criticism, depending on what was understood under this idea. While for most scholars in the *Nauka Polska* group it meant possibilities of expansion, linguist Andrzej Gawroński, for instance, equated it with communism and saw as endangerment of native science. In his articles he clearly took a pro-national stance and stated that 'in practice the ideals [of international science] lead to reign of Lenin and Trotsky and that the internationalization of science 'would at best weaken the unity of the nation'.[50] For him international science was to combine psychologically and linguistically defined styles of research, 'national sciences,' where he used the ideas of Wincenty Lutosławski on artistic (and scientific) styles and of Harald Høffding on the relation between cognition and feeling.[51] Referring especially to the soft sciences, Gawroński claimed that there is no such thing as a 'pure science' but 'the choice of questions, the manner in which they are approached, the interpretation of facts,'[52] is defined by emotive underpinnings (Pol. *podłoże uczuciowe*). In his eyes Polish science should first become a fully grown, autonomous science before it sets out to play with the scholarly superpowers. otherwise it would return to being dominated and exploited like in imperial times. But it is noteworthy that such positions were in the minority, and that the matter at stake was not if internationalization itself, but how internationalization could best be put to use for Polish science and scientists.

Searching for autarchy: Science and translation

In the direct aftermath of the First World War translation was one of the most vital issues discussed in *Nauka Polska*. It was also explicitly mentioned in the 1916 sur-

---

[48] *Organon*, vol. 2 (1938), 133–236. The 7-question survey was printed on p. 133.
[49] Paweł Rybicki: "Ankieta Organonu," *Nauka Polska*, vol. 24 (1939), 397–410: 406.
[50] Andrzej Gawroński: "Nauka narodowa czy międzynarodowa," *Nauka Polska*, vol. 4 (1923), 36–44: 44.
[51] Wincenty Lutosławski: *Volonté et Liberté*, Paris 1913, 166–168 (more detailed analysis of his thoughts on this issue can be found in Robert Zaborowski: "Sentiment, pensée et volonté dans les écrits de Wincenty Lutosławski (1863–1954)," *Organon*, vol. 31 (2002), 57–72); Harald Høffding: *The Outlines of Psychology*, transl. by Mary E. Lowndes, London, New York 1904, esp. ch. IV and V.
[52] Gawroński: "Nauka narodowa" (footnote 50), 43.

vey.[53] Below, I will use translation and language as markers of some of the questions addressed above since, firstly, translation does escape easy 'either–or' answers and presents differentiated views on prospects and pitfalls of internationalization. More importantly, language (and here also the question of its boundary, thus again translation) was increasingly becoming an identity issue and lead thinking about science leading to interesting theoretical considerations.

The 1916 survey asked about the need for translation in the context of deficiencies in Polish scientific literature and ways of filling them. This reopened a Pandora's box of old controversies,[54] and just as in the decades before, there were no straightforward answers. For some scholars, it was necessary to translate textbooks (including the specialist ones) to make science and scholarship more accessible in Poland. For others, it was just the opposite. In 1918 there had been a discord already. Many, like physicist Marian Smoluchowski demanded more intense translation efforts and included lists of books to be translated. For Smoluchowski, translations were especially important to students and early career researchers; they would popularize scientific thinking and unify terminology.[55] The last point is mentioned often, for example by philosopher Kazimierz Twardowski, one of most influential Polish scholars of the time and the father of the Lviv-Warsaw school of philosophy.[56] Once or twice translation was presented as necessary for Polish science to break free from 'German' science, and proponents of this idea stressed the need to diversify the source languages; here also translations from Russian were mentioned, a language that rarely came into discussion in other contexts of internationalization. Interestingly, the argument of diversification was to return after the Second World War and is until today linked with the idea of Poland 'between East and West' and a privileged place for pluralization of knowledge.

The projects to support Polish science through translations went beyond mere lists of publications, but included practical organizational ideas. In 1922 at the above mentioned First Congress on Organization and Development of Polish Science historian of law Stanisław Kutrzeba proposed that translations should not be copyrighted, and the government should buy the rights and allow texts to circulate freely (a proposal that seemed to earn him quite some applause).[57] The financial side of the translation business was indeed repeatedly brought up. Scholars agreed that scientific translations would not be profitable, that one cannot count on commercial publishers

---

[53] "Wstęp," *Nauka Polska*, vol. 1 (1918), I–XVI: VIII.

[54] See, for a well sorted overview on the controversies in the Enlightenment: Jadwiga Ziętarska: *Sztuka przekładu w poglądach literackich polskiego oświecenia*, Wrocław 1969.

[55] Marian Smoluchowski: "O potrzebach naukowych w zakresie fizyki," *Nauka Polska*, vol. 1 (1918), 19–42.

[56] Kazimierz Twardowski: "O potrzebach filozofii polskiej," *Nauka Polska*, vol. 1 (1918), 453–86. More ideas Twardowski's on this issue can be found in articles reprinted in Kazimierz Twardowski: *Rozprawy i artykuły filozoficzne. Zebrali i wydali* uczniowie, Lwów 1927, esp. "W sprawie polskich przekładów dzieł filozoficznych (1913)," 381–384.

[57] Stanisław Kutrzeba: "Nauka a państwo," *Nauka Polska*, vol. 3 (1921), 83–97: 92.

to cover the costs of translations, and called in the state to subsidize translators, who in their eyes were paid poorly even though they needed excellent academic qualifications.[58]

But there was also another side of the coin. The more students and scholars can rely on translations, the less they will read directly in foreign languages – a point Michael Gordin recently raised in the context of US-American scholars' diminishing linguistic competencies.[59] This danger was already mentioned by several scholars in their responses to the survey of 1916. Stefan Niementowski, a professor of chemistry in Lviv, published a rather aggressive statement in which he called translations of foreign authors 'harmful' because, first, every translation will take a Polish book's potential spot on the market and, second, it is necessity for scholars to know at least one 'world language' and they should consult such works in original.[60] Equally critical of translations of more recent literature was Jan Zawidzki, another chemist, who addressed the overvaluation of foreign science and Polish scientists' unnecessary inferiority complex. He also called for the publication of books by Polish authors. Nonetheless, he attached a list of foreign authors whose classical works should be made available in the Polish language.[61] Since translation takes time and also stabilizes certain knowledge (which by the time of publication is already outdated), this process would be also harmful to Polish science and makes it peripheral and belated in comparison to other nations. We must not assume that the critics of translations all belonged to the hard sciences, even though the question seems to have found particular resonance in this camp. The prominent modern historian Oskar Halecki, for instance, criticized the idea of translating specialist books in historiography, since 'without knowledge of foreign languages, historical work in the field of history is impossible' and such translations would only be relevant to popularize knowledge.[62] Ludwik Piotrowicz, historian of antiquity, added that specialists must consult original publications 'which change fast in new editions'.[63]

The question of translating from Polish into foreign languages was also raised, as well as the question of writing directly in a foreign language. This had been disputed on both local and international levels, especially after the war, as more languages aspired to be used in scholarly publications. Most scholars addressing this issue in

---

[58] Both issues are repeatedly discussed in Poland, most recently during the education reforms started by Barbara Kudrycka (Minister of Science and Higher Education 2007–2013), which heavily changed market for translations in Poland, linking translation subsidies with universities instead of publishers.

[59] Michael Gordin: *Scientific Babel: How Science Was Done Before and After Global English*, Chicago 2014, 159–185.

[60] Stefan Niemientowski: "Uwagi o potrzebach chemii," *Nauka Polska*, vol. 2 (1919), 30–41: 41.

[61] Jan Zawidzki: "O stanie chemii na ziemiach polskich oraz o środkach, zmierzających do jego podniesienia," *Nauka Polska*, vol. 1 (1918), 107–138.

[62] Oskar Halecki: "Potrzeby nauki polskiej w dziedzinie historii," *Nauka Polska*, vol. 10 (1928), 266–275: 270.

[63] Ludwik Piotrowicz: "Stan i potrzeby nauki polskiej w zakresie historii starożytnej," *Nauka Polska*, vol. 10 (1928), 276–285: 283.

Nauka Polska readily agreed as to the necessity of Polish scholars publishing more in foreign languages. The debate ranged from specialized journals published in Poland, to longer abstracts (and books of abstracts), to publishing in foreign journals etc., and all proposals were accompanied by different points of critique. Zoologist Jan Hirschler for instance criticized the idea of longer abstracts (to which most scholars agreed and which was commonly practiced), since 'there is no place in such résumés to describe the background, [...] to discuss methods of inquiry, to link the newly discovered facts to those already known'.[64] A foreign researcher reading such a summary would get the impression 'that something exists, but it is impossible to critically judge it,'[65] thus invalidating one of the basic points of reliable research practice. Meanwhile, Wacław Roszkowski called projects like the *Bulletin International* an 'omnibus edition' that collect dust in libraries, since the growth and diversification of knowledge creates a situation in which only specialist journals are taken notice of.[66] There were of course solutions at hand regarding the problem of Polish scholars' submissions to international specialized journals (this problem is commonly debated in smaller languages, while countries that use so-called 'world languages' ignored it for a long time in their discussions on internationalization; the situation changed recently when English started to be perceived as a threat to German and French). A translation office was to be established, to translate and correct articles for foreign journals.[67] Władysław Konopczyński and other supporters of this proposal were clear that it was impossible to require authors to write perfectly in foreign languages, but at the same time, a high linguistic standard was required to get an article successfully through peer review. Many Polish scholars agreed that scientist should not waste their time on correcting or translating articles. Apparently they had been affected by this problem themselves.[68]

But the most popular idea was to found Polish journals in foreign languages, as contributions would thus be distinguished as results of Polish science. As early as 1918 there were calls for a Polish book series in foreign languages. Most notably, Siedlecki proposed the foreign-language series of monographs *Fauna Poloniae* (the project was indeed realized, but only in 1973).[69] The pitfalls of this idea were also swiftly recognized: by 1929, *Fundamenta Mathematicae*, the oldest journal with international aspirations, did not print articles in Polish at all, which is why Jan

---

[64] Jan Hirschler: "Potrzeby zoologii polskiej," *Nauka Polska*, vol. 10 (1928), 198–205: 202.

[65] Ibid.

[66] Wacław Roszkowski: "O potrzebach polskie zoologii," *Nauka Polska*, vol. 10 (1928), 206–220: 217.

[67] Władysław Konopczyński: "Nauka Polska na terenie międzynarodowym," *Nauka Polska*, vol. 3 (1920), 197–203: 201.

[68] Parallel, similar offices for correcting Polish articles and unifying terminology were called for, obvious reaction to post-imperial situation with different terminological directions in different empires, and to influx of scholars socialized and working abroad.

[69] Siedlecki: "Potrzeby nauki" (footnote 37), 189–190.

Łukasiewicz called for a parallel Polish edition.[70] It was generally agreed that a balance was needed between articles in Polish and in foreign languages. But as in other cases, this was about drawing a thin line between nation and (universalist and international) science. Statistics for 1931–34 show that between 79 and 85 % of all non-periodical publications (books, brochures etc.) were printed in Polish, which can be called a balance for the interwar period.[71] Unfortunately, no statistics for periodical issues are available.

The most advanced analysis of translation came from Stefan/Stepan Baley, a Polish-Ukrainian philosopher and psychologist who was confronted with plurilinguality on a daily basis. His works at the time were already available in three languages – Ukrainian, Polish and German, all of which he spoke fluently. Writing on the needs of Polish pedagogical psychology, he discussed the structure of the scholarly world. In this, he divided the world into linguistically defined countries, which he put into hierarchies according to local scholars's need to know foreign languages. Scholars working in the US or in Germany could achieve an 'average' (średni) level in a given discipline working and reading only in their mother languages. Baley called such branches in these countries 'self-sufficient,'[72] though he also argued that any scholar who aims for a level higher than average must turn to internationality. By contrast, researchers in smaller countries must always work with foreign literature, even to attain that average level.[73] But the idea of self-sufficiency did not correspond well to the way science had worked in the past, i.e. its transnational ethos. Baley predicted that in the near future the national production of scientific knowledge would be so abundant that reading everything would become impossible, and new ways of international communication would emerge, such as 'translations of the most important works [or] an international office for reporting on the most important works.'[74] It is clear that Baley opposed the division of science into self-sufficient linguistic islands, but he still considered it a sign of his time that most sciences developed in this direction. Polish science must therefore manoeuvre between different sciences and be nourished by them (without depending on any one of them) until it becomes a self-sufficient science itself. To speed up this process, Polish psychology ought to have a central institution responsible for making foreign works available to 'scholars of the "middle category" who are not capable of linguistically mastering several areas of their work.'[75] The office would not only translate key works, make lists of recommended literature and also produce abstracts on demand. It would also offer grants for

---

[70] Jan Łukasiewicz: "O znaczeniu i potrzebach logiki matematycznej," *Nauka Polska*, vol. 10 (1929), 604–620: 619.

[71] Numbers after L.S.: "[Review] Statystyka druków 1935, ed. by Główny Urząd Statystyczny, Warszawa 1937," *Nauka Polska*, vol. 23 (1938), 360–368: 364.

[72] Baley used the expression *samowystarczalny*, which clearly links his ideas to one of the buzzwords of the interwar-period, economic autarchy.

[73] Stefan Baley: "Potrzeby psychologii pegagogicznej," *Nauka Polska*, vol. 10 (1929), 474–482: 476.

[74] Ibid., 479.

[75] Ibid., 481.

scholars to travel abroad to acquaint themselves with certain branches of knowledge and to support their Polish colleagues on returning.[76] Only in this way would Polish psychology – and other sciences – become self-sufficient by internationalization.

Science and language: On openness and closure

The problem of translation points to one last issue that I wish to mention (rather than analyse) in this paper: the influence of language on scientific production. As noted at the beginning, this issue undergoes various transformations and in many ways becomes a central topic in science studies – from Fleck debating contacts between styles of thought, to Kuhn speaking of incommensurability, and analytic philosophers claiming the need for a scientific metalanguage. Language – clearly a malleable term that signifies many things in these different projects – seems to have played a much more important role in the philosophy of science than is commonly acknowledged, both as topic of research and as a cultural context that influenced theories in the science studies.[77]

Polish scholars, rooted in traditions that reach back to imperial times, clearly saw language as having some influence on scientific production. In the nineteenth century the question of language was present in virtually every corner of Central Europe where educational imbalance lead to demands for teaching in the mother tongue at all levels, in order to develop thinking through language, which was impossible while using the 'imperial tongues.' Since the demise of Latin in the second half of the seventeenth century, nationalization or linguisticization, i.e. the use of the vernacular in all context of cultural life, had haunted scholars across Europe. But the imperial situation in which German (and partially Russian) attained the official status of 'language of education', lead to the question of 'native' language retaining a stronger presence in Czech, Polish or Slovenian scholarship than in France, for instance.[78]

The theoretical question of the relationship between language and science interested only a small number of scholars, mostly analytic philosophers, and was not explicitly discussed in the science of science. This is not to say that the need to theorize and analyse language was overlooked in *Nauka Polska*. In his programmatic article on the science of knowledge, Florian Znaniecki presented language analysis as an important component of his own research agenda, mentioning especially such influenc-

---

[76] Ibid., 482.

[77] The last part has been researched in particular for the Habsburg Science and its empire, following Ernest Gellner: *Language and Solitude: Wittgenstein, Malinowski and the Habsburg Dilemma*, Cambridge 1998.

[78] See on this topic, for example, Tomáš Hermann's analysis on the bilingual, yet pro-Czech bohemian philosopher Emmanuel Rádl, see Tomáš Hermann; "Originalita vědy a problém plagiátu. Tři výstupy Emanuela Rádla k jazykové otázce ve vědě z let 1902–1911," in: *Místo národnich jazyku ve výuce, vědě a vzdělání v Habsburské monarchii 1867–1918*, ed. by Harald Binder, Barbora Křivohlavá, Luboš Velek, Praha 2003, 441–468.

es on science as changes in flexion and the impact of writing, print and of building abstract terminologies.[79] In an article on the education of researchers, the pedagogue Zygmunt Mysłakowski similarly described language as a vital part of scientific socialization and of scientific conventions. Science was to be a permanent collective enterprise, a socializing process that would include learning language and terminology just as much as research techniques, ways of explication and presentation, as well as acquiring a 'feeling' as to which questions are of major and which of minor importance (here he referred to 'cultural factors'). He also mentioned the need to research psychobiological and individual factors, social factors (milieu, including language) and inner scientific factors (science is partly autonomous for Mysłakowski). This would give a clearer picture of how scholarly education should be perfected. His brief analysis – in which the primacy of psychologism and the relation to Émile Meyerson anticipate Kuhnian paradigms[80] – end with the remark that science cannot disappear behind a 'thicket of words'[81] but should remain concrete and realist – an implicit reference to language-obsessed Romantic philosophy, a tradition still alive among Polish scholars at the time.

Internationalization from the periphery: a summary

The last point mentioned above, remaining within the local tradition of Romantic concentration on language while simultaneously embracing pre-relativist thinking, shows the situatedness of debates on science in the interwar period. Analysing those debates we must keep in mind the dilemma of a linguistic community that perceived itself as a successor of the Polish-Lithuanian Commonwealth, one of Europe's superpowers of the early Modern period – a nation that in the 1920s wished to remain in a transitional zone just as it was beginning to regain its rightful place after having been wiped off the map for political reasons.

Situatedness included a constant push toward internationalism, often in its Olympic version. At the same time it included the ideals of better communication – a characteristic of a non-central scientific system.[82] This also meant that while the importance of manoeuvring between local and foreign was often recalled, it was not to depreciate scholarship from abroad, but to codify science in national terms. Polish researchers remembered from the imperial past that Polish scholarship was a part of global science and could only grow through constant contact, even with politically

---

[79] ▶ Florian Znaniecki: "The Subject Matter and Tasks of the Science of Knowledge" (1925), i. e. "Przedmiot i zadania nauki o wiedzy," *Nauka Polska*, vol. 5 (1925), 1–78.

[80] Zygmunt Mysłakowski: "Wychowanie pracownika naukowego," *Nauka Polska*, vol. 6 (1926), 24–48, esp. 29–31.

[81] Ibid., 48.

[82] See the analysis of peripherization-feeling among Polish scholars in Krzysztof Brzechczyn and Katarzyna Paprzycka (eds): *Thinking About Provincialism in Thinking* (= Poznań Studies in the Philosophy of the Sciences and the Humanities, vol. 100), Amsterdam, New York 2012.

inept neighbours and in times of crisis. Even if critics questioned the universality of science by accentuating its cognitive and emotional embeddedness, they tended to maintain a glimmer of hope about internationalization. Self-sufficiency through internationalization can be seen as the leading concept of the time, even while complete closure was obviously not a solution and was not even the purpose of most nationalist scholars. Interestingly, this characteristic also appears in other Central European countries that emerged after the Great War.[83]

Finally, almost every debate on internationalism turned into a theoretical discussion on international relations in science, even if this did not mean getting involved too deeply in the philosophy of interlingual communication. In most cases this was everyday theorizing, a search for models enabling communication with other scholars – an important point in the heated debates of the day, even if there was common agreement on basic problems. Besides traces of early professionalization of science of science, this is perhaps the most interesting point for historians of science in *Nauka Polska*: the Warsaw journal provided scholars with a forum to reflect on their work and to put their experience on paper, which gives invaluable insight into how they saw their agenda in the broader spectrum. Since a lot of early theorization of science was made precisely at this threshold,[84] *Nauka Polska* not only has much to offer in this regard, but can facilitate the turn from the big names of science of science to less researched scholars.

Two issues are conspicuously absent from the discussion on translation, which also points to the greater problem of how a periphery can participate in science that is defined internationally but remains limited to 'western' science in an almost imperialistic way. First comes the question which languages count as international – which ones should be allowed in international journals, for example. With few exceptions[85] there was no debate as to which languages are 'world languages' and in which languages Polish science should be promoted. This was at a time when Russian was still one of the three most used languages apart from Polish to print books in Poland.[86] One notable concession was for Italian for publications in mathematics. Given the anti-German resentments and political difficulties German encountered after the First World War,[87] the stability of 'world languages' is a phenomenon worth examin-

---

[83] Cf. the articles about interwar Czechoslovakia in Martin Franc, Antonín Kostlán and Alena Míšková (eds): *Bohemia docta. K historickým kořenům vědy v českých zemích*, Praha 2010.

[84] See for instance on theoreticization in Polish medicine Ilana Löwy (ed): *The Polish School of Philosophy of Medicine. From Tytus Chalubinski (1820–1889) to Ludwik Fleck (1896–1961)*, Dordrecht 1990. In broader context, especially true for the pre-Second World War Hans-Jörg Rheinberger: *On Historicizing Epistemology: An Essay*, Stanford 2010.

[85] The most notable exception was the question if other Slavic languages should be allowed. The answer tended to be negative. See for example Kazimierz Tymieniecki: "Uwagi o niektórych potrzebach historji w Polsce i o warunkach jej dalszego rozwoju," *Nauka Polska*, vol. 2 (1919), 148–174: 167.

[86] L.S.: "[Review] Statystyka druków 1935" (footnote 71), 364. These statistics include scholarly publications by minority groups, if they were printed in Poland.

[87] Roswitha Reinbothe: *Deutsch als internationale Wissenschaftssprache und der Boykott nach dem Ersten Weltkrieg*, Frankfurt am Main 2006 und id.: "Languages and Politics of International

ing. Second, it seems that the question of an international language of science was not discussed at all (not counting a few rather critical comments in reports on the logical positivists in *Nauka Polska* in the 1930s). In Poland a few organizations propagated international scientific languages, such as Esperanto. Białystok-born Zamenhof seemed an apt figure to be clad in the national colours. One organization, the *Societas Lingua Latinae usui internationali adaptandae*, established in 1933, even opted for international Latin. But it seems that the idea of a global scholarly idiom lost popularity once its main heroes like Wilhelm Ostwald turned to nationalism and the French government campaigned against Esperanto as a language of international relations.[88] Clearly, international science as imagined by Polish scholars cannot and does not even intend to counterbalance 'Western' science, even if remaining in tension with some of its proponents. The boundary of in-betweenness thus remained untouched and Polish scholars defined Polish science referring to the existing structures, not trying to propose or search for alternate modes of communication. In this, this very discussion of internationalization remained somewhat disappointingly noninnovative and derivative not leading to reconsideration of medial and lingual rigidity of modern sciences, something only post-colonial theory achieved decades later. The reason for it is perfectly obvious – Polish scholars felt already part of the international science and the question was the one of recognition and not the one of getting attention by questioning the norms. It was normal science, even if this normality had yet to be internationally acknowledged.

The writing of this article has been supported by the project "'Images of science' in Czechoslovakia 1918–1945–1968," financed by Lumina Quaeruntur fellowship of the Czech Academy of Sciences (Grant ID LQ300772201).

Scientific Communication in Central Eastern Europe after World War I," in: *Expert Cultures in Central Eastern Europe. The Internationalization of Knowledge and the Transformation of Nation States since World War I*, ed. by Martin Kohlrausch, Katrin Steffen and Stefan Wiederkehr, Osnabrück 2010, 161–177.

[88] Markus Krajewski: "Organizing a Global Idiom. Esperanto, Ido and the World Auxiliary Language Movement before WWI," in: *Information beyond borders*, ed by Rayward (footnote 2), 97–108; Ulrich Lins: *La dan̂ĝera lingvo. Studo pri la persekutoj kontraŭ Esperanto*, Gerlingen 1988, 61–70.

Strategies of Sincerity

Scientific Autobiographies in the Proceedings of *Koło Naukoznawcze*[1]

Marta Bucholc

A sociologist of knowledge, when confronted with the materials, which are collected under the title *Researching the Genesis and Development of Scientific Creativity* and in the proceedings of *Koło Naukoznawcze* (Science Studies Circle), is positively overwhelmed with a variety of research material. All this richness has been generated by an address of the editors of *Nauka Polska*, whose genesis is probably best explained by the greatest advocate of this initiative, Antoni B. Dobrowolski.

Dobrowolski's rich and in many ways programmatic paper *The Urgent Need for Mental Education in Poland* of 1918[2] contains a broad panorama of studies of science and their various social uses. Amongst those elements that the author finds crucial for satisfactory development of studies of science and, consequently, science in general, there are also conditions for truly original and excellent scientific work, the characteristics of which he encompassed in the semantically rich and philosophically ambiguous term "creativity". Dobrowolski's focus was scientific creativity and his goal – a new scientific methodology to study it empirically and systematically.

It is a justified course of action to examine Dobrowolski's ideas in more detail before plunging into the variety of answers. He was positive that "only a straightforward confession of the author (in the form of genuine and *sincere* notes" may be a reliable source of knowledge about the genesis of creativity.[3] I lay emphasis on the word "sincere", because it will be of primary importance in this chapter. Further elaboration of autobiographical sincerity in this chapter will include a distinction of three levels of sincerity, all of which I argue may be found both in Dobrowolski's paper and in the autobiographic accounts of his fellow scholars, together forming a *threefold sincerity standard*. On the *first level*, sincerity may be a requirement of even producing a statement. An autobiographer must expressly commit to the principle of sincerity as a guide in reporting facts, which brings about the necessity to restate this principle or

---

[1] This paper was prepared as a part of a research project financed by the Polish National Science Centre (2014/13/B/HS6/03741).
[2] ▶ IN THIS VOL. Antoni B. Dobrowolski: "The Urgent Need for Mental Education in Poland" 1918, i.e. "O pilnej potrzebie wychowania umysłowego w Polsce: o konieczności zasadniczej reformy nauczania w szkołach średnich oraz stworzenia w związku z ową reformą nowych placówek pracy naukowej," *Nauka Polska*, vol. 1 (1918), 489–502.
[3] Ibid., 297 (emphasis M.B.).

refer to a statement by another. On the first level, sincerity comes closest to genuineness of recollection and directness of reporting it. This in turn leads the autobiographer to the *second level*, where the mutual relation between sincerity and objectivity needs to be reflected upon. Sincerity on the first level may stand in the way of objectivity understood as a factual adequacy of a report. Sincerity on this level is not only a guiding principle of a reporter, but also a means of cognitive self-control juxtaposing the internally genuine recollection and – so to say – externally, publicly available knowledge or, simply, truth. Finally, on the *third level*, sincerity is also a meta-principle governing the selection of biographical material included in an autobiographical account. While all the experience available to an individual as a recollection may be reported sincerely on the first level (at least as long as we put any psychological reservations aside) and made subject to reflection on the second level, the final decision as to the report's contents is actually taken on the third level, where some things are labelled as irrelevant or superfluous. The latter point is particularly informative of an author's position to the autobiography as an intellectual product, but all levels on which sincerity operates as a reporting standard make it strategically useful, with different styles of autobiographical sincerity leading to different effects.

Autobiographical accounts of scientific creativity published in *Nauka Polska* demonstrate the work of sincerity on all levels: The authors explicitly declare their sincerity in relating their own experiences as creators, they appreciate the tension between objectivity and their personal, subjective reporting method applied, they also decide to cover some aspects of their lives and leave others untouched. Their choices are, at the same time, a result of their respective interpretations of sincerity as a feature of an autobiography and a strategic move determining their place in science and establishing a style of thinking about both scientific creativity and methods of understanding and researching it. It is in the interplay between these two dimensions that sincerity's role in positioning comes to the fore and its strategic aspects may be fully appreciated.

Semantic richness and complexity of the notion of sincerity, which allows for such a differentiation, also supports the thesis that sincerity was a vastly inclusive category for Dobrowolski. I argue that it was the insistence on the value of sincerity that made him doubtful of all second-hand efforts to study creativity, either in the aspect of the "ontogenesis" of a discovery, or its "psychogenesis"[4]; which were not based on autobiographical materials responding to all three levels of sincerity standard. That is why, in this chapter, I will use a concept of sociogenesis as developed by Norbert Elias, to analyze the basics of this autobiographical project, which will further bring me to a discussion of strategic effects of sincerity, based on Patrick Baert's positioning theory and a more detailed content analysis, aimed at discovering collectivity dynamics of a sociogenetic process expressed in the autobiographical fragments.

---

[4] Ibid., 298.

1. Sociogenesis of Science

Dobrowolski mentions ontogenesis and psychogenesis, which may be assumed to refer to emergence of a particular individual scientist or a particular product of scientific creativity and its birth out of a psychological setup of a particular individual respectively. However, the very operation of providing first-hand material for scientific study of scientific creativity is also a part of social process of creating science or, rather, making up relational, interpersonal and institutional conditions in which science can be made. It is also a remedy to a fault of even the "perfectly sincere"[5] firsthand notes and confessions. They will always be incomplete; "completeness" does not come into the semantic field of "sincerity" in Dobrowolski's programmatic paper. Nevertheless, the accumulative procedure to gathering and analyzing the auto-narratives of scientific creativity which he proposes is, in fact, an attempt to transcend the individualistic account of a social phenomenon by integrating the comparative, quantitative and relational aspect with individualistic.

It is rather doubtful whether this program was consciously conceived. Its formulation was fragmentary and its realization even more so. The editors asked their addressees to provide an account of their scientific work. The responses varied from short and relatively superficial statements of mixed facts and opinions to lengthy, ten and more pages mini-treatises. Serious attempts at self-analysis were from time to time accompanied by philosophical excurses and generalized remarks on methodology of science. To use such means of extracting information about the most intimate and personal aspects of scientific creativity and biography of a scientist was in accordance with the basic tenets of the Science Studies Circle. For them, science was a social, collective enterprise or, at least, could be studied as such. It was not a view unrepresented among their contemporaries and indeed it stemmed from 19th-century ideas, originating in a somewhat older, Enlightenment view of science as a product of the individual mind educated by a society in order to serve it and solve its problems. Nevertheless, what was new and novel to the Science Studies Circle, was an attempt to trace social mechanisms which make science and in particular innovative science possible, be it the organizational structure of institutions, educational curricula or the life paths of researchers. All these problems were constantly present on the pages of *Nauka Polska* and formed a context for the autobiographical accounts, a meaningful if somewhat disparate one.

Early insights into the sociogenesis of scientific achievement in all probability resulted in subsequent developments of individual theorists active in this circle, notably Florian Znaniecki and Stanisław Ossowski.[6] Nevertheless, it will not be a genealogy of later and usually more prominent efforts which will be the focus of this

---

[5] Ibid., 299.

[6] See Florian Znaniecki: *The Method of Sociology*, New York 1968; Stanisław Ossowski, *O osobliwościach nauk społecznych*, Warsaw 2001.

chapter. Instead, a seemingly minor issue internal to the autobiographical works analyzed here will be brought to the fore: the tension between sincerity as an ideal in reporting biographical experience and the positioning stratagems.

This tension brings us to the very core of a problem which haunts sociologists exploring the background of social production of knowledge. On the one hand, there is always a person, to whom a sociologist habitually refers to as an individual, who is a carrier of knowledge. To focus entirely on the individual would be a wrong approach altogether, as Norbert Elias famously noted in his account of "thinking statues"[7]. Elias, born in 1897, which makes him a contemporary of Znaniecki, Ossowski, Maria Ossowska and many of the members of the Science Studies Circle, shared many of their ideas and concerns, also those regarding scientific creativity and its social origins. This is evidenced by some of his works published after the World War II, including in particular *Involvement and Detachment*. In his social ontology and epistemology, Elias argued against Neo-Kantianism in which he received his philosophical education and claimed that there is no such thing as an individual, disentangled and separated from its neighbors. There are only human beings involved in networks of interdependence, which Elias himself called "figurations".[8] However, there is undoubtedly an individual element in all human beings, which is why Elias insisted on distinguishing two aspects of genetic account of all social phenomena: a sociogenetic and a psychogenetic one. The ontogenesis, as an emergence of an individual being, would be redundant to the two, as nothing can be born either human psyche or in society without coming to existence as such, in a more basic philosophical sense which Elias might discard as "post-philosopher".[9] Be that as it may, Elias only used these terms in order to explain "social generation" or "social production" of concepts, practices and organizational structures. Incidentally, Elias would not have been content with the notion of "social construction" because, as Eric Dunning and Jason Hughes noticed, he opposed the overly voluntaristic and rationalistic visions of human agency which the notion of construction conveys.[10]

The dilemma of explanatory priority between the individual and the collective may be pushed further. This problem was also featuring a lot in social science of the time, and was elaborated on particularly by Émile Durkheim and his circle in the late 19th and early 20th century. Durkheim's reasoning, especially in *The Rules of Sociological Method* and *Suicide*, reflects the tension between two extreme interpretations of individual actions: either they are deliberate attempts to produce certain predesigned effects or they are simply epiphenomena of interpersonal processes beyond the

---

[7] See Norbert Elias: *The Society of Individuals*, ed. by Michael Schröter and Robert van Krieken, transl. by Edmund Jephcott, Dublin 2010, ch. 2.

[8] See Marta Bucholc: *A Global Community of Self-Defense. Norbert Elias on Normativity, Culture and Involvement*, Frankfurt am Main 2015, 37–39.

[9] See Richard Kilminster: *Norbert Elias: Post-philosophical Sociology*, London 2007.

[10] See Eric Dunning and Jason Hughes: *Norbert Elias and Modern Philosophy. Knowledge, Interdependence, Power, Process*, London 2012, 81.

agents' control. But, of course, no human action if just one or the other: it is inevitably both, and none of the extreme explanations could satisfy a sociologist. Nevertheless, the Durkheimian school failed to work out a convincing alternative to the dichotomy of free will and social determination and drifted towards what would later be called "an oversocialized conception of man"[11]. Elias's notion of sociogenesis was meant as a device to overcome the problem of social entanglement of individual psychological development, used to express the connection between the personal existential experience and its social frame.

Science is socially useful and functional, structurally conditioned, but deeply personal and profoundly idiosyncratic. Hence the Polish term *twórczość*, for which the English word "creativity" is but a very pale equivalent (though undoubtedly the best there is), may be applied to scientific activity as well as to artistic one. *Twórczość* is something original, coming from *tworzyć*, a verb, which – as one may still hear in Catholic religion lessons – essentially means "to make something out of nothing". It is a verb used to describe what God did when the world was created, and it can only apply to an action insofar it bears the sign of originality, uniqueness and identifiable personal agency. It is a term used today in intellectual property laws, in aesthetic discourses and in art criticism, always in the same meaning: making something new and unreducible to what has been. But *twórczość naukowa* (scientific creativity) is not really *creatio ex nihilo*.

Social embeddedness of scientific creativity, which the Science Studies Circle took for a theoretical premise, opens a way for very delicate reflection on the balance between the two elements: the social and the individual, in the self-reflection of scientists who responded to the appeals of the editors of *Nauka Polska* (those appeals to be repeatedly formulated and, it may be indirectly inferred, not always readily followed). In order to tackle this problem, I propose to use the positioning theory proposed by Patrick Baert. I will first present the basics of this theory, which I will use to analyze selected positioning moves of the scientists. The analytical part will be followed by a revision of interplay between sincerity and strategic self-representation, which I interpret as an attempt to rework the basic sociological dilemma of individuality versus social determination. Scientific creativity in the inquiries of the Science Studies Circle is thereby sometimes raised to an emblem of socialization, much as science would be in the later writings of Ossowski.[12] To endow science with such significance is undoubtedly a relic of positivism. But it also bears a mark of neo-Kantian scientific romanticism reverberating in Weber's lectures on science and politics[13], coming

---

[11] See Dennis H. Wrong: "The Oversocialized Conception of Man in Modern Sociology," *American Sociological Review*, vol. 26/2 (1961), 183–193.

[12] See Anna Matuchniak-Krasuska: "Estetyka i socjologia sztuki Stanisława Ossowskiego. Ilustracje 'U podstaw estetyki' – studium wartości artystycznych," in: *Ossowski z perspektywy półwiecza*, ed. by Antoni Sułek, Kraków 2014, 79–116; Stanisław Ossowski: *The Foundations of Aesthetics*, transl. by Janina and Witold Rodzinski, Boston 1978; id.: *Wybór pism estetycznych*, Kraków 2004.

[13] See Max Weber: "Politik als Beruf," in: id., *Gesammelte Politische Schriften*, ed. by Johannes

from a time when science was still, and very self-consciously, an art, but quite aware of its imminent aesthetic decline.

In this sense, the approach of the Science Studies Circle is an heir to a long tradition of philosophy of science, but also a portent of new times of 20th-century scientific imperialism. Positioning theory is a device to account for strategic uses of sincerity in autobiographical narratives, thus contributing to a better understanding of difficulties encountered in using personalized self-narratives as a source of knowledge about social creation of creativity. At the same time, it also allows the philosophical context of an early 20th-century scientific self to be adequately represented as a product of positioning in a field delineated by former theoretical and literary traditions.

2. Positioning and threefold sincerity standard[14]

Baert's positioning theory is meant as a novel approach in sociology of intellectuals, different from many other contemporary perspectives, including Pierre Bourdieu's reflexive sociology, Randall Collins's sociology of philosophy or new sociology of ideas developed by Charles Camic.[15] Baert's purpose is essentially to find a way out of dilemmas of classical history of ideas, ricocheting between Arthur O. Lovejoy[16], the Cambridge school (at which university Baert is currently based)[17] and, sometimes, German contributions like Reinhart Koselleck's historical semantics[18]. At the same time, he seeks a more sociological approach to the dynamics of intellectuality as an intelligible social phenomenon whose interpretation is not entirely dependent on and reducible to factual findings related to individual biographies. To accomplish this task, he turns towards the domain of rhetoric of intellectual products, which in his case are mostly (though not only) texts. Baert's starting point is the following:

[T]he reception, survival and diffusion of intellectual products [...] depends not just on the intrinsic quality of the arguments proposed or the strength of the evidence provided, but also

---

Winckelmann, Tübingen 1988 [1919], 505–560; id.: "Wissenschaft als Beruf", in: id., *Gesammelte Aufsätze zur Wissenschaftslehre*, ed. by Johannes Winckelmann, Tübingen 1985 [1919], 581–613.

[14] A more extensive review of the main tenets of positioning theory can be found in Marta Bucholc: "Balansując na marginesach. O strategii intelektualnej Norberta Eliasa," *Kultura i społeczeństwo*, 59/1 (2015), 1–26.

[15] See Patrick Baert and Joel Isaac: "Intellectuals and Society. Sociological and Historical Perspectives," in: *Routledge International Handbook of Contemporary Social and Political Theory*, ed. by Gerard Delanty and Stephen P. Turner, London 2011, 200–211: 203; Charles Camic and Neil Gross: "The New Sociology of Ideas," in: *The Blackwell Companion to Sociology*, ed. by Judith R Blau, Malden (Mass.) 2001.

[16] See Arthur O. Lovejoy: *The Great Chain of Being*, Cambridge 1976.

[17] See Quentin Skinner: "Meaning and Understanding in the History of Ideas," *History and Theory*, vol. 8/1 (1969), 3–53.

[18] See Reinhart Koselleck: *Futures Past. On the Semantics of Historical Time*, transl. by Keith Tribe, New York 2004.

on the range of rhetorical devices which the authors employ to locate themselves (and position others) within the intellectual and political field.[19]

Baert argues that a particular valor of his concept lies in the shift of attention to the strategic aspect of scientific work, which is directly present in any scientist's activities and not only in accompanying actions. A consequence of this preference for the strategic is, unexpectedly enough, a rejection of intentionalism characteristic of the Cambridge school of history of ideas and a very strong focus on effects of actions and not on the agents' desires, schemes or plans. Strategy does not necessarily, as Michel Foucault proved, require a strategist[20]. Much in this line, although not drawing directly on Foucault in this respect, Baert focuses on the reconstruction of life trajectories and professional careers generated by a specific use of rhetoric devices in what he defines as "positioning moves", made in a form of "producing" intellectuality by writing books, publishing articles, giving lectures and talks and, last but not least, writing biographies and autobiographies.

Intellectual products as moves have more unintended consequences than those which could have been anticipated. Nevertheless, they form a pattern which under given circumstances results in the actor achieving a certain position; whether or not according to his or her intentions, is irrelevant.

However, what do the actors compete for? Usually, in sociology of intellectuals and in sociology of science, the goal of all positioning moves would be to occupy a position of power in the field, maximizing the efficiency of all resources which an actor has at his or her disposal. To become an important, powerful figure in the field is a goal easily fitting into the game logic of winners and losers. Examples of such an approach may be found in many scientific biographies, but also in sociology of science and of professions.[21] A central position in the field seems to prevail as a theoretically assumed ultimate goal of all scientific endeavor[22], and the strategic aspect of positioning moves is construed accordingly. Nevertheless, I believe this assumption should be qualified, and in more than one way. In my recent article on Elias's intellectual strategy I insisted on the importance of alternative strategies, which would not maximize centrality versus marginality, but just staying in the field versus falling out of it altogether.[23] An outsider's strategy does not emerge out of a wish to become an outsider, but out of a conglomerate of moves and their subsequent proximate and remote effects. But it is one of the many possible strategic choices of a scientist.

---

[19] Patrick Baert: "Positioning Theory and Intellectual Interventions," *Journal for the Theory of Social Behaviour*, vol. 42/3 (2012), 304.

[20] See Hubert L. Dreyfus and Paul Rabinow: *Michel Foucault: Beyond Structuralism and Hermeneutics*, Chicago 1982, 109.

[21] See e.g. Izabela Wagner: *Becoming Transnational Professional. Kariery i mobilność polskich elit naukowych*, Warsaw 2011.

[22] See Baert and Isaac: "Intellectuals and Society" (footnote 15), 203.

[23] See Bucholc: "Balansując na marginesach" (footnote 14).

According to Baert, actors position themselves by referencing other actors in the field, as well as their respective intellectual products. References to names, schools, theories, book titles, institutions etc. operate as markers of relative positions of actors, as does "intellectual coldshouldering", not taking notice or failing to quote, adhere to or criticize.[24] Actors not only position themselves by referring to others in the field, but at the same time also position others and are themselves used as positioning resources. This interdependence of positioning strategies in any given field is particularly evident in science, with its strict rules for referencing other people's work and increasing number of references in scientific works, resulting – among other things – from rigorous standards of protection of intellectual property rights. This may tempt scholars, especially those enjoying an established position, to use other means of positioning, including biographies and autobiographies. These genres impose specific and very stringent demands on their reader. Baert writes:

I take it as essential to establish a critical distance vis-à-vis the way in which most intellectuals portray themselves to their audience. Indeed […], intellectuals have a tendency to depict their own intellectual trajectory as untainted by these material, symbolic and institutional constraints. For instance, there are remarkably few intellectual autobiographies that acknowledge the full extent to which considerations of this kind interfered with the intellectual choices that were made. This is because autobiographies too – just like other intellectual products – position their authors, their allies and opponents.[25]

For these precise reasons, autobiographies are one of the most efficacious modes of scientific positioning (provided, of course, that they are read). They are not, strictly speaking, scientific texts and escape the limitations usually framing scientific writing. They are by definition subjective and have an appearance of direct personal expression coming with the first-person narrative, which an impersonal scientific style usually lacks. On the other hand, they usually refer to intellectual products of the author, mention his or her friends, teachers and competitors, enumerate books and institutions and thereby contain a powerful set of positioning moves. They are apparently informative and genuinely persuasive at the same time: they serve self-creation and make the creator vulnerable to critical close-reading. However, biographies and autobiographies are also excellent tools for a student of positioning understood more broadly and in more general terms as a way of finding one's way round in the field by referencing some fundamental discussions and distinctions. The intellectual products used as landmarks for positioning purposes would in this case be concepts and general classifications. By applying them, adhering to them or failing to notice them, an actor may produce a positioning effect and join a club of thinkers in a certain style. Such act of intellectual allegiance may also be rendered in autobiographical form.

---

[24] For an example of this positioning moves by Elias, see ibid.
[25] Baert: "Positioning Theory" (footnote 19), 309.

3. A sincere autobiography?

Let us consider what standards and strategies of sincerity are emergent in the accounts published in *Nauka Polska*. They are "scientific autobiographies", reports of an individual's path to the status of a scientist, reflecting "the exact relationship between the work and the author".[26] The demanding character of this method was universally recognized, both by its adversaries and proponents. It was commonly felt that in a scientific autobiography introspection, extraspection and circumspection had to be reunited, in order to produce not a poetized self-account, but workable data for a student of science. Therefore, a standard of sincerity appears to crystallize around the notion of guaranteeing a certain dose of objectivity, which – as Lorraine J. Daston and Peter Galison demonstrated – incidentally leads its proponents to commit to a cultivation of a distinctive scientific self[27]. The ideal of objectivity is, however, rather puzzling since sincerity, on the first level of the threefold standard, is also commonly related to an accurate and unartful rendition of inner moods, feelings and memories which are inherently subjective.

Czesław Białobrzeski, a prolific story-teller, in his vibrant self-narrative of 1927 thus summarized the difficulties of submitting oneself to the exigencies of autobiographical method:

Will it suffice to point out in what way a certain manifestation of scientific creativity had been prepared and carried out, or ought I to reach deeper and attempt to outline the development and formation of these creative faculties? [...]

I will try to carry out my task from the broader perspective, that is to say by attempting to outline the type of mentality I represent. Every type possesses traits that are characteristic of the individual, and therefore I must stress them, too. I will, however, disregard unduly minute details that might put to the test the reader's patience, and whose existence only seems justified when portraying an individual of exceptional calibre.

Aside from the difficulties related to introspection, the task as defined above demands great impartiality of judgement. At any rate one must, as it were, step out of oneself and observe one's psyche like an object to be studied scientifically.

It is possible that in the great majority of cases an astute outside observer is much better placed to satisfy this demand for impartiality than the subject studying himself.

And yet, accounts of personal experience have a unique value that is not called into question even by a certain partiality or self-love, or by the lack of perspective that results from too close a proximity to the object under investigation. [...]

I will try to limit to a minimum the external details of my life. But we cannot ignore the backdrop against which the individual develops.[28]

---

[26] ▶ Stefan Błachowski: "The Problem of Scientific Creativity" 1928, 192, i.e. "Zagadnienie twórczości naukowej," *Nauka Polska*, vol. 9 (1928), 1–67.

[27] See Lorraine J. Daston and Peter Galison: *Objectivity,* New York 2007.

[28] ▶ Czesław Białobrzeski: "An Autobiographical Sketch and Remarks on Scientific Creativity" 1927, 306, i.e. "Szkic autobiograficzny i uwagi o twórczości naukowej," *Nauka Polska*, vol. 6 (1927), 49–76.

A major problem of interpretation is clearly manifest in this passage: what is and what is not crucial for an examination of scientific creativity? Białobrzeski clearly chooses to try and look at himself impartially in order to produce a description of his mental processes, thus following a doubly safeguard function of sincerity, which defends objectivity by being truly subjective. But, Białobrzeski at the same time notices that an impartial external observer might stand a better chance to produce an objective picture of another's mental capacities, thus contributing to the second level of sincerity standard, on which the tension between objective and subjective narrative is thematized as a methodological dilemma. Still, he also values the effect of engagement (or, as Elias would put it, involvement) of the subject with his or her own narrative as a source of knowledge about creative work. This purely introspective aspect prevails over what might have been at least equally important for a sociologist of Eliasian persuasion: a description of social conditions of life and scientific development. Białobrzeski declares that to put the facts of his life in the foreground would not be in accordance with the autobiographical genre. Thereby, he defines it as a psychological tool rather than a sociological one: his sincerity is a sincerity of a penitent, not of a witness.

This idea of penitent-like introspection is, naturally, bound to cause criticism on methodological grounds, as in the following passage by Antoni Dobrowolski:

I am not convinced that "scientific biographies" of this kind will actually prove useful as material for *scientific* research on creativity. They might be "interesting" to those who are interested in the human mind and soul; they might even be "instructive" for young researchers or teachers, by providing certain guidelines, based on personal experience, about what ought to be done *perhaps* and what one should *perhaps* avoid. But collecting such descriptions is unlikely to yield anything accurate, because the descriptions are neither accurate nor complete or systematic; they lack *immediate* observations and contain too many interpretations, distortions and stylisations, since facts are reflected in the waves of memory. In my opinion the only method that might yield more serious results is not the "scientific biography" of a given person, but the biography of a given *work*: the researcher's immediate confession about the birth of a given idea, its development and maturation, as well as of the circumstances that influenced the research process in one way or another. This kind of declaration ought to be made in real time: fresh, full, undistorted.[29]

Interestingly enough, authorial perspective is deconstructed in this passage and the scientific work in its immediacy is juxtaposed to memory-mediated, systematic and linear biographic account. This is also a crucial epistemological point: an undistorted picture can hardly be achieved in narration procured by one believing that a personal perspective is necessarily a distortion, a point which Znaniecki challenged in his *Method of Sociology*.[30] Personal bias of sincerity and sincerity as immediacy (so the two elements of the first level of sincerity standard) also clash in this passage. Stylization is a usual suspect, if we take style as an aesthetic surplus to the idea itself and not,

---

[29] ▶ Antoni B. Dobrowolski: "My 'Scientific Biography'" (1928), 320, i. e. "Mój 'życiorys naukowy,'" *Nauka Polska*, vol. 9 (1928), 68–216.

[30] See Znaniecki: Method of Sociology (footnote 6).

as Ludwik Fleck, amongst others, would argue, as its inherent element[31]. But to denounce stylization is in itself a rhetoric device employed to style the author as objective and sincere.

Apart from these very general methodological doubts there are minor misgivings. Should the research address the exact and idiosyncratic context of origin of a specific work, including its sometimes illicit genealogy, or just focus on its internal "creative potential" and leave the context aside? Stefan Błachowski's answer to that is clear and unequivocal.

> Smelling out the irregularities of a work's origin, a Holmesian chase after its spiritual father, is neither pleasant nor does it benefit science. [...] But we are not concerned with that. Let us assume honesty and examine individual scientific products for the degree of scientific creativity manifest in them.[32]

It is not a very promising approach for a sociologist, who would gladly know the context as fully as possible and would tend to relate the "degree of scientific creativity" to the conditions. Here, yet another limitation on the standard of sincerity is introduced: on the third level, when we decide which things are best left out of biographical readings, genealogies of creative works are narrowed down to what is manifest and expressed. Sincerity and critical thinking do not seem to go hand in hand in this case. Granted, critical thinking need not always seek truth in intentional omissions. But, if the research material for examining scientific creativity is an autobiography, in assuming honesty and focusing on the creative product instead of its origination, a limit is set on the criticism exercised on the third level of sincerity standard to the author, for the opportunity to exercise critical judgment never extends to the reader. Therefore, on the third level, the sincerity standard works as an author's weapon against the reader.

Taking into account these and other difficulties of sincere writing, the authors frequently feel obligated to give their reasons to take up this exercise, which also forms a part of sincerity standard. They do it for the greater good:

> The task before me is not only rife with difficulties, but it is also a delicate and thankless one. By writing this I must, as it were, exhibit my intellectual individuality for public viewing; I am sure to meet with a great variety of accusations, be it of conceit or of false modesty and so forth. But I undertake to produce this autobiographical sketch about my creativity in the conviction that if other researchers, more worthy than I, follow in my footsteps, this will unquestionably be of use for a great cause, namely the effort to understand the development of scientific creativity – the foundation of modern civilization.[33]

Rhetoric tactic in this passage is to foreclose all accusations by humbly admitting inadequacy of one's own effort, its pioneering character and potentially great use in

---

[31] See Ludwik Fleck: *Genesis and Development of a Scientific Fact*, ed. by Thaddeus J. Trenn and Robert K. Merton, transl. by Fred Bradley and Thaddeus J. Trenn, Chicago 1979.
[32] ▶ Błachowski: "The Problem of Scientific Creativity" (footnote 26), 192.
[33] ▶ Białobrzeski: "An Autobiographical Sketch" (footnote 28), 304.

the task of understanding science better – and also of being able to better understand the civilization which rests on the foundation of scientific creativity. A worthy task indeed, and one which adds to the sincerity standard a grave weight: only those autobiographies which stand the trial may contribute to achieving this goal.

4. Positioning by sincerity styles

Let us now take a closer look at the kind of positioning effected by using the sincerity standard as a style of autobiographical story-telling. The style will develop around key concepts and distinctions, which in turn relate to the sincerity standard for their justification as a proper subject-matter of an autobiography. The main concept, an axis to the style of autobiographies published in *Nauka Polska* was, of course, science.

Science is, firstly, a particular form of human creativity, as evidenced by all authors. However, they differ as to the exact relationship between science and other forms of creativity. This makes the distinction between science and other human creative efforts a major positioning landmark: whereas there is naturally more than just creativity to science, amongst other products of human spirit aspiring to the status of science only those endowed with a creative element may be regarded as truly scientific. Creativity thus connects science to other fields of human creative activity, while scientificity becomes a *differentia specifica*, one supposedly better known to the narrators than the *genus proximum*.

> A scholar's work is scientific only when it contains an element of creativity. The term "scientific" ought to be reserved for such articles, discourses and products that are the results of creative work.[34]

Which also means that some works may aspire to the status of scientificity, but never achieve it: they are not truly scientific.

> Certainly, some characteristics are common to all manifestations of creativity, which is the highest form of human mental activity. A particularly close relationship should exist between scientific and philosophical creativity. And yet the differences in thought content and method across the sciences are immense, and the diversity of the creative processes at work in them is evident from the variety of skills required for their successful practice. What is more, within each science we can single out a handful of distinct types of creative ability on whose concurrence the rapid expansion of knowledge is contingent.[35]

Scientific creativity is generic, a subgenre of a general human capacity for original and relevant work. What is the purpose of this work in which a *differentia specifica* for the whole genre may be found?

---

[34] ▶ Błachowski: "The Problem of Scientific Creativity" (footnote 26), 190.
[35] ▶ Białobrzeski: "An Autobiographical Sketch" (footnote 28), 304.

[S]cience is the constantly changing product of the efforts of the human mind, efforts that aim to explore and to master reality. These sustained efforts of the human mind in the field of science express themselves in opinions that represent real progress in relation to the state of research at a given moment.[36]

Exploring and mastering reality is a Baconian and Comtean paradigm of *savoir pour prévoir pour pouvoir* revisited: it refers to a general social function of science and thereby puts the autobiographical narrative in a larger philosophical context. Another highly efficient positioning landmark, and one which seems to link to many more categories, is related to localizing agency in science. It is a creative activity serving human progress and command of reality, but whose activity is it?

"[…] [I]s it worth the effort to accumulate material about the creativity of that multitude of scholars who have followed in the masters' footsteps, making only modest contributions to the repository of knowledge?" This complaint is unjust for several reasons. Above all, science has become second nature to humanity. A great number of people work on its development and on the transmission of scientific accomplishments from one generation to the next. The fruits of this labour are colossal and the group effort serves to bolster and to provide material for the geniuses who produce the great syntheses.

Furthermore, it is not as if ingenious scientists resembled lonely peaks rising high above the plain. There is an intangible gradation of talent. Each individual, endowed with abilities in this or that domain, possesses a certain amount of creative intuition. Epoch-making discoveries, at least the experimental ones, have often occurred to people whose mentality had none of the characteristics of greatness. Given how scientific work is organized today, it would be no easy task to determine if science owes more to ingenious minds or to the shared labour of talented or simply capable people.[37]

Science as a paradigm of socialization, the trope which I mentioned in the introduction to this chapter, is present in this passage: science is second nature to humans, but it is a nature which only actualizes within a certain social organization. Science is strongly institutionalized and collectivized in this fragment of Białobrzeski, but we would be in the wrong to overlook an important tension which appears simultaneously with a statement that science "today" is a domain of collective effort. "Geniuses", "ingenious minds" are one category of scientific workers and the "talented or simply capable people" are another. Historically, this suggests that it has not always been that way; the role of the geniuses might have been more prominent in the past. On the other hand, they are still the consumers of group efforts of the mediocre crowd pursuing Kuhnian "normal science", but it seems that it is their work and their work only which is really original and creative. The answer to the question of scientific agency is ambiguous: a necessary condition of collective organization does not in itself make the collectivity an agent, even if, as Białobrzeski admits, "[a] lack of this influence" may be a major handicap in scientific work.[38]

---

[36] ▶ Błachowski: "The Problem of Scientific Creativity" (footnote 26), 184.
[37] ▶ Białobrzeski: "An Autobiographical Sketch" (footnote 28), 305.
[38] Ibid., 314.

Once science is established as a paradigm of socialization and a problem of scientific agency is set in terms of collective effort versus individual genius, all axial concepts are outlined and the positioning strategy based on sincerity standards is launched. Each and every author taking part in this positioning game by telling his autobiography must apply the sincerity standard strategically to answer the following question: How do I as a scientist locate myself among the agents of scientific creativity?

The first and the most sociological answer would be to name science a product of formation, a derivative of group belonging. Favorable conditions are crucial[39] and conditions of work may even determine local varieties of scientific work[40], but the tension between individual agency and collective effort continues:

> Being naturally endowed with no exceptional abilities, one can still cultivate and prepare one's mind for creative work, and one can skilfully put to use the potential for intuition supplied by nature.[41]

Natural ability and disposition are a prerequisite: a creative scientist cannot be collectively produced, even though the environment may undo his or her chances of development and career.

> The fact that an uninterrupted and regular course of study is relatively unimportant for people of the highest abilities is also apparent in the fact that a great number of ingenious individuals can be listed who did not study in their youth, and yet came to achieve exceptional results in later years. [...]
> But Galton's views have met with opposition from a great number of psychologists. It has been observed that the environment in which a highly gifted individual is raised often determines the appearance and development of his ability.[42]

Science is, therefore, a product of a right kind of mind, and one of the strongest voices positioning the author vis-à-vis this problem is that of August Krogh:

> I shall begin this autobiographical note by stating my firm belief that the essential traits in the intellectual make-up of a person are inherited and can be modified only to a comparatively slight extent by environmental influences.[43]

Here we find a strategic application of the threefold standard of sincerity in a particularly succinct form. The genre of autobiography requires the author to state what he believes about scientific agency, because it is in the light of this declaration that the whole account will be shaped (level one). He needs to state how objectivity will be guaranteed in his report and what facts (truths) will be included as a testimony (level

---

[39] See e. g. ibid., 314–315.
[40] See ▶ Emile Borel: "Contribution" (1936), 331–332, i. e. "Contribution (Documents sur la Psychologie de l'Invention Dans Le Domaine De La Science)," *Organon*, vol. 1 (1936), 33–42.
[41] ▶ Białobrzeski: "An Autobiographical Sketch" (footnote 28), 305.
[42] ▶ Błachowski: "The Problem of Scientific Creativity" (footnote 26), 196.
[43] ▶ August Krogh: "Visual Thinking" (1938), 334, i. e. "Visual Thinking," *Organon*, vol. 2 (1938), 86–94.

two). Thereby, the scope of report is limited by directing attention to innate properties, leading to a reduced attention to the author's environment (level three).

A minor positioning point, on which the authors differ, which seems to indicate that it was lacking strategic potential at the time, was the question about what kind of mind the right kind of mind actually was. The authors diverged on this point, some believed the right mind to be very narrow and specialistic[44], whereas the others preferred to think that a general, philosophical turn of mind is more useful to a scientist[45]. Among the innate gifts of the right kind of mind one is particularly singled out: a right kind of imagination: for a specific science or for science in general[46], sometimes though not always associated with genius[47]. Attempts at sincere self-classification are quite frequent and usually follow a similar line:

> The philosophical mind possesses no sharply defined abilities at all. Its curiosity is open to the whole world. The designation is based on the fact that philosophers usually possess such a universal inquisitiveness. This is not to question that creative work in philosophy requires certain special intellectual qualities. To be sure, even the discipline of philosophy is currently experiencing increased specialization, a narrowing of scope, but universality still characterizes most philosophers.
>
> The wealth of aptitudes and interests of an expert mind is focused on a given conceptual field, while for a philosophical mind it is dispersed across a number of fields that can be even be distant from each other. [...]
>
> I unquestionably belong to the second category, the philosophical mind. This is evidenced by my keen interest in fields outside of physics and related subjects such as mathematics and chemistry; as I already mentioned above, my interest spans the entirety of knowledge, and philosophy in particular.[48]

The tension between individual genius and collectivity leaves little place for other, alternative explanations of scientific creativity. A notion of serendipity appears, especially in the form of a lack of any obvious psychological and motivational 'deal-breakers'[49], but it is a lesser motif. Any psychoanalytical (or quasi-psychoanalytical) attempts are incredibly modest and I have only managed to spot one clear reference to science as a product of sexual drives in the account of Błachowski[50]. However, the rhetoric strategy here is distance and mediation through another author: Błachowski clearly does not believe that the self-narrative of a sexual background to scientific creativity might form a part of sincerity standard to be applied in scientific autobiography.

Surprisingly, certain elements are missing in the positioning strategy reconstructed here, which we might expect in the writings initiated by the Science Studies Circle.

---

[44] See ▸ Błachowski: "The Problem of Scientific Creativity" (footnote 26), 189.
[45] See ▸ Borel: "Contribution" (footnote 40), 113.
[46] See ▸ Białobrzeski: "An Autobiographical Sketch" (footnote 28), 313.
[47] See ▸ Błachowski: "The Problem of Scientific Creativity" (footnote 26), 195.
[48] ▸ Białobrzeski: "An Autobiographical Sketch" (footnote 28), 307.
[49] See ▸ Białobrzeski: "An Autobiographical Sketch" (footnote 28), 312.
[50] See ▸ Błachowski: "The Problem of Scientific Creativity" (footnote 26), 189–190.

Notably, any mission statements which would make a definite ethical stance internal and inherent to science are absent. Science has its goals and these goals are (as may naturally be assumed) cherished by eminent scientific creators. Nevertheless, scientific creativity as such does not seem to require a right kind of heart apart from a right kind of mind. No moral formation is necessary for a creative scholar, unless we perceive sincerity as a moral value, but sincerity is, arguably, necessary to produce an account of one's work, it is not a requirement of work as such. What is indispensable is a training in the right ways of thinking. It is a small but significant point, which in my opinion indicates to a strong domination of the genius-mediocrity dichotomy not only in narrative self-accounts, but also in the vision of science as a social phenomenon.

Another missing element – or, rather, an element which does not appear systematically and which is, therefore, of little strategic relevance – is the role of teachers and mentors in individual scientific development and in creative work. Leaders, teachers or colleagues are mentioned frequently, but they are part of the environment, a stimulus, in the absence of which creative powers might wane. However, this does not make them important individually, as person to person: a scholar in these accounts is invariably a lone star. His intellect does not emerge in a Platonic discourse with another mind of the right kind, which would make the partners in the dialogue important for each other as anything but catalyzers or facilitators. No Aristotelian friendship is necessary in scientific community.

Relative absence of this motif[51], a well-established and traditional one in debates on science and its ways since the classical antiquity, may be due to the fact that the Science Studies Circle attached great importance to the organization of science. The same reason may have acted against mission statements related more explicitly to the ethical formation of a scientist. Ethos and organization are two concepts coming from very different philosophical agendas. Ethos is about ways, habits and attitudes which form and reform practices and institutions, whereas organization is a way of procuring the right ways, habits and attitudes by means of precluding opportunities for their alternatives to come into being. There is a suggestion of a paradox in that a focus on subjective sincerity as a guarantee of objectivity was accompanied in these narratives with a focus on organization of science favoring creativity, of which ultimately one thing only could be said for sure: that it cannot be collectively produced.

5. A collective of sincere individualists

In the autobiographical accounts collected by the editors of *Nauka Polska* a group style of thinking about science is clearly manifest, despite the differences in the au-

---

[51] Nevertheless, see on star pupils ▸ Tadeusz Kotarbiński: *On the Skills of a Researcher* (1929), esp. 252–253, i.e. "O zdolnościach cechujących badacza," *Nauka Polska*, vol. 11 (1929), 1–10.

thors' worldviews, professional careers and – to an extent – their interpretation of the goal of the biographical exercise they took. This style was construed around a threefold standard of sincerity, defining autobiography as a narrative exercise in introspection, accompanied by an extraspection, but only insofar as to ensure an adequate and objective accounting for facts of creative biography. Circumspection, which might consist in examining the social surrounding of creative work and forming hypotheses regarding the external influences on internal process of creative work, was limited to few causal remarks. Quite on the contrary, the introspection was carefully carried out in full trust in its cognitive value and productive potential. Sincerity concentrated on defining the turn of mind necessary to report a creative mind's work and to create science, and this aspect was the most prominent. The third level, of delimitating facts which had and which had no relation to creative biography, was served by perfunctory or arbitrary remarks which usually derived their justification from a second level: the level of which objectivity and subjectivity collided by means of a juxtaposition between the ingenious, creative mind and the mediocre one.

The second level is also the ground for very intensive positioning work which demonstrates strategic uses of sincerity. Various rhetoric devices were used in order to preserve the picture of an objective narrator who describes himself as a creative scholar which at the same time implies that his mind is of more ingenuity than those of his peers and that this situation depends mostly on innate functions and dispositions of mind and only to a limited extent on organizational capability of society whom science is meant to serve.

This position resulted – a good example of a probably unintended consequence – in a preclusion of certain ethical issues. Some of them found their place in other discussions, including those regarding the role of university and professional versus liberal education. Others feature relatively little on the pages of *Nauka Polska*. Another consequence, and a graver one, is a rather a-sociological character of resulting accounts. It does not only lie in the weakness of circumspection. A more likely reason for the lack of sociological drive among the authors is the fact that they fail to see broader schemes of conditioning circumstances which might not only have an impact on their work but which might even challenge the idea that it is them, the individual scholars, who are the agents of scientific creativity.

I began this chapter with a metaphor of "thinking statues" introduced by Elias in order to depict the state of sociological and philosophical imagery which he had struggled all his life to undermine and dismantle.[52] *Homines clausi*, individuals set apart from their surroundings, influenced by it and influencing it in turn, but always individualized and complete in their mental setup, cognitive chances and goals, are not only a sad, but an inaccurate picture of what humans are like, including those humans whose occupation is science.[53] However, to be able to see that and to be able

---

[52] See Elias: Society of Individuals (footnote 7), ch. 2.
[53] See Bucholc: Global Community (footnote 8).

to put it into an autobiographical account, a very different strategy of sincerity would be required, which would position the narrators closer to sociological perspective of interdependent moves in constant interplay with each other. This more active, dynamic and network-like vision of science is absent in auto-biographies published in *Nauka Polska*, even though it is very much reverberating in the papers dealing with problems of scientific organization, where the collective perspective comes to the fore. However, the programmatic vision of science as a collective enterprise and a social phenomenon characterized not only by a psychogenesis, but also by a sociogenesis was not shared by the authors gathered under this unconsciously groundbreaking research agenda. One possible reason for that is their conviction that without a right institutional framework it hardly makes sense to analyze environmental determination of creativity. Where there is no rational organization of work, any systematic research striving to connect social causes and scientific results would inevitably fail in the face of diverse and heterogeneous social factors, potentially meaningful but supposedly much harder to classify than psychological variables. It may well be said as a concluding point that the most important aspect of sociogenesis of science which stems out of the material referred to in this chapter is the inherent individualism of scientific practitioners. Their ideas are more related to each other than their creators themselves, which supports Dobrowolski's argument about advantages of researching biographies of scientific achievements rather than those of individual scientists.

Of course, the authors of those autobiographical accounts were not a collective in any theoretically meaningful sense. Only some of them belonged to a regular network of exchange of ideas, be it on the subject of their professional interest or of their scientific vocation in general. They were united by various interests, personal links and philosophical connections and few of them ever worked together as scientists. Even Ludwik Fleck's conceptual framework[54], despite all its generosity, would be very difficult to apply systematically to the group of authors and correspondents of *Nauka Polska* before 1939. Nevertheless, marks of a style of thinking about science and scientists, which are marked by individualism and prevalence of psychogenetic over sociogenetic approach, are too distinct to be disregarded, even though the outlines of their social carrier remains blurry. The project of studying scientific creativity based on autobiographical narratives of scientists turned into a very interesting exercise in sociology of knowledge, debunking visions underlying scientific activities of people whose philosophical self-consciousness was, it seems, much more potent than their sociological imagination.

---

[54] See Fleck: *Genesis and Development* (footnote 31).

A Grammar of Actual Thinking

Antoni B. Dobrowolski within and beyond the Warsaw Science Studies Circle

Friedrich Cain

"Don't You have something in Your rich drawer that You could present at our science studies circle in November, something about scientific culture, science, or scientific creativity?" It was late in October 1929 that Stanisław Michalski (1865–1949) scribbled these words on a sheet of writing paper of the *Józef Mianowski Institute for the Promotion of Science* (Pol. Kasa im. Józefa Mianowskiego – Instytut Popierania Nauki), the largest science funding organization in Poland, to which he served as general secretary. "You could certainly produce it just like that, for You lectured about it […], and there is also Your new, mysterious work!"[1] The addressee, however, was reluctant. About one and a half months later, Antoni Bolesław Dobrowolski (1872–1954), a well known geophysicist and theorist of pedagogy received another letter from Michalski, yet this time under the letterhead of the director of the *National Culture Fund* (Pol. Narodowy Fundusz Kultury). "Antek, deserter," Michalski started one day after the 6th session of the *Science Studies Circle* (Pol. Koło Naukoznawcze) at which Dobrowolski had not presented, "how could You lay down arms yesterday! […] The majority demands that You present Your ideas […] in March".[2] Another day later, a follow up letter raised the stakes:

Give us <u>Your</u> evening of discussion […]. Today, it is necessary to promote a cult for science in our "faithless" society that longs for nothing; I feel this in every step of my activities, fighting with the opportunism of the ordinary people […]. Do provoke a discussion on the foundations of this "religion of science" at the sess. of the Science Studies Circle – it does not necessarily have to be a lecture. Just a plan for a discussion You would lead. If it was a lecture – high day.[3]

Michalski, a central figure in Polish science funding, who filled several posts at the same time, had one more ace to play.[4] He offered the immediate publication of the

---

[1] Muzeum Ziemi PAN: Materials of Antoni Bolesław Dobrowolski, box 262 (Korespondencja przychodząca, litery M–Z, 1902–1953), letter from St. Michalski, 30 Oct. 1929, no pagination. Translations from Polish by F.C.

[2] Ibid., letter from St. Michalski, 13 Nov. 1929, no pagination.

[3] Ibid., letter from St. Michalski, 14 Dec. 1929, no pagination. Accentuations follow the original text.

[4] For the most detailed biography of Stanisław Michalski see Jan Piskurewicz: *W służbie nauki i oświaty. Stanisław Michalski (1865–1949)*, Warszawa 1993.

text (and 10.000 reprints) in the upcoming volume of *Nauka Polska* – one of the Mianowski Fund's journals of which he was editor in chief: "It is necessary to ignite this question with a thunderbolt, as You are capable to do, to teach searching god within mankind, and not beyond."[5]

Alas, several more letters needed to be sent over the course of eight years before Dobrowolski would finally give a talk at the Science Studies Circle.[6] Although he had been a frequent attendant and discussant from the very beginning, he did not deliver a paper before the last session of the Circle of which a report was published.[7] On 27 April 1938, he presented his ideas on the role of learning in democratic societies as "Introductory Considerations on Higher Schools, and Especially Academic Schools (theory, questions, demands)". The "prescientific essay," as he called it, should launch a series of presentations to streamline discussions.[8] Strengthening the Science Studies Circle, however, was only an intermediate goal of Dobrowolski's "mysterious work" to which Michalski had alluded. Ultimately, Dobrowolski sought to describe a "Grammar of Actual Thinking" that would not only make scientific work most effective, but also render any other activity intellectual, too. He thus designed it as a vehicle for what Michalski had referred to as a "cult for science".[9]

This article traces the history of Dobrowolski's "Grammar of Actual Thinking" and its history between the Mianowski Fund and pedagogical discussions in interwar Poland. The project was "mysterious" in Warsaw, as Dobrowolski had never been reluctant to announce a solution to what he called an intellectual-civilizational crisis without ever presenting the monumental leap. While his first ideas date back to the late 1890s, Dobrowolski only managed to cast it into bigger frames in the 1920s. However, though he published smaller, highly polemic sketches over the years, he would leave behind an uncompleted manuscript when he died in 1954, partly due to losing greater parts of his materials during the German occupation in the Second World War. Although Dobrowolski was an eccentric figure in the Science Studies Circle, the scope that he was aiming for stands emblematic for the mindset around

---

[5] Muz. Ziemi PAN, Mat. of A.B. Dobrowolski, box 262 (footnote 1), letter from St. Michalski, 14 Dec. 1929, no pagination.

[6] Cf. ibid., several letters, no pagination, as well as ibid., box 228 (Kasa im. Mianowskiego. Korespondencja dotycząca współpracy A.B.D. z Kasą, 1923–1929, 1939), no pagination.

[7] For an overview of the discussions at the Science Studies Circle cf. ▶ **IN THIS VOL.** *Sessions of the Science Studies Circle (1928–1938)* with detailed bibliographical information. For an overview see also Stefan Zamecki: *Problematyka naukoznawcza na łamach periodyku "Nauka Polska. Jej potrzeby, organizacja i rozwój,"* Warszawa 2016.

[8] "Sprawozdanie dwunaste z działalności Koła Naukoznawczego," *Nauka Polska*, vol. 24 (1939), 187–241: 230–241. The original title of the talk was "Wstępne rozważania o szkołach wyższych, w szczególności akademickich (teoria, zagadnienia, postulaty)."

[9] The Polish term *nauka* has a broader scope than the English *science*. As an umbrella term, it can cover the whole specter of disciplines from the natural sciences to the humanities. Next to research, it can allude to the realm of teaching and learning or studying. Michalski's mentioning of the cult for science reflects the latter, broad idea, which was widely spread at the turn of the centuries but also met opposing voices. For the Polish case, see Magdalena Micińska: *Inteligencja na rozdrożach, 1864–1918*, Warszawa 2008, 115–122.

the Mianowski Fund – nothing less was at stake than modernizing the Polish nation and the young II. Republic, which was framed as a successor to the democratic tradition of the 18th century Polish-Lithuanian Commonwealth, a constitutional monarchy granting the nobility participation in decision-making.

The technological and infrastructural organization of the state was the paramount project to many government agencies and academic circles. Especially the latter often linked the blossoming of a new state and democratic culture to a broad intellectual development of the people. Cultural conservatives as well as leftist reform theorists often embraced education as a key moment to their ideas. An analysis of Dobrowolski's interventions helps to grasp the scope of ideas and political attitudes within the Science Study Circle and to embed it in contemporary pedagogical discussions.[10] It also adds another aspect to the academic self-figuration in public discourse in the young II. Republic – next to technocratic expertise and conservative politics of memory, Dobrowolski started from cultural criticism.[11]

Based on his specific critique of civilization, Dobrowolski proposed a holistic approach to education that included all layers of society. As he told his audience in the Science Studies Circle right away in April 1938, a future theory of higher education would "demand" and "build" a "democratic society" at the same time.[12] A sane society could only rely on consciously structured minds of individuals, who would follow

---

[10] In the literature on the history of Polish pedagogy the most comprehensive descriptions of Dobrowolski's concepts are probably to be found in Stanislaw Michalski: *Koncepcje systemu edukacji w II Rzeczypospolitej. Studium z pedagogiki porównawczej*, Warszawa 1988, 151–178 and Wincenty Okoń: *Wizerunki sławnych pedagogów polskich*, Warszawa 1993, 58–83. Dobrowolski's unpublished texts were edited within a decade after his death in 1954: Antoni B. Dobrowolski: *Ustrój oświatowy* (= Pisma pedagogiczne, vol. I), Wrocław 1958, *Nowa dydaktyka* (= Pisma pedagogiczne, vol. II), Wrocław 1960 and *Moralność umysłowa (materiały)* (= Pisma pedagogiczne, vol. III), Wrocław 1964.

[11] Literature on Polish interwar pedagogy is vast, yet a comprehensive history of political and cultural contexts is still missing. For specific perspectives cf. Danuta Krześniak-Firlej, Stanisław Majewski and Waldemar Firlej: *Organizacja szkolnictwa w II Rzeczypospolitej 1918–1939*, Kielce 2014, Maciej Fic, Lech Krzyżanowski and Miłosz Skrzypek (eds.): *Dwa dwudziestolecia Rzeczypospolitej. Oświata, religia, kultura i społeczeństwo. Próba bilansu*, Katowice 2010, Mirosław S. Szymański: *Pädagogische Reformbewegungen in Polen 1918–1939. Ursprünge – Verläufe – Nachwirkungen*, Köln e.a. 2002, Joanna Sadowska: *Ku szkole na miarę Drugiej Rzeczypospolitej. Geneza, założenia i realizacja reformy Jędrzejewiczowskiej*, Białystok 2001, Władysława Szulakiewicz: *Historia oświaty i wychowania w Polsce 1918–1939. Studium historiograficzne*, Toruń 2000, Danuta Drynda (ed.): *Inspiracje dla współczesnej edukacji w dydaktyce Drugiej Rzeczypospolitej*, Katowice 2000, Leonard Grochowski: *Studia z dziejów polskiej szkoły i polskiej pedagogiki lat międzywojennych w kontekście europejskim*, Warszawa 1996, Karol Poznański (ed.): *Oświata, szkolnictwo i wychowanie w latach II Rzeczypospolitej*, Lublin 1993, Danuta Drynda (ed.): *Studia z historii polskiej pedagogiki. Koncepcje pedagogiczne w Drugiej Rzeczypospolitej*, Katowice 1993, id.: *Pedagogika Drugiej Rzeczypospolitej. Warunki, orientacje, kontrowersje*, Katowice 1987, Franciszek Bereźnicki: *Innowacje pedagogiczne w Polsce (1918–1939)*, Szczecin 1984, Feliks W. Araszkiewicz: *Ideały wychowawcze Drugiej Rzeczypospolitej*, Warszawa 1978.

[12] "Sprawozdanie dwunaste z działalności Koła Naukoznawczego" (footnote 8), 230. There is some linguistic ambiguity to this quote. It rests on the unequivocal use of the Polish verb „zakładać", which translates as „presuppose" and „establish" at the same time.

a cult of science rather than religious or other "pre-scientific" modes of reasoning, among which he also listed philosophy. He would never miss a chance to point out such shortcomings, and criticize them even more fiercely if he detected them among 'well-educated' people. Thus, he did not hold back when spotting alleged intellectual flaws in critcized as too loose and random. He frequently urged to narrow down topics and methods.[13] Though his own remarks could not yet draw on strong research insights, as Dobrowolski introduced his talk in 1938, he would refrain from simple "philosophical considerations". Instead, he announced to present a plan for the reorganization of universities that would entangle research, application, and education and save culture and civilization.[14]

Both the quoted correspondence and the long awaited talk elucidate some dynamics within and around the Circle. They also help to comprehend the greater framework in which the Circle was working. All its members aimed at the well-being of the new Polish state that had to administrate territories formerly governed by no less than three different powers and to integrate citizens educated without any official Polish curriculum for over a century. However, the biographies of the Circle's members were as diverse as the specter of academic disciplines and political standpoints they represented.[15] Dobrowolski was no exception. His career had taken remarkable intellectual and professional shifts and its geographical trajectory spanned vastly, from Warsaw across Tiflis, Zürich, Liège, Uppsala, and Antarctica. Starting from philosophical studies, he had become a star in Polish biology and geology by the end of the First World War, and later one of the most eminent meteorologists in interwar Poland. At the same time, he worked his way up from a clerk's position in the *Ministry of Religious Beliefs and Public Education* (Pol. Ministerstwo Wyznań Religyjnych i Oświecenia Publicznego, MWRiOP) to professorial honors. When Michalski asked Dobrowolski to speak to the Circle, the latter had only shortly been appointed professor of theoretical pedagogy at the *Free Polish University* (Pol. Wolna Wszechnica Polska) in Warsaw.

Although Dobrowolski was deeply unsatisfied with the diversity of subjects in the Circle's discussions, he did attend most sessions. He never hesitated to discuss and even if he found it defective at times, Dobrowolski kept considering the Circle an

---

[13] See for example "Sprawozdanie dziesiąte z działalności Koła Naukoznawczego," *Nauka Polska*, vol. 22 (1937), 191–218: 204.

[14] "Sprawozdanie dwunaste z działalności Koła Naukoznawczego" (footnote 8), 231–234. Dobrowolski divided his talk into two sections: *I. Higher Schools as Centers of Research* and *II. Higher Schools as Colleges.*

[15] Prussia, Imperial Russia, and Austria-Hungary divided the territories of the Polish-Lithuanian Commonwealth in the late 18th century to incorporate them in their realms until the First World War. Recent historiography has suggested to turn away from national to imperial biographies in this context: Martin Aust and Frithjof Benjamin Schenk (eds): *Imperial Subjects. Autobiographische Praxis in den Vielvölkerreichen der Romanovs, Habsburger und Osmanen im 19. und frühen 20. Jahrhundert*, Köln 2015 or Tim Buchen and Malte Rolf (eds): *Eliten im Vielvölkerreich. Imperiale Biographien in Russland und Österreich-Ungarn (1850–1918)*, Berlin 2015.

important forum for his project. It was a unique institution in interwar Poland, where many leading figures of Science Studies in Poland met. The Circle was closely connected to the Mianowski Fund and its publications, where the proceedings were published.[16]

For more than fifty years, Dobrowolski ordered his humanistic œuvre around two core projects of civilizational critique. Both his *Catalogue of Mental Actions* and the *Code of the Intellectual Morality*[17] were originally drafted as interventions into scholarship, and as such ever connected to the well being of society and, of course, the Polish state. In fact, they were to be two cornerstones of a new intellectual civilization that would rely on the "Grammar of Actual Thinking" Dobrowolski aimed to describe. He planned to design it by the way of induction and for this he relied on the Circle, the Mianowski Fund, and especially its two journals *Nauka Polska* and *Organon* as a repository for sources. Michalski collaborated closely when Dobrowolski suggested establishing a series of *biographies of scientific thoughts* in the two journals. Dobrowolski dreamt to categorize a set of basic intellectual acts by dismantling actual processes of human thought and uncovering the complex structures of intuitive sparks, creative dead ends and reckless shortcuts. Describing what he called "intellectual sins," he hoped to extract the basic virtues of scientific and artistic creativity.

After some notes on Dobrowolski's life in the second section, the following section will present his cultural critique of historical stages of thinking. Section four will shed light on Dobrowolski's attempt to study creativity and the morals of thinking. The fifth section will reconsider Dobrowolski's ideas about how these virtues, once described, would help to reconstruct a network of institutions to foster a "higher intellectual culture" and thus a good social life in a democratic society. Based on the protocol from the discussion after Dobrowolski's talk, the final section analyzes the political array of the Science Studies Circle.

---

[16] Michalski had organized intellectual circles earlier, where many of the regular attendants of the Science Studies Circle had met before. Cf. Piskurewicz: *W służbie nauki i oświaty* (footnote 4), 79 or several diary entries of the young Stanisław Ossowski: *Dzienniki (vol. 1 1905–1939)*, ed. by Róża Sułek, Warszawa 2019, 136, 153, 170, 177.

[17] Fragments of both works were published posthumously from Dobrowolski's materials. Cf. Dobrowolski: *Moralność umysłowa (materiały)* (footnote 10). For the Polish typescript and materials of the *Catalogue of Mental Action* see Muz. Ziemi PAN, Mat. of A.B. Dobrowolski, box 116 (Katalog czynności umysłowych. Klasyfikacja czynności, materiały ogólne. Notatki, 1949–1953). The materials also contain an uncompleted English translation of the text. Cf. ibid., box 33 (Catalogue of the Mental Action). The Polish typescript and materials of the *Code of the Intellectual Morality* can be found in box 99 (Kodeks moralności umysłowej. Tłumaczenie, interpretowanie, teorie. Notatki rękopiśmienne, 1906, 1934, not dated), an also incomplete translation in box 32 (Code of the Intellectual Morality).

1. The life of Antoni Bolesław Dobrowolski

When Dobrowolski spoke at the 37th session of the Science Studies Circle, he presented a provisional conclusion of the forty years that he had worked on his pedagogical endeavor. Yet, although he was reluctant to deliver this talk, he had not been shy to tell his own story before. The most detailed narrative of his life is a lengthy autobiographical text that was published in Nauka Polska in 1928. After his death in 1954, a reworked and supplemented version was published from his materials.[18] Born from a peasant family in 1872 in the Russian partition of Poland, Dobrowolski left for Warsaw to attend a lyceum after primary school. From the age of 12, he had to provide for himself as a private tutor, which was not uncommon at the time. He was also part of various reading circles that Polish students organized to expand the official Russian curriculum.[19] Studying in Warsaw, Dobrowolski became part of the Polish *inteligencja*, a specific social stratum that consisted of people with higher education, but very different vocations and a wide array of political views.[20]

At that time, Dobrowolski must have become acquainted with Polish positivism, a philosophical and literary movement close to the ideas of Auguste Comte, John Stuart Mill, and Herbert Spencer. It promoted scientism and empirical studies against metaphysics, speculation, and mysticism in search of a *Religion of Humanity* (Comte). It gained influence after the failed January Uprising in 1863/64 led to a crisis of romanticism. The concept of "organic work", promoting economic and intellectual development, replaced heroic fight and martyrdom as central keywords. From now on, social life should be based on (scientific) reason.[21] In 1881, the Mianowski Fund was founded to support Polish scientists with scholarships and printing subsidies.[22] At the same time, women, who were excluded from official higher education, organ-

---

[18] ▶ Antoni B. Dobrowolski: "My 'Scientific Biography'" 1928, i.e. "Mój 'życiorys naukowy,'" *Nauka Polska*, vol. 9 (1928), 68–216 and for the posthumous extended text id.: *Mój życiorys naukowy*, Wrocław 1958.

[19] For the most comprehensive biography of Dobrowolski see Stanisław Michalski: *Antoni Bolesław Dobrowolski Prekursor Współczesnej Edukacji*, Warszawa, Poznań, 1984. The most detailed CV can be found in Dobrowolski's archive: Cf. Muz. Ziemi PAN, Mat. of A.B. Dobrowolski, box 256 (Curriculum vitae Antoniego Bolesława Dobrowolskiego). So far all biographical work heavily builds on Dobrowolski's autobiography, his collections have rather been neglected.

[20] For a broad history of Polish *inteligencja* see the volumes in the series *Dzieje inteligencji polskiej*: Maciej Janowski, *Narodziny Inteligencji, 1750–1831*, Warszawa 2008, Jerzy Jedlicki, *Błędne koło, 1832–1864*, Warszawa 2008, as well as Micińska: *Inteligencja na rozdrożach* (footnote 9).

[21] Cf. Jan Surman: "The Contexts of Early Polish Positivisms, 1840s–1900s," in: *The Worlds of Positivism. A Global Intellectual History, 1770–1930*, ed. by Johannes Feichtinger, Franz L. Fillafer, id., London 2018, 239–272. For a broad overview on Polish national activities in the late 19th century see Brian Porter: *When Nationalism Began to Hate: Imagining Modern Politics in Nineteenth-Century Poland*, Oxford, New York 2000.

[22] A concise English history is told in Piotr Hübner, Jan Piskurewicz and Leszek Zasztowt: *A History of the Józef Mianowski Fund*, Warszawa 2013. Generally on funding science and culture in 19th century Warsaw see Jan Piskurewicz: *Warszawskie instytucje społecznego mecenatu nauki w latach 1869–1906*, Wrocław e.a. 1990.

ized in clandestine circles and became even more influential within the *inteligencja*. Since 1885, these circles formed a densely meshed network, henceforth offering systematic education on a large scale, soon known as *Flying Universities*. One of the central figures was Zofia Daszyńska-Golińska, who had studied in Switzerland and was the first woman to attain a doctorate at the university of Zürich in 1891. The best-known graduate from these courses undoubtedly is two times Nobel Prize winner Marie Skłodowska-Curie. When Tsarist restrictions were loosened after the Russian Revolution in 1905, the clandestine activities could be made official, and soon the *Society for Scientific Courses* (Pol. Towarzystwo Kursów Naukowych, TKN) was established in Warsaw. After 1918, in independent Poland, the society founded the private Free Polish University, which gained official recognition in 1929. Another important educational initiative was the *Handbook for Autodidacts* (Pol. Poradnik dla Samouków) written by well-known Polish scholars. Distributed from 1898 onwards, these well received textbooks adjusted their subject matter to the needs of autodidactic learning with subsequent levels of complexity. Extensive bibliographical data paved the way for further individual studies.[23]

Coming to Warsaw, Dobrowolski also got in touch with the conspiratorial activities of the *Polish Social-Revolutionary Party "Proletariat"* (Pol. Polska Socjalno-Rewolucyjna Partia "Proletariat"), also known as the *Second Proletariat* (Pol. Drugi Proletariat), which did not refrain from violent action. Shortly after finishing school in 1891, he was arrested along with many of the members of group. After spending parts of a three years prison sentence in the infamous Warsaw citadel, Warsaw's Pawiak prison, and St. Petersburg, he was exiled to the Caucasian Tiflis Governorate in 1892. With the help of Armenian revolutionaries he finally escaped across the Black Sea. For a year, Dobrowolski tried to settle in Zürich, but left for Belgian Liège in 1895 when he could not attain a permanent residence permit in Switzerland. In Zürich, he attended various scientific classes, but also the philosophical lectures of Richard Avenarius, whose empiriocriticism related to Positivist ideas. Eventually, Dobrowolski figured that the natural sciences granted more systematic knowledge and finally concentrated on physics and biology when moving to Liège, where he earned a degree in 1897.[24] When the famous Belgian Antarctica Expedition left Antwerp in August 1897 on board the *Belgica*, such eminent figures as Roald Amundsen (1872–1928) and Frederick A. Cook (1865–1940) were accompanied by Dobrowolski,

---

[23] The *Poradniki* series was a very dynamic project. Five parts of a first edition were published between 1898 and 1911. Part I covered mathematics and natural sciences, part II philology and history, part III social sciences and jurisprudence as well as philosophy, part IV pedagogical problems, and part V the relationship of mankind and the world, especially theories of evolution. Part VI described the history of thought. Ten further parts were published between 1915 and 1932 in a second edition. These parts covered philosophy, logics, theory of science, but also physics, geophysics, meteorology, mathematics, crystallography, mineralogy and petrography, botanics, zoology, paleozoology and others.

[24] Dobrowolski: *Mój życiorys naukowy* (footnote 18), 34–57.

who was hired as an assistant by the Polish geophysicist Henryk Arctowski (1871–1958).[25]

Upon his return, Dobrowolski was granted a scholarship to publish the meteorological data he had gathered in Antarctica.[26] He only left Liège in 1907, when an amnesty allowed many Poles to return from exile after the Tsarist autocracy was shaken in 1905. Until the First World War, Dobrowolski worked as a teacher near Warsaw. After publishing a popular book on Antarctica in 1914, the Mianowski Fund granted him a scholarship sending him to Sweden for three years. Eventually, he finished a crystallographic monograph on the *Natural History of Ice* in 1923. This book made him famous as the (Polish) father of *kriologia*, i.e. the science of ice, a subfield of hydrology. In 1918, when Warsaw was still under German military control at the end of the war but civil administration had already been reactivated, Dobrowolski returned to the city without ever moving again.[27]

In the young Polish state, Dobrowolski first took a post in the planning department of the newly founded Ministry of Education. He was entrusted with strategic and structural planning of educational politics and mostly with the re-organization of the school system. However, integrating three systems that had functioned in the partitioned Polish lands before 1918 was not easy, as the ideas on the final outlook differed widely. In fact, these discussions were never settled until 1939. Important problems to resolve were connected to the system's overall structure: Which minimum of education was necessary for all pupils? When could ideal candidates for different careers (e.g. future scholars and craftsmen) be separated without wasting material and intellectual resources? Which pedagogical concepts should be applied? Dobrowolski opted for a broad system of basic, general, and specialist education including institutions that should trigger intellectual interest within the masses. However, he left the ministry in 1922, allegedly after severe conflicts with higher officials.[28] Having worked at a bank for some months, he joined the *State Meteorological Institute* (Pol. Państwowe Instytut Meteorologiczne), where he was appointed vice-director in 1924 and director three years later. Poor health conditions forced him to retire in 1929.[29]

Next to his inhibited career as a clerk and his success in geophysics, Dobrowolski also managed to climb the ranks in another field: In 1927, he accepted the chair for

---

[25] Ibid., 87–96, 309–311.

[26] For a bibliography of Dobrowolski's publications on the expedition see ibid., 402–412.

[27] Antoni B. Dobrowolski: *Historia naturalna lodu*, Warszawa 1923. For the stay in Sweden cf. id.: *Mój życiorys naukowy* (footnote 18), 127–141.

[28] Ibid., 146–153. Cf. also Muz. Ziemi PAN, Mat. of A.B. Dobrowolski, box 256, 7–8.

[29] Dobrowolski did not completely retire from geophysical and related activities. In 1929, he co-founded the *Association of Geophysicists* (Pol. Towarzystwo Geofizyków) and in 1934 he initiated a seismological observatory. He was patron to the 2nd Polar Year in Poland and the Polish exhibition to Spitzbergen in 1934. In 1938 he founded the *Polar Circle* (Pol. Koło Polarne). For later popularization activities see Joanna Popiołek: "'Polar Action' of Antoni Bolesław Dobrowolski in the Interwar Period," *Polish Polar Research*, vol. 19/1–2 (1998), 31–36.

theoretical pedagogy at the Free Polish University, where he would work among many reform oriented scholars until 1939.[30] At the same time he was an important figure to the Society of Democratic Education "New Tracks" (Pol. Towarzystwo Oświaty Demokratycznej "Nowe Tory").[31] Dobrowolski survived the German occupation during the Second World War in seclusion while continually working on pedagogical matters.[32] In 1945, he first helped to establish the new university in Łódź and later became a professor of pedagogy at Warsaw University before retiring for the second time in 1949. Dobrowolski went on to publish in all his fields of interest, yet never finished the two central pedagogical projects before he died.[33]

The profound heterogeneity of Dobrowolski's biography owes to the partitions of Poland and its threefold periphery before the First World War. The breakdown of classical imperialism left a void that was filled by the idea of an integrate national state and the search for a corresponding political system, infrastructure and (democratic) culture. In 1918, old careers were as precarious as the situation of many individuals and ethnic groups, who had to re-establish themselves within the new borders of the II. Republic, but also needed to reconceptualize in relation to the new state. The post-imperial situation forced to re-narrate life stories and to integrate past experiences in the new order, which was greeted emphatically but also seemed fragile to many.[34]

2. A New Man, a New Conscience, a New Religion, and a New Church

In the same year that Dobrowolski was appointed professor of theoretical pedagogy, he published the first version of his lengthy autobiography in Nauka Polska, in which he also sketched his way into pedagogy. He traced his first teaching experiences back to his teenage years as a tutor for younger pupils and himself when much of his education relied on autodidactic study and attending secret circles to read books censored at the time. He recalled devouring the relatively well equipped prison library of the Warsaw Citadel, and henceforth referred to this time as his "First University".[35] Theoretical interest rose when he acquainted himself with progressive pedagogy in

---

[30] Michalski: *Koncepcje systemu edukacji w II Rzeczypospolitej* (footnote 10), 93.
[31] Ibid., 105–121.
[32] The editors of Dobrowolski's autobiography state that he worked out his basic theses in the Humanities during the War. Janusz Ostrowski, Andrzej Brożyna: "Nota Redakcyjna," in: Dobrowolski: *Mój życiorys naukowy* (footnote 18), VIII–XII: IX.
[33] For comprehensive résumés of Dobrowolski's pedagogical projects see Michalski: *Antoni Bolesław Dobrowolski* (footnote 19) and Okoń: *Wizerunki sławnych pedagogów polskich* (footnote 10), 58–83.
[34] Cf. Martin Müller-Butz: *Blicke zurück nach Osten. Erfahrungen des Imperialen in Lebenserzählungen der polnischen Intelligenz im 20. Jahrhundert*, München, Wien 2019.
[35] Dobrowolski: *Mój życiorys naukowy* (footnote 18), 20. See for similar accounts Padraic Kenney: *Dance in Chains. Political Imprisonment in the Modern World*, New York 2017, ch. 1.

Zürich. Later, when in Belgium, experimental, cooperative, and personalized learning attracted his attention.[36] Dobrowolski recalled installing his very own "pedagogical laboratory" upon his return to Warsaw, where he put to test the "whole variety of traditional and new 'methods,'" with an emphasis on "'heuristic methods' [that grant] a maximum of autonomy to the child".[37] Unfortunately, there are no descriptions of a physical space and Dobrowolski very likely used the term metaphorically.

This narrative forms the background to Dobrowolski's drop out from the ministry of education a few years later. It was not without pride that he recalled his reform ideas being rejected as too radical. He criticized what seemed all too pragmatic to him, namely approaches that only aimed to produce work force. Instead, he aimed to exchange teaching by a "cultivation of thought".[38] He wanted to foster a culture of learning, leading every individual to autonomous decisions and the perfectly fitting vocation. No talent should be lost due to unfavorable social factors. Yet, instead of knitting a tighter net of institutions providing educational guidance, Dobrowolski wanted to design structures that would train individual interest and curiosity. He was eager to equip individuals with a set of mental techniques and bring forward a general intellectual mindset fond of democracy instead of sketching sets of facts to be known for specific occupations or positions.

One basic assumption was the distinction of four historical stages of thinking. Only the final stage could bring forth a society of free people, namely "proper scientific reasoning". The worst of the three antecedent stages was a "misleading way to think," with dangerous deductions from certain theoretical corsets that could not meet life's necessities: philosophy.[39] Not much better, but less suspicious was magi-

---

[36] Many of these theorists came from medical disciplines, such as Edouard Claparède, Maria Montessori or Ovide Decroly, whereas John Dewey, Helen Parkhurst, Rudolf Steiner, Robert Baden-Powell, or Georg Kerschensteiner had a background in philosophy, psychology, or classical pedagogy. The most prominent progressive pedagogue in Poland was physician Janusz Korczak. He gained international fame with his orphanage, before being killed in a German death camp in 1942, together with a group of children.

[37] Dobrowolski: *Mój życiorys naukowy* (footnote 18), 115–225: 119.

[38] Id.: *Wstęp do wydawnictwa protokóły lekcyj*, Lwów, Warszawa 1921, 6.

[39] Id.: *Moralność umysłowa (materiały)* (footnote 10), 11. Dobrowolski's archives contain several preliminary English translations of his Polish texts that have not been published. They are used as guidelines for translations from the Polish original here. In the "Introduction to the Catalogue of Mental Actions" Dobrowolski wrote concerning philosophy: "If it is a fact that philosophers as philosophers have certainly not contributed to the explanation of the structure and dynamics of mental work, but have rather obscured the problem (I repeat: as philosophers!), it is none the less a fact that philosophers were also many of the most powerful brains which were able not only to succumb to that natural weakness of the mind, but also to observe and draw conclusions in a fruitful manner. In philosophical systems of all times can everywhere be found though scattered and very fragmentary – interesting and valuable remarks on the human mind [...]. But all [the philosopher's] sharp observations, deep remarks, and pertinent conclusions have nothing in common with their philosophies as philosophies; they have only been interwoven, uselessly and unfortunately, into their philosophical cobwebs which have played here only the role of a serious obstacle on the solely efficient road of systematic empirical research." Muz. Ziemi PAN, Mat. of A.B. Dobrowolski, box 33, 3–4. Accentuations follow the original text.

co-mystical (or religious) thinking in ancient societies, explaining the unexplainable with myths and forms of transcendental will. The third, still insufficient mode was pre-scientific or practical reasoning, which could make life easier, yet still left space for social asymmetry.[40] According to Dobrowolski, the four modes of reasoning overlapped historically, inhibiting democratic life in modern societies. Their analysis, description and practical separation became subject to an inductive, pedagogical theory.

Again, Dobrowolski was not a solitary with this approach, but deeply embedded in the history of thought. Auguste Comte, whose work was well known in Positivist Warsaw, had formulated a similar critique of philosophy several decades earlier. According to Comte, a theological era, in which personified deities granted for an explanation of the world, had been superseded by a metaphysical age when impersonal abstract concepts formed the basis of knowledge. Yet the highest, modern stage was the era of scientific method that would thrive to explain relationships of cause and effect based on scientific method, and thus experiment and method driven comparison.[41] Ideas about practical thinking based on empirical observation were known in contemporary progressive pedagogy, but also left their mark in Pragmatism.[42] In 1938, Gaston Bachelard refined Comte's three stage model of the formation of the scientific mind.[43]

Deeming himself a radical thinker, Dobrowolski was confident enough to not climb any giant's shoulders explicitly. Although he referred to several authors, his œuvre mentions only one name with unconditional praise. "Proper scientific reasoning", as he called it, could only be possible in the integrate realm of humankind and the world. This clearly recalls Dobrowolski's Zürich contacts with Avenarius' empiriocriticism, and, given his socialist past in 1890s Warsaw, it would also make plausible references to Alexander Bogdanov's *Empiriomonism*.[44] Yet, interestingly enough, in the inaugural lecture to his professorship of theoretical pedagogy, Dobrowolski neither referred to Avenarius nor Bogdanov. Instead, his talk gravitated towards Rudolf Maria Holzapfel (1874–1930), a Cracow born student of the former, whom Dobrowolski might have met in Switzerland. Holzapfel had published a "psychology of social feelings," the *Panideal*, in 1901. None other than Ernst Mach contributed a pref-

---

[40] Id.: *Moralność umysłowa (materiały)* (footnote 10), 11–12.

[41] See for example Auguste Comte: *Discours sur l'esprit positif*, Paris 1844, 2–5.

[42] Cf. Julian Bauer: "Kann man lernen, mit Gedanken zu experimentieren? Ernst Machs Vorstellung des Gedankenexperiments im Kontext der zeitgenössischen Pädagogik," *Berichte zur Wissenschaftsgeschichte*, vol. 38/1 (2015), 41–58: 44–46.

[43] Gaston Bachelard: *The Formation of the Scientific Mind. A contribution to a Psychoanalysis of Objective Knowledge*, Manchester 2002, 18–23.

[44] Cf. Aleksandr A. Bogdanov: *Empiriomonizm: Stat'i po Filosofii* (3 vol.), Moscow 1904–1906. Although very much committed to political discussion, Dobrowolski did not comment on any ideological conflicts between different currents of Marxism or in the Soviet Union. However, he did praise the Soviet school reforms when he inveighed against the Polish educational system in a later article.

ace, praising Holzapfel's introspective abilities as a challenge to the methods of modern psychology.[45] Now mostly forgotten, Holzapfel was a star for some time in the early 20th century. Romain Rolland and others even nominated him for the Nobel Prize in Literature several times.[46]

Holzapfel's largely aphoristic *Panideal* identifies morals, conscience, fight, solitude, prayer, longing, and hope as basic human feelings. They are conceived as motors of social life that need to be equilibrated by educating the soul to develop itself to the full in order to enhance human culture in all important fields, namely the conduct of social life, education, politics, economy, art, religion, science.[47] Dobrowolski praised Holzapfel's analysis of contemporary life – "today's chaos"[48] – and his idea to develop a "New Man" with a "New Conscience", who would orient himself along the lines of a "New Religion". A "New Church" of art and science would guard the new morals and ethics developed in this spirit.[49] The "New Man" would finally leave behind belief and magico-mystical thinking to turn to knowledge. Dobrowolski's inauguration lecture shows his full admiration for Holzapfel's investigations, especially in the psychology of scholars, inventors, artists, founders of religions – Holzapfel also mentioned "cultural creators" ("Kulturgestalter"). Adding the pedagogue to this order, Dobrowolski finally formulated a special interest for his future investigations in theoretical pedagogy. He announced to develop a new, genuine pedagogical method

---

[45] Rudolf Maria Holzapfel: *Panideal. Psychologie der sozialen Gefühle*, Leipzig 1901. See pages VII–VIII for Mach's preface.

[46] For nominations cf. nobelprize.org. Wladimir Astrow, a follower of Holzapfel, mentioned Thomas Mann and further admirers. Wladimir Astrow: *Rudolf Maria Holzapfel – Der Schöpfer des Panideal. Ein neues Leben*, Jena 1928, 72. Astrow's book on Holzapfel was prefaced by Romain Rolland. In 1903, Holzapfel completed a philosophical dissertation "Wesen und Methoden der sozialen Psychologie" with Ludwig Stein in Bern.

[47] Holzapfel's Panideal was strongly influenced by Avenarius, but also Friedrich Nietzsche. Dobrowolski later reported that Hermann Lotze had also been an important inspiration for Holzapfel. Dobrowolski: *Mój życiorys naukowy* (footnote 18), 39–40.

[48] Antoni B. Dobrowolski: *Rudolf Maria Holzapfel i podstawy naukowe wychowania uczuć*, Warszawa 1927 (= reprint form the monthly *Droga*, vol. 5/1–3 (1927), 22–39), 15. The reprint announces a Polish edition of Holzapfel's book for the same year. However, the two volumes of *Wszechideał. Życie duszy i jego nowe postacie społeczne* translated by Stanisław Vincenz and Izydor Blumenfeld were only published in 1936 in Warsaw.

[49] Ibid., 19. This is very close to the *Catalogue of the Mental Action* as it was published posthumously. Dobrowolski: *Moralność umysłowa (materiały)* (footnote 10), 9. The translation from Dobrowolski's archive reads as follows: "Consequently, the primary postulate of the new man, the first cardinal condition of the new life, is – beside a certain indispensable amount of general education – the general training of the mind […]. This […] means acquisition of one's own justifiable judgement in general problems that should be the concern of every one. Among these problems the most important are those which pertain to the so called weltanschauung [sic!], in particular the view of the problems of mankind, since it determines, truly or falsely, what we are to deem to be of paramount importance, and thereby provides us with a guide to life. The problem of weltanschauung [sic!] comes to-day to the forefront with an elemental force in connection with the universal crisis of values: creeds, ideals, ideologies, [normative world views, religions] which is clearly noticeable for all those who are able to see." Muz. Ziemi PAN, Mat. of A.B. Dobrowolski, box 33, 1. The insertion in the last sentence was left out in the translation.

granting a "maximum of true autonomy" to thought. With Holzapfel, he wanted to finish Nietzsche's religious dream of the "Übermensch" and Jean-Marie Guyau's evolutionist attempt to create a moral system on a "scientific basis".[50]

Though quite in favor of progressive models, Dobrowolski was reluctant to attach to any specific pedagogical theory or method. Based on his own teaching experience in his Warsaw 'laboratory', he dismissed the greater part of pedagogical literature as far too general and denounced most concepts of autonomy as "masked suggestion". Also the psychological and sociological literature that he would read later did not convince him. Only parts of Hugo Münsterberg's *Psychotechnik* (1914) made an exception.[51] Apart from Holzapfel there was only one other reference to receive similar honors from Dobrowolski, namely Ovide Decroly, a Belgian pedagogue, for whom he wrote an enthusiastic obituary in 1932.[52]

Dobrowolski attested a very poor condition to (Polish) pedagogy, which he also linked to wrong political decisions and administrative mistakes. In fact, in most of his polemics he did use a narrative of a pedagogical tabula rasa. Theoretical pedagogy, as he diagnosed, had been neglected systematically in the formative years of the II. Republic. It was only upon his personal initiative as a ministry official that two relevant projects had been promoted, namely Józefa Joteyko's intelligence tests and a publication series of lecture protocols by outstanding teachers.[53] Dobrowolski was convinced that it was now up to him to collect and systematize material, and to develop the first scientifically grounded, synthetic handbook for teachers, which would be based on an inductive study of creative processes. How he combined his early political socialization and philosophical readings in order to realize the New Man of intellect that he dreamt of with Holzapfel under his pillow will be subject in the following section.

---

[50] Dobrowolski: *Rudolf Maria Holzapfel i podstawy naukowe wychowania uczuć* (footnote 48), 15–18. In 1945, Dobrowolski again referred to Nietzsche as a precursor, and also added André Gide to this list. Id.: "O potrzebie katalogu czynności umysłowych", *Kuźnica*, vol. 1/13 (1945), 5–6: 6.

[51] Id.: *Rudolf Maria Holzapfel i podstawy naukowe wychowania uczuć* (footnote 48), 7.

[52] Id.: "Owidiusz Decroly (1871–1932)", *Przegląd Współczesny*, vol. 11/43 (1932), 399–403.

[53] Cf. id.: *Mój życiorys naukowy* (footnote 18), 154–156, as well as Muz. Ziemi PAN, Mat. of A. B. Dobrowolski, box 256, 8–9. Dobrowolski wrote an introduction to the series: Dobrowolski: *Wstęp do wydawnictwa protokóły lekcyj* (footnote 38). For the tests and protocols see Józefa Joteyko: *Poziom inteligencji uczniów gimnazjum niższego. Badania eksperymentalne*, Warszawa, Lwów 1922, Wanda Haberkantówna: *Protokóły lekcyj przyrodoznawstwa odbytych w r. szkolnym 1918/19 w klasie pierwszej Gimnazjum p. Kowalczykówny w Warszawie (Cz. 1)*, Lwów, Warszawa 1922, id.: *Protokóły lekcyj przyrodoznawstwa odbytych w roku szkolnym 1919/20 w klasie drugiej Gimnazjum p. J. Kowalczykówny w Warszawie (Cz. 2)*, Lwów, Warszawa 1922, id.: *Protokóły lekcyj przyrodoznawstwa odbytych w roku szkolnym 1919/20 w klasie trzeciej Gimnazjum p. J. Kowalczykówny w Warszawie*, Lwów, Warszawa 1923, and Zofia Bohuszewiczówna: *Lekcje botaniki, w IV klasie szkoły średniej ze wstępem metodycznym*, Lwów, Warszawa 1925.

3. A proper method for studying creativity: codifying intellectual morality

When presenting at the Science Studies Circle in 1938, Dobrowolski heartfully embraced Holzapfel's all-encompassing approach. He used it as a utopian horizon for his own pedagogical theorizing, yet not without paying tribute to a critique by Ernst Mach. Though favoring Holzapfel's introspective approach, Mach's preface had urged for more clarity to support readers with less experience in the respective matters. And Mach seemed to be right. Although Holzapfel attracted a lot of attention in a certain social stratum, he neither influenced a broad social movement nor did he have lasting effects on actual pedagogical theory. Dobrowolski, however, focused on the practice of teaching, which he finally wanted to stand on solid ground. Thus, in 1918, he introduced an inductive research program in the first volume of Nauka Polska that would help to reform pedagogical practice in a scientific way. Rather than being presented with facts, students should be enabled to think properly. A teacher, as Dobrowolski wrote,

[…] must *know the art of thinking*: he must know the *catalogue of the faculties of the mind*; he must know their applications […]. [He] must be familiar with the mind's tendencies, its age-related development, its errors and faults. It will not do to be versed in logic, epistemology and general psychology of the mind. The teacher of thinking must be a *practical* psychologist of the mind; he must know the *grammar of actual thinking*.[54]

This was very close to the pedagogical ideas of Mach, which came from the background of a specific epistemic culture that tried to include the experiment in pedagogical practices to activate creative thinking.[55] Dobrowolski related closely to this, and wanted to find the smallest parts of these processes and hence the basic mental mechanisms of any creative activity. Yet, such "Grammar of Actual Thinking" still awaited proper description in a so-called *Catalogue of Mental Action*. According to Dobrowolski, none of the academic disciplines in charge of examining human thought had reached a scientific level so far. His earlier attempts from when he worked in the Ministry of Education had been cut off.[56]

To study all aspects of creativity systematically, a broad material basis was needed in order to observe, compare and classify creative processes. For the same reason Dobrowolski had turned to the natural sciences during his studies, he aimed to reduce anything that seemed close to philosophical speculation. Scientific method and a strong body of sources were paramount to his pedagogical endeavor. However, Do-

---

[54] ▶ Antoni B. Dobrowolski: "The Urgent Need for Mental Education in Poland" 1918, 291 (Italics as in the original), i. e. "O pilnej potrzebie wychowania umysłowego w Polsce: o konieczności zasadniczej reformy nauczania w szkołach średnich oraz stworzenia w związku z ową reformą nowych placówek pracy naukowej," *Nauka Polska*, vol. 1 (1918), 489–502.

[55] For Austria, Great Britain, the USA, and France cf. Bauer: "Kann man lernen, mit Gedanken zu experimentieren?" (footnote 42), 42–50.

[56] ▶ Dobrowolski: "The Urgent Need for Mental Education in Poland" (footnote 54), 295.

browolski diagnosed a severe lack of sources and thus was eager to find more material to understand "how [we] really think". So far, only very few works seemed to meet his demands, although it is hard to establish how systematically and exhaustively Dobrowolski did (or could) actually read in the libraries of Tsarist Warsaw.[57] He frequently quoted certain passages of John Dewey's *How we think*[58], and also Henri Poincaré's description of working on Fuchs' functions.[59] There was an autobiographical text of Herman von Helmholtz, as well as a handbook of the German psychologist and pedagogue Otto Schultze.[60] Quite interestingly, Dobrowolski often mentioned Edgar Allan Poe's account on the genesis of his poem *The Raven*, and later added also Thomas Mann's report on the process of writing his Faust novel.[61]

Dobrowolski lined out six fields to draw from, namely (1) experimental psychology, (2) past intellectual achievements as well as (3) research reports and law suits that display "intellectual strategies in different circumstances". He further planned to (4) analyze successfully finished scientific projects that had faced particularly fierce conditions, and (5) systematic behavioral observation of children's minds during school lessons, which was related to the first point.

Yet, the core of Dobrowolski's program was a (6) collection of biographies "of the thought of creative individuals".[62] In order to end "more or less ingenious [speculations]" on the "creative faculty of the mind" he wanted to base any future analysis on "adequate *first-hand material*," and hence genuine reports about the "ontogenesis of the attainments of the intellect," which would preferably be produced by actual researchers themselves:

They who transform, improve, discover, invent – in one word, they who do original work in the field of science, art, technics – shall initiate the world not only into the results of their creative thought, but also into its "biography," this last consisting of notes, arranged in chronological succession, and giving details about the origin of their thought, its development and its coming to maturity, and about everything that has, somehow or other, contributed to the creative process.[63]

---

[57] During his stay in Uppsala, Dobrowolski was impressed by the well-equipped library and the quick inter-library loan system. Dobrowolski: *Mój życiorys naukowy* (footnote 18), 133.

[58] John Dewey: *How We Think*, Boston e.a. 1910. A Polish translation by Zofja Bastgenówna was published in the Series *Bibljoteka Przekładów Dzieł Pedagogicznych* (Eng. Library of Translations of Pedagogical Works) with a foreword by Zygmunt Mysłakowski: *Jak Myślimy?*, Lwów, Warszawa 1934. Dobrowolski mentioned especially chapter VI. *The Analysis of a Complete Act of Thought*.

[59] Henri Poincaré: *Science et Méthode*, Paris 1908.

[60] Cf. Hermann von Helmholtz: "Erinnerungen: Tischrede gehalten bei der Feier des 70. Geburtstages, Berlin 1891", in: id.: *Vorträge und Reden* (vol. 1), Braunschweig 1903, 1–22 and F. E. Otto Schultze: *Systematische und kritische Selbständigkeit als Ziel von Studium und Unterricht für Lehrende und Studierende*, Leipzig 1914.

[61] Cf. Edgar Allan Poe: "The Philosophy of Composition," *Graham's Magazine*, vol. 28/4 (1846), 163–167 and Thomas Mann: *Die Entstehung des Doktor Faustus. Roman eines Romans*, Amsterdam 1949.

[62] ▶ Dobrowolski: "The Urgent Need for Mental Education in Poland" (footnote 54), 295–300.

[63] Ibid., 298 (Accentuations follow the original text).

By 1939, this call had been published three times in Nauka Polska and the Organon, twice in Polish and once in English.[64] The responses were published in both journals and should build the core of a pioneering "Archive of Materials for Research on Creativity".[65] This archive should grow until one day "each published piece of original research" would be accompanied by "an addendum containing a 'biography' of the creative scientific thought of that work, its origin and development".[66]

The initiative raised much attention, especially with the help of the journals' editorial boards and their head figure Stanisław Michalski, who helped convince contributors.[67] In fact, it picked up on a series of questionnaires that had been published in the *Handbooks for Autodidacts* around 1900. Already then the editors – among them Michalski – had hoped to raise "statistical material on self education" and asked for "*autobiographies of [the readers] intellectual development*".[68] Until 1939, about ten biographies were published in Nauka Polska and Organon, usually under the headline "Documents concerning the Psychology of Creativeness in Science and Scholarship".[69] Whereas Michalski and others were fascinated, Dobrowolski was disappoint-

---

[64] Cf. ibid., esp. the editorial introduction. For the texts see Antoni B. Dobrowolski ["A.B.D."]: "'Biografja' myśli twórczej," *Nauka Polska*, vol. 6 (1927), 136–140, and id.: "Letters to the Editor," *Organon*, vol. 1 (1936), 290–294. The English text was a partial translation of the first text from 1918.

[65] ▶ Dobrowolski: "An Archive of Materials for Research on Creativity" (1927), 35, i.e. "Archiwum materjałów do badania twórczości," *Nauka Polska*, vol. 6 (1927), 140.

[66] Ibid.

[67] As editor in chief of both journals, general secretary to the Mianowski Fund, and director of the National Culture Fund, Michalski had the contacts and the influence to convince contributors. Cf. Piskurewicz: *W służbie nauki i oświaty* (footnote 4), 91–93, 101–102. Piskurewicz quotes extensively from the materials of Zygmunt Lubicz-Zaleski (Bibliothèque Polonaise de Paris), a historian of literature, publicist and Polish consul general in Paris, who organized international contacts and distribution of both journals. Lubicz-Zaleski also arranged for Emile Borel's contribution to Dobrowolski's archive project (▶ Emile Borel: "Contribution" (1936), i.e. "Contribution (Documents sur la Psychologie de l'Invention Dans Le Domaine De La Science)," *Organon*, vol. 1 (1936), 33–42).

[68] Aleksander Heflich and Stanisław Michalski (eds.): *Poradnik dla samouków* (Cz. IV), Warszawa 1902, 472–474. See also eid. (eds.): *Poradnik dla samouków* (Cz. II), Warszawa 1899, IX and eid. (eds.): *Poradnik dla samouków* (Cz. III), Warszawa 1900, VII.

[69] The following sometimes anonymous texts were published in Nauka Polska until 1939: Stanisław Ossowski ["Student"]: "Kartka z życia studenta," *Nauka Polska*, vol. 4 (1923), 63–75, ▶ Dobrowolski: "An Archive of Materials for Research on Creativity" (footnote 65), ▶ Czesław Białobrzeski ["C.B."]: "An autobiographical sketch and remarks on scientific creativity" (1927), i.e. "Szkic autobiograficzny i uwagi o twórczości naukowej," *Nauka Polska*, vol. 6 (1927), 49–76, Franciszek Bujak ["F.B."]: "Drogi mojego rozwoju umysłowego," *Nauka Polska*, vol. 6 (1927), 77–136, ▶ Dobrowolski: "'Biografja' myśli twórczej" (footnote 64), cf. editorial introduction to ▶ Dobrowolski: "The Urgent Need for Mental Education in Poland" (footnote 54), ▶ id.: "My 'Scientific Biography'" (footnote 18), anonymous ["X.Y."]: "Życiorys II-gi," *Nauka Polska*, vol. 9 (1928), 217–245, anonymous ["J.Z."]: "Wspomnienia o drogach do pracy naukowej," *Nauka Polska*, vol. 9 (1928), 246–259, anonymous: "Życiorys I," *Nauka Polska*, vol. 15 (1932), 241–259, anonymous: "Życiorys II," *Nauka Polska*, vol. 15 (1932), 260–272, anonymous ["J.O."]: "Szkic autobiograficzny," *Nauka Polska*, vol. 16 (1932), 39–58. In Organon, one French and one English contribution were published, while the latter was later also translated for Nauka Polska: ▶ Borel: "Contribution" (footnote 67), ▶ August Krogh: "Visual Thinking" 1938, i.e. "Visual Thinking," *Organon*, vol. 2 (1938), 86–94. A Polish translation was published as "Myślenie wzrokowe," *Nauka Polska* vol. 24 (1939), 35–42.

ed. Though the archive grew and cross-references occurred in *Nauka Polska*,[70] he criticized the majority of the texts. The last preserved version of his analysis quotes this material only once.[71] Instead of biographies of their ideas, most authors had submitted personal autobiographies containing detailed stories about how they had entered academic life. Such a "'scientific biography' of a given person" seemed of no worth to his 'Grammar of Actual Thinking'.[72] In fact, most contributions simply contained the life stories of scholars, very close to other collections that had great success in other countries at the same time, for example in Germany and Great Britain.[73]

It is ironic that Dobrowolski failed to meet his own demands when he gave in to Michalski and finally submitted his own scientific curriculum vitae to the journal. On the 148 pages of the by far longest contribution, he presented a strictly autobiographical narrative. However, he did not forget to tell the early history of his project in his recollections: Returning from Antarctica, he had made the acquaintance of the engineer Siméon Gelblum, with whom he started theoretical discussions on the process of invention in Liège.[74] In early 1905, Gelblum gave two talks to the local *Association des Ingénieurs*, during which he promoted experimental investigations in the "mécanisme de l'invention". He claimed that in the age of high industrialization with its extremely complicated machines, only precise knowledge about the process of invention could help to develop even more detailed solutions. Otherwise mankind would be doomed to live with the imperfect machines.[75] Referring to discussions

---

[70] Stefan Błachowski for example worked with the texts collected in the *Archive of Materials for Research on Creativity* in Nauka Polska: ▶ Błachowski: "The Problem of Scientific Creativity" (1928), 213–215, i.e. "Zagadnienie twórczości naukowej," *Nauka Polska*, vol. 9 (1928), 1–67. Cf. Maria Ossowska: "[Review] 'A History of Psychology in Autobiography' Vol. I, red. C. Murchison (1930), Worcester, Mass., London," *Nauka Polska*, vol. 24 (1931), 363–365 and id.: "[Review] 'A History of Psychology in Autobiography' Vol. II, red. C. Murchison (1932), Worcester, Mass., London," *Nauka Polska*, vol. 28 (1933), 351–352.

[71] See Dobrowolski: *Moralność umysłowa (materiały)* (footnote 10), 163, where he quotes the anonymous ["J.O."]: "Szkic autobiograficzny" (footnote 69).

[72] ▶ Dobrowolski: "My 'Scientific Biography'" (footnote 18), 320.

[73] Cf. Jeremy D. Popkin: *History, Historians, and Autobiography*, Chicago e.a. 2005 or id.: "The Origins of Modern Academic Autobiography: Felix Meiner's Die Wissenschaft der Gegenwart in Selbstdarstellungen, 1921–1929," *Rethinking History: The Journal of Theory and Practice*, vol. 1 (2009), 27–42. For the Polish connection see Friedrich Cain "Jak pisać biografie pojęć i myśli? O technikach pracy twórczej," *Stan Rzeczy*, vol. 10 (2016), 376–397. On the Polish tradition of autobiography writing see Katherine Lebow: "The Conscience of the Skin. Interwar Polish Autobiography and Social Rights," *Humanity*, vol. 3 (2012), 297–319.

[74] ▶ id.: "My 'Scientific Biography'" (footnote 18), 322–324. Dobrowolski had mentioned this already before, cf. ▶ id.: "The Urgent Need for Mental Education in Poland" (footnote 54), 300.

[75] As Dobrowolski recalled, he had met Gelblum around the turn of the centuries over technical questions, when he had to adjust a microscope for special needs. They must have begun to discuss epistemological questions, soon. Not much is known about the engineer Siméon (or Szymon) Gelblum. Apart from being born in Poland and of a Jewish family background he must have come to Belgium for higher education. According to Dobrowolski, Gelblum was the first to invent a caterpillar vehicle for civil use. When Gelblum committed suicide in 1906, the joint project came to an end. A brochure with the title *Par l'historique à la science des idées*, which Gelblum still wanted to revise,

with "Monsieur A. Dobrowolski"[76], Gelblum called all authors of original ideas working in the arts, applied arts, and science to write detailed empirical accounts on the birth, development and maturation of their ideas and add them to their actual publications. Thus, in 1927, Dobrowolski framed the origins of his idea in a context of engineering, which survived in the third and fourth fields of sources to his project and which are closely connected to patent law.[77]

However, this early call of Gelblum, which was published in the proceedings of the Liège *Association*, differed from Dobrowolski's later works in Warsaw. Around 1900, the project was still limited to the enhancement of technological and scientific processes, whilst later versions appealed more generally to the wider social and cultural horizon so passionately drawn from Holzapfel's work. Though Gelblum had also referred to the general progress of humanity,[78] he introduced the "science des idèes" as a means to enhance the industrial capacities of nations.[79] Industries could no longer rely on supernatural, intuitive sources or imaginative speculations of philosophers, but instead needed to know about intellectual mechanisms and should be based on first hand material. Gelblum expected the rough shape of an idea to be imposed on the inventing individual by the "milieu", which would also trigger the birth ("naissance") of a thought. He postulated a mutual relationship between newly invented objects and their milieus ("fécondation réciproque de l'objet et du milieu") in which both constantly modify each other by the rules of chance ("hasard").[80] Said reciprocity relied on human intermediation, and therefore "comparative studies of such histories [would] allow perceiving modifications in [the] manners of seeing, under-

---

was never published. ▶ Dobrowolski: "The Urgent Need for Mental Education in Poland" (footnote 54), 300, ▶ id.: "My 'Scientific Biography'" (footnote 18), 322–324 and id.: "W rocznicę wynalazku czołgu," *Odrodzenie*, vol. 1 (1946), 5.

[76] "Séance du 8 janvier 1905," *Bulletin de l'Association des Ingénieurs sortis de l'École de Liége. Nouvelle série*, vol. 29 (1905), 18–28: 26. N.B.: The spelling „Liége" was officially changed to „Liège" only in 1949.

[77] Writing his call, Dobrowolski sticked closely to Gelblum's French version: "Que tous ceux qui modifient perfectionnent, inventent, autrement dit, que tout auteur d'une idée *personelle* dans l'art, l'art appliqué ou la science la livrent á la publicité en l'accompagnant d'un relevé exact *empirique et dans l'ordre chronologique* de tout ce qui, d'une manière ou l'autre, a contribué à sa naissance, à sa naissance, á son développement et à sa maturation." See "Séance du 5 février 1905," *Bulletin de l'Association des Ingénieurs sortis de l'École de Liége. Nouvelle série*, vol. 29 (1905), 163–177: 167.

[78] "Séance du 8 janvier 1905" (footnote 76), 26.

[79] "Séance du 5 février 1905" (footnote 77), 165. Gelblum had started to think about creativity from an engineer's viewpoint. However, though well informed by discussions from international patent law which was thriving since the middle of the 19th century in order to determine authorship (Cf. Monika Dommann: *Authors and Apparatus. A Media History of Copyright*, Ithaca 2019), Gelblum and Dobrowolski shifted from determining the inventor to investigating the actual process of invention in order to enhance it. One reference was the then famous collection on inventions, trade and industry edited by Franz Reuleaux, a protagonist of international patent law: *Das Buch der Erfindungen, Gewerbe und Industrien. Rundschau auf allen Gebieten der gewerblichen Arbeit* (8 vol.s), Leipzig 1884–1892. Dobrowolski later also referred to other authors. Cf. for example ▶ Dobrowolski: "The Urgent Need for Mental Education in Poland" (footnote 54), 298.

[80] "Séance du 5 février 1905" (footnote 77), 165–166.

standing, thinking, etc." This, in turn, would render processes of inventing more effective.[81] Hence, Dobrowolski's holistic, social program was clearly informed by classical positivist motives.

Although the responses coming in to fill the "Archive of Materials for Research on Creativity" did not meet his expectations, Dobrowolski would not give in. For the eleven years that he held the chair for theoretical pedagogy at the Free University, he taught two classes that helped him to order the material for two empirical works that should become the basis for a truly scientific pedagogy, namely the aforementioned *Catalogue of Mental Actions* and the *Code of the Intellectual Morality*. However, neither one was put to print during Dobrowolski's lifetime, although even an English translation was well advanced. It remained up to his students to posthumously publish "materials" on *Intellectual Morals* from his notes and other pedagogical writings.[82] The fragmentary *Catalogue* consists of twelve chapters with several subsections (chart 1). The *Code*, in its latest form, was divided into two parts, collecting intellectual and moral shortcomings (chart 2).[83]

Catalogue of the Mental Action	*Code of the Intellectual Morality*
– Introduction	1. Intellectual shortcomings
	– Intellectual vulgarity
– Awareness	– Thoughtlessness
– Capturing differences and similarities	– Generalization
– Analogies	– Mental blindness
– Separations and bonds	
– Comparisons and classifications	2. Moral shortcomings
– Identities	– Mental sloppiness
– Generalisations	– Mental dishonesty
– Mental feelings	– Evaluation mistakes
– Intuition	
– Posing questions	
– Evaluations	

Chart 1   Chart 2

---

[81] Ibid., 167.

[82] Cf. For the posthumously published texts cf. the three volumes of *Pisma pedagogiczne* (footnote 10).

[83] History of science has rediscovered this field lately, yet not so much from the angle of pedagogical theory, but under the auspices of moral economies (Lorraine Daston: "The Moral Economy of Science," *Osiris*, vol. 10 (1995), 3–24.) or *epistemic virtues*. Cf. Andreas Gelhard, Ruben Hackler and Sandro Zanetti (eds): *Epistemische Tugenden. Zur Geschichte und Gegenwart eines Konzepts*, Tübingen 2019.

The *Catalogue* presents the basic elements to creative mental action as Dobrowolski had identified them in his data. He tried to keep up with his inductive – or "experimental" – project by collecting a great amount of texts, excerpts, newspaper articles, larger and smaller pieces of paper etc. Comparing and systematizing the pieces, he tried to identify basic similarities and differences, based on which he would come to describe fundamental mental processes. Based on these, future education should lead pupils and adult people to think properly and avoid the fallacies of mental sloppiness, intellectual vulgarity etc. The bigger the archive, thus Dobrowolski, the more precise his catalogue would become, enabling to establish an "ontogenesis of the attainments of the intellect"[84], and to identify and fill the conscious or unconscious voids in other "biographies of ideas". Eventually, Dobrowolski composed rather short, dictionary-like articles presenting the identified basic actions and their functioning. Larger parts of the posthumous *Intellectual Morals* have remained fragmentary, while the *Catalogue* contains longer, seemingly finished passages.[85]

Whereas the *Catalogue* seeks for scientific soberness, the *Code of the Intellectual Morality* is rather a moral pamphlet. It is based on exemplary "intellectual" and "moral shortcomings" that Dobrowolski found in the collected materials. Hence, the *Catalogue* fulfills one part of the program from Dobrowolski's inaugural lecture, namely the search for a scientifically underpinned human culture. The *Code*, instead, seems to follow the second appeal, which is a radical critique of magico-mystical, pre-scientific, and especially philosophical thought.

Dobrowolski spared neither opponents nor friends with critique. Uncovering "intellectual vulgarity" or "mental sloppiness" but also "dishonesty" was undisputable when creating the new morals and ethics, the New Conscience of the New Man living only in and for science and art. As the proceedings from the Science Studies Circle show, Dobrowolski was a fierce discussant who did not hold back with direct personal critique if he found it necessary. Aversion rather accelerated than stopped his moral project and he did not hesitate to tell the history of his own project as one of fight and sacrifice. He never refrained from mentioning that the project had irritated many scholars from the very beginning on. The idea of "[taking] ideas to confession" had not been welcomed in Liège, where some commentators even tried to block the publication of a report on Gelblum's presentation – probably those, who had something to hide, as Dobrowolski speculated somewhat maliciously.[86] Though not including this episode in the unfinished manuscript of the *Code*, he did not fight shy of naming colleagues and their alleged intellectual sins.[87] Although Dobrowolski was not the only one to take interest in obstacles or (alleged) mistakes when studying intellectual

---

[84] ▶ Dobrowolski: "The Urgent Need for Mental Education in Poland" (footnote 54), 296.

[85] For the complete preserved materials cf. Muz. Ziemi PAN, Mat. of A.B. Dobrowolski, boxes 94–119.

[86] Dobrowolski recalled colleagues explaining such behavior with probable theft of ideas. ▶ Dobrowolski: "My 'Scientific Biography'" (footnote 18), 323.

[87] Cf. for many examples id.: *Moralność umysłowa (materiały)* (footnote 10).

processes, others did not take his accusing stand. Ludwik Fleck for example, who also lived and worked in interwar Poland but had no connection to the Science Studies Circle, attributed a central role to the "resistance aviso" in the genesis and development of scientific facts. Alexandre Koyré pointed out the epistemic potential of mistakes several years later.[88]

Dobrowolski did not get tired of promoting his project. In 1937, he wrote a theoretical introduction, which he must have circulated and discussed at least with Tadeusz Kotarbiński and Czesław Białobrzeski, two other regular attendants of the Science Studies Circle.[89] Right after the Second World War, he published a final call in the leftist magazine *Kuźnica*.[90] Admitting that his *Catalogue* would be a never-ending project, always preliminary and subject to change, he hinted at Carl Linné's project of classification, which was rather basic, but still effective. Therefore, any attempt seemed worthwhile as long as it was sincere. Under such preconditions also a shorter version, a "Small Code" containing only the "main sins" of the mind could prove to be important when educating the "New Man". This final goal received a new twist in 1945, shortly after the War had torn a void into the moral costume of societies.[91]

4. Universitas Rediviva

From the very beginning Dobrowolski designed his own idea for a Science of Science with strong inclinations to application. Opting to train as a natural scientist, he was a disciple of a Positivist objectivity, which rooted in the modern natural sciences and had to be spread not only over other academic fields, but society in total. This is where his attempt to overcome non-scientific, and especially what he suspected to be transcendental modes of reasoning were linked to his studies in theoretical pedagogy. And yet, Dobrowolski used a language that was overloaded with religious vocabulary. His effort to codify the moral economy of science doomed sins to be punished. The anthropological and social project did not stop at the education of a New Man

---

[88] Ludwik Fleck: *Genesis and Development of a Scientific Fact*, transl. by Fred Bradley and Thaddeus J. Trenn, Chicago, London 1979 (1935), 98 and Alexandre Koyré: "Perspectives sur l'histoire des sciences," in: id.: *Études d'histoire de la pensée scientifique*, Paris 1966, 352–361: 360–361. For a broad perspective on this cf. Giora Hon, Jutta Schickore and Friedrich Steinle (eds): *Going Amiss in Experimental Research*, Dordrecht 2009. Christoph Hoffmann links this to the social imagery of scientific research: *Die Arbeit der Wissenschaften*, Zürich, Berlin 2013, 124–132.

[89] Cf. Dobrowolski: *Mój życiorys naukowy* (footnote 18), 122 and Muz. Ziemi PAN, Mat. of A.B. Dobrowolski, box 263 (Korespondencja wychodząca, 1907–1954), letter to T. Kotarbiński, 12 Nov. 1937, no pagination.

[90] Dobrowolski: "O potrzebie katalogu czynności umysłowych" (footnote 50), 5–6. Dobrowolski actually reedited older texts without much adaption of Postwar situation. At some point he wrote that his *Catalogue* was "currently" in publishable shape, yet added the year 1937 in brackets (p. 6).

[91] On the Polish case see Marcin Zaremba: *Wielka Trwoga. Polska 1944–1947. Ludowa reakcja na kryzys*, Kraków 2012 and Andrzej Leder: *Prześniona rewolucja. Ćwiczenie z logiki historycznej*, Warszawa 2014, but also earlier Czesław Miłosz's classic *The Captive Mind*, London 1953.

but pushed for a New Religion of proper thinking. Such religion, of course, needed a New Church to be promoted. This was quite provocative on many levels. With Catholicism extending its great influence in Polish interwar society, Dobrowolski must have reckoned to stir emotional upheaval. Then again, his numinous wording deterred other commentators who were actually on track with his ideas of rationalization, although Auguste Comte had framed it similarly, before.[92]

In the 1930s, Dobrowolski tried to institutionalize his vision of an intellectual society. His ecclesiastic language culminated in a project of a network of schools. Having been hindered to organize a sound system in the early 1920s (as he might have said) from within the ministry, Dobrowolski now opted to create core institutions that would be an inspiration for further school reforms. When minister of education Janusz Jędrzejewicz launched a school reform in early March 1932, Dobrowolski and many other scholars, teachers, and intellectuals protested against elitist elements and the extensive political influence the government could now exert on elementary, middle and higher schools.[93] A special concern was the plan to introduce different levels within the compulsory school attendance of seven years. About a week after the reform bill had been published, Dobrowolski diagnosed a triple crisis with effects in economy, education, and – much worse – "bankruptcy in ideas of leadership" leading to a division of two castes. The contemporary school system would foster a small privileged group opposed by the "masses of billions" of uneducated people, practically "illiterates that can read, write, and count" but could not cope with civilization. The people were required to free themselves from their own benightedness, which could only be triggered by the creation of a "New Religion" with science and art as "human rights" and "obligations of civilization" at the same time.[94]

Two months later, Dobrowolski followed up on this article claiming that the Soviet school system had succeeded in granting broad education to vast masses, and thus overtake Poland in terms of civilization.[95] Two more weeks later, Dobrowolski lined out a sketch for a school system that would offer the opportunity (and duty) to a dec-

---

[92] The religious imagery was not uncommon at the time. In this respect, Dobrowolski's and also Holzapfel's publications were in no way special against the contemporary plethora of Monist texts, which, in fact, were often specifically directed against the Catholic Church. Cf. the essays in Johannes Feichtinger, Franz L. Fillafer, Jan Surman (eds): *The Worlds of Positivism. A Global Intellectual History, 1770–1930*, London 2018.

[93] Cf. Drynda: *Pedagogika Drugiej Rzeczypospolitej* (footnote 11), 47–49 and Sadowska: *Ku szkole na miarę Drugiej Rzeczypospolitej* (footnote 11), esp. chapters I and II. A comprehensive history of the reform, its effects and contemporary reactions has not been written yet. See, however, Stanisław Mauersberg: "Reformy szkolne w Drugiej Rzeczypospolitej (1918–1939)," *Kwartalnik Pedagogiczny*, vol. 40/4 (1995), 19–37. For a recent description of the situation at Warsaw University cf. Marek P. Deszczyński: "Uniwersytet Warszawski w II Rzeczypospolitej," in: *Dzieje Uniwersytetu Warszawskiego 1915–1945* (= Monumenta Universitatis Varsoviensis 1816–2016), Warszawa 2016, 325–427: 362–369.

[94] Antoni B. Dobrowolski: "Minimum oświaty szkolnej. O trzech kryzysach," *Wiadomości Literackie*, vol. 9/12 (20 March 1932), 1.

[95] Id.: "Minimum oświaty szkolnej! Post-scriptum Rządu Sowietów do moich artykułów," *Wiadomości Literackie*, vol. 9/20 (15 May 1932), 2.

ade of school education and vocational training for every individual. The central concern remained a general rise in intellectual culture that should not spare any social (or vocational) class. Only a human right to education would secure a common basis for civilized life in a democratic system.[96] In this respect, it was very important to train the ability and the will to self education in every single individual, an aspect which leads back to Dobrowolski's pedagogical research.[97]

During these years, the "intellectuals" were a central category to Dobrowolski's texts. In fact, he aligned his whole educational project with strengthening what he called an "inteligencja". This is paradoxical at first sight only, as Dobrowolski did not use the term for a specific social group with academic education, but often as a model for future democratic citizenry. Educating the masses meant to teach and promote autodidactic methods and to encourage a drive to educate oneself. Dobrowolski envisioned institutions where people would study and learn even after finishing ten mandatory years at school and vocational or academic training. A "cult of learning" should be established that would drive intellectual citizens to deepen and broaden their knowledge throughout their life.[98] In 1936, Dobrowolski reported on a model institution, which he and "about 15 colleagues" had organized at the Free Polish University. This so called *Universitas Rediviva* should foster "intellectual [...] or general education on the highest level" and revive a "general intellectual education" along the tradition of medieval colleges – much different from contemporary models of specialization.[99]

Dobrowolski was eager to spread the idea throughout Poland and beyond, for example when he visited his alma mater in Liège in 1937. The manuscript of his talk was published in French, and an English version was printed in *The Sociological Review*.[100]

---

[96] Id.: "Minimum oświaty szkolnej. Nowy postulat społeczno-oświatowy i jego umotywowanie," *Wiadomości Literackie*, vol. 9/22 (29 May 1932), 2. Dobrowolski made a similar argument in: "Zagadnienie szkoły powszechnej jako zasadnicze zagadnienie naszej cywilizacji," *Przegląd Współczesny*, vol. 11/40 (1932), 205–217.

[97] Id.: "Sprawa szkoły powszechnej jako zasadnicza sprawa naszej cywilizacji," in: *II Kongres Pedagogiczny Związku Nauczycielstwa Polskiego w Wilnie od 4–8 lipca 1931 r.*, ed. by Związek Nauczycielstwa Polskiego, Warszawa 1932, 342–377: 344–345.

[98] Cf. for example id.: "Sprawa oświaty inteligenckiej, czyli wykształcenia ogólnego na wyższym poziomie (projekt wyższej szkoły ogólnokształcącej, czyli organizacji samokształcenia inteligentów)," *Kultura i Wychowanie*, vol. 2/3 (1935), 133–149: 142–144.

[99] For the programmatic report cf. id.: *Universitas Rediviva. Sprawa oświaty inteligenckiej czyli wykształcenia ogólnego na wyższym poziomie (Zapoczątkowanie organizacji samokształcenia inteligentów)*, Warszawa 1936. A shorter English version, in which Dobrowolski reported first experiences was published one year later: id.: "A Liberal Education: A Plea for a General-Knowledge Faculty in Every University," transl. by Walter H. Dawson, *The Sociological Review*, vol. 29/4 (1937), 355–369: 359–360, 364–366.

[100] Id.: "Sur la nécessité de créer, dans le Universités, une faculté d'instruction générale supérieure pour hausser le niveau de culture de l'esprit des intéllectuels", *Bulletin trimestriel de l'Association des Amis de l'Université de Liége*, vol. 9/1 (1937), 3–19. This text, which was introduced under the heading "Le problème de l'education intellectuelle générale" (p. 2) was, in fact, the text that Walter Dawson translated into English as: "A Liberal Education" published in *The Sociological Review* in 1937.

His vision was that every university would create courses, schools, study programs, or even a "Faculty of Higher General Instruction"[101] that would not only help students "*to find appropriate reading-matter and make use of it in the most efficacious way, but, above all, to exercise themselves in the art of observing and thinking.*"[102] Thus, the "Subject-matter of the Programs" needed to be as broad as possible:

1. An intellectual should, before all, be practiced in the art of thinking: should learn to think honestly, correctly, skillfully, instead of dishonestly, faultily, clumsily. It is a mistake to suppose that this necessarily involves the study of philosophy. Such study might be too much and would certainly be too little. There are other ways, independent of philosophy, which may be recommended:
    (a) To read, mark, learn, and inwardly digest masterpieces and compare them with inferior works dealing with the same subject.
    (b) Critically analyze appropriate passages and participate in the discussion of questions suitably chosen or relating to the subject matter of the courses.
    (c) Note the mechanism and methods of various kinds of intellectual work, including the difficulties to be overcome and conditions to be satisfied. First of all the constituents of this work: the elementary intellectual functions with the conditions for their efficacy; then the tendencies and natural inclinations of the mind, its vagaries and shortcomings and their correction – in short, the technique of mental effort. These things are not to be found in the manuals of logic old or new (logistics), nor even well catalogued. But there are already some first essays.
2. General history of human experience and thought. [...]
    (a) *Pre-scientific and scientific.* [...]
    (b) *Philosophic and magico-mystic.* [...]
3. Of all these lines of approach *Science* is selected for closer study as the most recent, with *Technology* (in the widest sense) linked integrally with it. Science (with technology) as the most characteristic mark of our great epoch, as the new historic factor, creating a new era, a new value, as it were. Sciences pure and applied, their reciprocal relations. Unity and diversity of science, differences in nature and in stage of evolution between the different branches; the interdependence of these. Organization and functioning of science, its differing laboratories, ways and methods of work, difficulties and obstacles, external and internal, successful and unsuccessful efforts in the struggle to overcome them. Epic of science against the background of the history of labor, of social economy and thought, with a closer view of certain chosen moments of this epic; comparison with the history of religions and philosophies. Ever-growing progress and its more and more explosive and unexpected character, as characteristic of modern science, especially its vanguard, – the mathematical and physical sciences. Research and the researcher, discovery and invention. Science as an individual and, at the same time, a social enterprise. Rôle and influence of science.
4. *Art*, as the second cultural value, elder sister of science. [...] Conditions of work, difficulties, obstacles. [...] Evolution, long and slow [...], studied against the background of the history of labor, of social economy, and of thought. [...]
5. *Literature.* Its infinite diversity. [...] Emphasis on the essay as a literary form for the expression of thought, parallel, therefore, to philosophy.

---

[101] Dobrowolski: "A Liberal Education" (footnote 99), 359, 367.
[102] Ibid., 367 (Italics as in the original.).

6. *Psychological and sociological sciences.* Importance and primitive state of these "sciences." Efforts to elevate them to a truly scientific level; difficulties and obstacles. History and actuality, different strains of tendency, watchwords, schools. Treasures of pre-scientific psychology to be found in some masterpieces of belles-lettres.
7. *Economic and social problems*, as fundamental problems of humanity in the past and present. Scientific and quasi-scientific researches and theories, doctrines, watchwords.
8. *Problems of education and public instruction*, past and present, considered against the background of life as a whole, as of the rights and duties of man, past and present. Ever-growing importance of these problems. Scientific and quasi-scientific researches and theories, doctrines, watchwords.
9. *Problem of the use of leisure.* Ever-growing importance. Diversions, amusements, games, in different countries and epochs. Amateurs and dilettantes; scientific, technical, artistic, literary; their utility, too often unrecognized; the conditions of efficacy of their labors. Ever-growing use-fullness of non-specialists who, guided by specialists, indirectly come to the aid of different branches of science or technology, offering their services systematically, e.g. explorers.
10. *Objects of life and criteria of values.* Conceptions of the world and ideals, ideologies and watchwords, religions and "philosophies of life"; their actuality and their past considered against the background of economic, social, and political life.[103]

This 1937 outline for Universitas Rediviva summed up forty years of Dobrowolski's conceptual work. Over the course of four semesters, attendants would have to study three days a week with different lecturers – twice three, and once two hours – and embrace "the whole of civilization" to create unity out of greatest diversity. General self-education required absorbing the entirety of knowledge instead of piecemeal.[104] Dobrowolski's personal archives do not seem to hold much material on the project. However, a fragmentary program for a holiday course in the academic year 1938/1939 gives an impression of the wide range of topics that has indeed been covered.[105]

---

[103] Ibid., 361–364 (Italics as in the original.). In a footnote to point 1(c) Dobrowolski alludes to John Dewey and F.E. Otto Schultze again, but also to his own project: "A.B. Dobrowolski. *Catalogue of Intellectual Functions, etc.*, based on observations made since 1900" (p. 362).

[104] Ibid., 365, 366.

[105] This fragment lists six departments of the project: Dep. I: Basic Natural Sciences, Dep. II: Concrete Natural Sciences, Dep. III: Psychological Sciences, Dep. IV: Social Sciences, Dep. V: Sciences of Normative Character (e.g. Pedagogy), Dep. VI: Arts and Literature. The fragment further tells about nine courses and teaching staff for the period between 2 and 31 July 1938: *On Mathematics* (Adolf Lindenbaum, 15 hours), *Basic Natural Sciences* (Władysław Kapuściński, 15 hours), *From and About Sociology* (Józef Chałasiński, 20 hours), *On Intellectual Work and Self Education* (Antoni B. Dobrowolski, 10 hours), *Contemporary Polish Literature and Theatre Against the Background of European Cultural Trends* (Bohdan Korzeniowski, 20 hours), *Music* (Julian Pulikowski, 20 hours). Three more courses were planned during Christmas Break: *Geological, Geographical and Related Sciences* (Zb. Bujkowski, 10 hours), *On Intellectual Work and Self Education* (Antoni B. Dobrowolski, 4 hours), and *From and About Logic* (Janina Hosiasson-Lindenbaum, 10 hours). Muz. Ziemi PAN, Mat. of A.B. Dobrowolski, box 225 (Instytut Wyższej Kultury Umysłowej. Rozkład godzin i program 1938/1939), no pagination.

5. Koło Naukoznawcze – Science Studies Circle

When Dobrowolski finished his "Introductory Considerations on Higher Schools, and Especially Academic Schools" at the Science Studies Circle in late April 1938, four out of 30 members of the audience raised questions. Most of the questions either commented on Dobrowolski's remarks on the relationship of pure and applied or younger and older scientific disciplines, or touched upon the specifics of linking research and education in contemporary universities. However, two questions also investigated Dobrowolski's greater concern. Responding to a Dr. Truszkowski, who had asked about the perspectives of raising public science funding, Dobrowolski said that governments of democratic countries should finally understand the importance of science "propaganda" in order to change the public opinion in a way that a rich funding of science would become a natural public aim.[106] In response to Czesław Białobrzeski's question of how such propaganda should be organized, Dobrowolski referred to his Universitas Rediviva project. Instead of only educating specialists for narrow tasks, such "institutes for higher intellectual culture" should promote an all-encompassing worldview and raise a new consciousness for learning.[107]

Interestingly enough, Dobrowolski shut down the questions posed by writer and literary critic Artur Górski, an influential figure in interwar Poland, who had been a frequent opponent of Dobrowolski during the discussions in the Circle. Górski criticized that the talk rested on highly constructed theses with odd anthropological presuppositions. Human beings and their interactions, he stated, would not become favorable if only the environmental conditions served that condition. Instead, true virtues would rather (and especially) come to the surface in times of hardship, when the individual was in a condition of fight. A wide range of political philosophies and ideologemes came to clash when Górski, who leaned towards a neo-romanticist messianism, opposed Dobrowolski and his positivism informed approach.

In the Polish context positivism and messianism are generic terms combining artistic and political positions from various backgrounds, which in turn are heterogeneous in themselves. Górski is especially interesting, as he was famous for coining the term *Młoda Polska* ("Young Poland") in 1898, which was adapted by a wide range of modernist artists connected to decadence, symbolism, and art nouveau. They opposed the positivists' ideas and their concept of "organic work" that based social life on science and reason. However, neither did a certain artistic choice precondition specific political opinions, nor the other way around. In other words, the Science Studies Circle was rather heterogeneous when it came to political attitudes and it is hardly possible to inscribe a specific label in this respect. Obviously, Dobrowolski could not accept Górski's messianism. He made this clear by stating that the intervention was based on a false and mischievous interpretation of his thoughts and that

---

[106] "Sprawozdanie dwunaste z działalności Koła Naukoznawczego" (footnote 8), 187–241: 236.
[107] Ibid., 238.

he therefore would not bother to answer. He clearly based his Universitas Rediviva on classical enlightenment and positivist initiatives. He had introduced it at the Free Polish University, which in turn had been founded by the Society for Scientific Courses, and thus the legal successor to the Flying Universities from Tsarist times. It was in this milieu, which cherished the ideal of self-education, that Dobrowolski had started his aforementioned – metaphoric or not – laboratory for pedagogical concepts.

Against the backdrop of this ideological variety, the Science Studies Circle surely was a fair of conflicting ideas and thus Dobrowolski diagnosed an ineffectiveness of discussions. To his taste the Circle talked too much about science without proper reflection. This encouraged him to go back to the basic operations of human creativity in order to substantiate notions of genius and scientific greatness. He did not convince the entire audience in early 1938, but must have touched a nerve. As the logician Jan Łukasiewicz, who presided the session, stated in a closing remark, the discussions had excited the audience very much. Thus, future sessions should treat on "what science actually is" and on the question about the relationship between science and morale, for example.[108] However, the Second World War ended these discussions abruptly, and brought a whole new significance to the latter question.[109]

During the war, Dobrowolski made contacts with clandestine groups that fought the German occupants. He also spent a lot of time working through his materials on creative action and must have gotten far with both the *Catalogue* and *Code*. In vain, yet, as the manuscripts got lost in the late days of occupation. Thus, in early 1945, he once again suggested to organize an archive for biographies of intellectual activities, encouraged by the civilizational catastrophe of the highly industrialized war. A backdrop to magico-mystical and religious speculation needed to be prevented by promoting honesty, sincerity, and critical thought more than ever before.[110] A couple of weeks later, Dobrowolski also renewed his interwar calls to reorganize the educational system by embedding schools in a higher culture of reasoning. Otherwise a "new benightedness" governed by "half-" and "quarter-intellectuals" would emerge to threaten civilization. Full or "hundred percent intellectuals" should be able to handle the three highest achievements of civilization, and namely science (in a broad sense), technology, and the arts. Conceited smart alecs, obsessed with books, with assumingly encyclopedic knowledge, and only thinking for the sake of thought were a threat to a civilized world as they would neither push criticism nor mental honesty to limits.

These categories were an obvious outflow from Dobrowolski's work on the *Catalogue of Mental Actions*, and especially from the collections of mental sins. However, an even bigger threat, and, in fact, "more horrible and dangerous than complete

---

[108] Ibid., 241.

[109] The 26th volume of *Nauka Polska*, which was freshly printed in September 1939, was burned completely during the German siege of Warsaw. Thus, it is not clear how many sessions actually followed. Cf. Piskurewicz: *W służbie nauki i oświaty* (footnote 4), 100.

[110] Dobrowolski: "O potrzebie katalogu czynności umysłowych" (footnote 50), 6.

benightedness" seemed to be a certain attitude of high school graduates, who – as Dobrowolski feared in 1945 – did not understand graduation as a rite of passage but as the step into a realm of wisdom without any significant loopholes:[111] "The expensive, […] careerist character of the high school, and the ever growing, unreflected popularization […] of science and art among the intelligence at school, but mostly via books, the press, lectures, radio, exhibitions, excursions – and the parallel […] 'general education,' bring forth a fast growing mass of polymorph and deformed individuals: half- and quarter-intelligents, a threat to civilization."[112]

The reflecting institutions of a Universitas Rediviva as well as similar "Institutes of Higher Intellectual Culture," and courses for self-education had become necessary more than ever before, as Dobrowolski diagnosed. Once again, he called to bundle experience from various Polish and indeed existing international research on science popularization in order to formulate a program for general intellectual education.[113] However, a Science Studies Circle in its interwar shape did not seem to be important for this – if at all, then only as a forum of self reflection for Scientists that would support and stabilize intellectual culture.

---

[111] Id.: "Nowa ciemnota, czyli nowe niebezpieczeństwo grożące cywilizacji," *Twórczość. Miesięcznik literacko-krytyczny*, vol. 1/5 (1945), 76–93: 76–77, 79–81.

[112] Ibid., 76.

[113] Dobrowolski cited for example Aleksander Hertz ("Książka popularno-naukowa a kultura," *Wiedza i Życie*, vol. 10/1 (1935), 45–51) and Stanisław Loria ("Jeszcze o popularyzacji," *Wiadomości Literackie*, vol. 10/39 (10 Sept. 1933), 7) from Poland and referred to Henri Berr's French *Centre international de synthèse*, which organised an annual *Semaine international de synthèse* and the weekly magazine *La Science*. Though Dobrowolski had doubts about the *World Encyclopedia* project of H.G. Wells', he favoured the general idea. Ibid., 76, 88.

Part II:
Sources

Translator's Note

Tul'si (Tuesday) Bhambry

I translated ten of the seventeen texts that make up this reader, texts by about half a dozen different writers and published in the interwar period. Scholarly writing from the early twentieth century poses unique translation challenges. In this case, however, I realised that my strategies needed to be tailored to every single writer. Let me address this last point before I give an account of my choices more broadly.

The style and tone of the originals differ greatly. Some of the Polish scholars represented here must have carefully weighed every word, others seemed more nonchalant about the craft of writing, privileging spontaneity. This alone does not indicate who was a "better" writer, but it is worth bearing in mind that this reader represents a broad spectrum from classic coherence (Białobrzeski) to the energy and fluency of spoken language (Kotarbiński) or an experimental or idiosyncratic style that can seem a bit haphazard (Dobrowolski). A writer who was no natural born stylist always poses the greatest difficulties to the translator. I was least faithful to the letter where the style was most unconventional, where I felt that repetitions, unorthodox punctuation and emphases in italics distracted from the writer's argument.

Looking at the choices I made across the spectrum, the most important one was to determine to what extent the translation should be modernised. I knew that my target audience would expect a fluent rendition that uses the discipline's current terminology and conventions. However, an aggressively updated translation can distort the original's historic grounding, which risks making the author sound naïve or behind their time. The word "naukoznawstwo" is a case in point. Modern readers might baulk at the pompous construction "science of science," which appears in the original *Organon*. The unpretentious "science studies," which appears in an English translation by Maria Kantor 2014, seems like a more attractive alternative. But "science studies" did not come into its own as a discipline until the 1970s and 80s and was not directly related to the "science of science" as we have here. The editors of this volume have offered me their expert advice on appropriate translations of specialist terms, saving this English version from a few inappropriate anachronisms.

My translation generally ended up reflecting the sensibilities of the era, i.e. honouring the historical context over more reader-friendly alternatives. Ideally, I believe, the translation would create the impression that the text could have been written before World War II, but without sounding antiquated. I would consider it a bonus if my translation also managed not to perpetuate biases and stereotypes, for instance relating to gender. But sometimes, that bonus just cannot be had and we modern readers just have to face the facts of history and let them inspire us to become more open-minded and inclusive in our own work.

Another challenge I faced relates to the use of non-English words in the translation. First, there is the question of proper nouns such as the title of the Polish-language journal *Nauka Polska* (est. 1918), which addressed the needs, organisation and development of the sciences in Poland. When the multilingual journal *Organon* was launched in 1936 to disseminate Polish scholars' work on an international platform, *Nauka Polska* was translated as "Science and Letters in Poland"; an alternative rendering would be "Polish Science". The editors of this volume and I agreed that we should explain the term in a footnote when it first appears in the respective texts and then use the original. The same applies to other proper names, such as *Koło Naukoznawcze*. Next, I needed to settle on the presentation of foreign-language quotations. The Polish original texts are peppered with words and phrases in French, German, English and even Latin. This is not surprising – the members of the *Koło Naukoznawcze* saw themselves as belonging to a thoroughly European elite. The problem is that there is no consistency – even within a single author's work – in the way non-Polish quotations are presented. Błachowski, for instance, sometimes quotes foreign authors without giving a Polish translation. At other times, he paraphrases the quotation, or provides an already published Polish translation, or even translates the passage into Polish himself. So how do we deal with this inconsistency? The most straightforward and generally accepted approach would be to translate into English whatever he translated into Polish, leaving French, German or Latin quotations in their respective originals. But this would leave many modern readers at a loss, as few of us are as well-versed in the same foreign languages as these writers and their contemporaries. I have adopted the broadest, most inclusive approach, tracking down originals and either hunting down existing English translations where possible, or providing my own. My hope is that this selection of texts will be accessible to readers today and tomorrow.

What would a translator be without her readers? In this case, the first readers were the participants of the workshop "A New Organon: Science Studies in Poland between the Wars" at the University of Konstanz on 20–21 February, 2015. The organisers, Friedrich Cain and Bernhard Kleeberg, and their international guests offered valuable feedback on the first version of this translation. While any shortcomings are mine, the final version would not have been as solid without their helpful comments. I also wish to thank Friedrich Cain's for his tremendous support. He helped me contextualise the material, provided source texts, and was often the sounding board for my decisions. It is a rare privilege for translators to be part of a team that takes our work so seriously and thus makes it so rewarding.

Berlin, February 2021

Editorial Introduction

Starting with vol. 4 in 1923, the journal Nauka Polska changed its outlook. Stating a need for systematic knowledge about science, its theory, its foundations, its organization, and its psychology, this "Editorial Introduction" presented the new and much broader programmatic approach of the journal for the years to come. It was authored by editor in chief Stanisław Michalski.

Before, vol. 1 and 2 had assessed the state of affairs in Polish sciences and letters and vol. 3 had been a collection of materials from the I. Congress designated to Questions of the Organisation and Development of Polish Science and Letters (*I. Zjazd poświęcony sprawom organizacji i rozwoju Nauki Polskiej*), which was held at Warsaw Univ. 7–10 April 1920. The new program was only interrupted twice. Vol. 8 contained the materials from the II. Scientific Congress held in Warsaw 2.–3. April 1927 (*II Zjazd Naukowy*) as well as vol. 7 and 12, which gave an extensive overview on scientific institutions and learned societies in Poland. Translation by Tul'si (Tuesday) Bhambry.

ORIGINAL REFERENCE
"Wstęp redakcyjny," *Nauka Polska*, vol. 4 (1923), VII–IX.

The materials so far published in the volumes of Nauka Polska have highlighted a set of issues concerning the life of science as a social phenomenon; a separate department of *knowledge about science* has formed. Science is an object of research, similarly to other products of human culture, for example art and religion.

To certain extent, this separation of knowledge about science is connected to life itself in our country. Incentives of practical character – e. g. considering today's needs of science, and questions of supporting and organizing it, questions on the relation of science to other branches of culture and to life in general – force us to think about scientific creativity and conditions of its development. Wherever a certain activity starts, the need for theoretical foundations of this activity will come with time: the need to study science. That is why among the many articles in Nauka Polska, which were devoted to the needs of separate sciences or matters of organization, some are also theoretical in nature, for example those from the field of sociology of science.

What we have said so far points to the programme and direction of the journal. Its main task is to draw public attention to the pertinent, yet still not very popular question of science; for this purpose we desire to separate this question from others, with which it is interwoven, in order to deal with it separately, both from the theoretical and the practical standpoint.

Several articles in the present volume concern more general problems pertaining to research on science. Next, while previous volumes have broadly (though not exhaustively) catalogued individual sciences' prerequisites, this volume considers the requirements of the entire body of science and groups of sciences (e. g. scientific work in the provinces, the lack of a scientific workforce and research workshops); most of the necessities of the individual sciences should anyway be reflected in the numerous specialist journals that are currently being established. A couple of pages deal with the history of the organization of science in Poland. There follows a chronicle of the latest scientific events in Poland as well as information on the state of science abroad.

The ideal development of science requires an appropriate atmosphere across society as a whole. This atmosphere must be nurtured by spreading not only education but also knowledge about science.

It is not uncommon nowadays to be faced with a lack of understanding of the nature and needs of creative scientific activity, which is often seen as a luxury when it has no utilitarian goals. This is due to the current problematic situation, which has marginalized the issue of pure science. Contemporary life reflects the entirety of society's needs like a curved mirror, distorting actual relationships between things.

In earlier days we needed pioneers of education to combat the superstitions that hampered the advancement of learning in society. Today, *it is science that needs such pioneers.*

A new branch of social engagement is emerging. Participants are required to have both education and practical skill, especially since specialized establishments, run by [private] societies as well as the state, already exist. Members of staff are in demand. They, like anyone who is interested in science and who wishes to join the effort to support it, must nowadays be equipped with theoretical knowledge about science as well as practical know-how about what is needed for its organization and development.

The wartime and postwar activities of the Mianowski Foundation indicate that interest in science is gradually increasing. Even if it is slow, this process suggests that perhaps in the near future our sciences will fare better, thanks to awareness raising and self-reliance in society. The facts listed in the Chronicle in the second volume of our publication and the Chronicle in the present volume both corroborate this expectation. Ever-broader circles exhibit a dedication to scientific goals; they include individuals, private institutions, schools (the assiduity of teachers and students), as well as self-governing organizations (local parliaments [*sejmik*], municipalities).

The efforts of society as a whole will determine the accelerated development of our own scientific culture.

The efficiency of Poland's defense against co-optation by foreign culture hinges on the rapidity of this development.

Warsaw, 28 April 1923

Preface

In the English preface to vol. 1 of Organon, editor Stanisław Michalski lines out the intent of the new journal. It is very clear about the strong connections to Nauka Polska and sets the task of making accessible the materials published there in Polish for a wider, anglo- (and franco-)phone audience. The edition reproduces the original text.

ORIGINAL REFERENCE
"Preface," *Organon*, vol. 1 (1936), V–VI.

The ORGANON, the first volume of which we are offering to our readers, is to be devoted to the »science of Science«. The term »Science« is used here in its broadest sense, including both Natural Science and Humanities, covering practically the whole field of organized knowledge.

The idea of this publication owes its origin to Mianowski Institute for the Promotion of Sciences and Letters; it is to be issued as an international organ corresponding to a Polish periodical, published by this Institute, since 1918, under the title »Nauka Polska« (Science and Letters in Poland). In the first stage of its existence the chief concern of »Nauka Polska« was to deal with practical questions of the organization of Science in Poland, In the course of time, however, it came to include questions of theory as well, as practical activity generally gives rise to the need of a corresponding theory which is indispensable for the extension of this activity and for testing its usefulness. Thus »Nauka Polska« has gradually undertaken the treatment of various psychological problems, connected with creative research work, of the social conditions under which Science has developed, and also of the question of the connection of Science with other branches of civilization. (The table of Contents annexed to the present volume gives the subject matter of twenty volumes of »Nauka Polska« published till now).

The first article of the ORGANON shows the manifold character of the problems concerning the »science of Science«. Hence the need of a wide collaboration which prompts the Editor of »Nauka Polska« to look for contacts with those who are interested in these studies outside Poland. He hopes such contacts will be established through this periodical, published in languages which are more widely understood.

The matters which this publication proposes to treat have so far been touched upon by many other periodicals, devoted to philosophy, psychology, or sociology. It has, however, been deemed useful to single out all the questions pertaining to the »science of Science« and bring them together on a special organ.

The Editor

Report on the activities of the Science Studies Circle

From vol. 11 onwards, Nauka Polska published consecutive reports on the meetings of the Science Studies Circle. The reports were printed in the section Polish Chronicle (*Kronika Polska*) together with information on latest legislation concerning science, prizes etc. While relatively short at first, the reports grew longer when the discussions were printed in detail (for an overview see ▶ **IN THIS VOL.** List of Sessions). The papers themselves were usually not reiterated, as they were mostly printed in Nauka Polska.

While this edition of the first report concentrates on the passages about the third and fourth meeting with presentations by sociologist ▶ Paweł Rybicki (1929) and philosopher ▶ Tadeusz Kotarbiński (1929), it skips accounts of the sessions with historian ▶ Franciszek Bujak (1938), pedagogue Bogdan Suchodolski, and Wojciech Przybyłowicz, an advisor from the Science Department at the Ministry of Religious Affairs and Public Education (*Ministerstwo Wyznań Religijnych i Oświecenia Publicznego*). Translation by Tul'si (Tuesday) Bhambry.

ORIGINAL REFERENCE
"Sprawozdanie z działalności Koła Naukoznawczego," *Nauka Polska*, vol. 11 (1929), 353–355.

FURTHER TEXTS MENTIONED
Franciszek Bujak: "Działacz i badacz," *Nauka Polska*, vol. 11 (1929), 11–23.
▶ — "The Man of Action and the Student," *Organon*, vol. 1 (1938), 20–32.
▶ Tadeusz Kotarbiński: "O zdolnościach cechujących badacza," *Nauka Polska*, vol. 11 (1929), 1–10.
Wojciech Przybyłowicz: "Uwagi o stosunku państwa do nauki," *Nauka Polska*, vol. 11 (1929), 65–91.
▶ Paweł Rybicki: "Nauka a formy życia społecznego. Kilka zagadnień z pogranicza socjologii i teorii nauki," *Nauka Polska*, vol. 11 (1929), 24–64.
Bogdan Suchodolski: "Review of Max Scheler, Die Wissensformen und die Gesellschaft. Probleme einer Soziologie des Wissens, Leipzig 1926," *Nauka Polska*, vol. 11 (1929), 379–381.

The *Science Studies Circle* was established in June 1928 by the Research Division [*Dział Naukowy*] of the Mianowski Fund. Its creation responds to the perceived need for theoretical foundations for the activities undertaken by institutions that deal with the administration and management of research, in particular the Mianowski Fund. Invitations to the Circle's meetings are extended to research workers interested in the science of science as well as employees of social and government institutions in charge of research. During the meetings participants present papers in the science of science and research administration, which are then discussed.

Between June 1928 and May 1929, the Circle held five meetings, where the following papers were presented (including in parentheses the number of participants and the number of individuals taking part in the discussion following the paper):

14. July 1928:	Dr B. Suchodolski's paper on Max Scheler's *Die Wissenschaftsformen und die Gesellschaft (Probleme einer Soziologie des Wissens)* – (11, 5)
10. December 1928:	Prof. Dr F. Bujak's paper "The man of action and the student" – (14, 7)
7. February 1929:	Dr P. Rybicki's paper "Research and the forms of social life" – (22, 7)
7. April 1929:	Prof. Dr T. Kotarbiński's paper on "On the skills of a researcher" – (20, 8)
27. April 1929:	W. Przybyłowicz's presentation "Remarks on the state's relationship to research" – (22, 10)

Besides, three smaller meetings took place, where discussions concerned the method and technique of performing ongoing research and the means of implementing the activities of the Mianowski Fund's Research Division.

[...]

The Circle's third meeting focused on Dr P. Rybicki's paper "Research and the forms of social life", which outlined the sociological trend in the theory of science. A lively discussion followed the presentation. The following remarks stood out: 1) studying the genesis of scientific activity we must not overlook the moment where teaching inspires an inquisitive relationship to the material; 2) specialization in research is not rooted in social concerns but in a proliferation of material; 3) the boundaries between certain disciplines have recently become completely blurred, giving rise to great syntheses in some fields (physics); 4) there is no fundamental difference between the thinking of primitive and contemporary humans; there has merely been an evolution from animism towards objectivism, though remnants of that animism are still felt in many sciences; 5) we are aware today of the birth of a new discipline, the science of science, which aims to uncover the truths governing science as a phenomenon. A vocation for research in this field is most common among people who do scientific work; in order to make foundations of the science of science as solid as possible, we ought to begin by examining the most concrete issues within the field of sociology of knowledge, such as the recruiting of workers or the influence of social conditions on the researcher's work.

During the fourth meeting of the Circle, prof. Kotarbiński gave a paper entitled "The abilities characterizing a researcher", setting out the skills of a researcher, which he illustrated with an analysis of the intellectuality of a leading Polish logician. While the presenter argued that we cannot pin down any specific characteristics common to all researchers, during the discussion the following ensemble of qualifications was proposed as necessary to researchers of all kinds of temperaments and preferences: a) a capacity for abstract thinking, b) an ability to identify similarities and to synthetize phenomena, c) a penchant for theoretical problems, d) a tendency to search for the causes of phenomena, rather than just their effects, e) there wherewithal to experience scientific "revelations". There are, however, sciences on the lower rungs of development (descriptive ones), where it is possible to do successful work without the above-mentioned conditions; there are also highly intellectualized practical fields that require thinking that is just abstract as in research. A methodological indicator for studies on these issues might be to examine not the researcher but the product of the research.

[...]

Second Report on the activities of the Science Studies Circle

Starting with vol. 11, so-called reports on the activities of the Science Studies Circle were printed in Nauka Polska. They appeared in the section Polish Chronicle (*Kronika Polska*) alongside information on recent science legislation, prizes etc. Rather short in earlier years, later reports were longer, when the discussions were documented in detail (for an overview see ▶ **IN THIS VOL.** List of Sessions). Usually, the papers presented were not reiterated exhaustively, as most of them were published in the journal.

The Second Report printed in vol. 13 informed about presentations by sociologist Stanisław Ossowski, physicist Czesław Białobrzeski as well as sociologist and economist Stanisław Rychliński, whose text was not printed at length, however. This translation focusses on Ossowski's presentation (166–167), which was an earlier version for the seminal text on science studies that he wrote together with Maria Ossowska (▶ Ossowska/Ossowski 1936). Translation by Tul'si (Tuesday) Bhambry.

ORIGINAL REFERENCE
"Sprawozdanie drugie z działalności Koła Naukoznawczego," *Nauka Polska*, vol. 13 (1930), 166–169.

FURTHER TEXTS MENTIONED
Czesław Białobrzeski: "Religja i nauka," *Nauka Polska*, vol. 13 (1930), 1–15.
Maria Ossowska, Stanisław Ossowski: "Nauka o nauce," *Nauka Polska*, vol. 20 (1935), 1–12.
▶ — "The Science of Science," *Organon*, vol. 1 (1936), 1–12.

Between June 1929 and June this year the *Koło Naukoznawcze* held three meetings. Presentations included the following papers:

25. November 1929: Dr S. Ossowski on "The Problematics of the Science of Science" (12, 6)[1]
12. December 1929: Prof. C. Białobrzeski's paper on "Science and Religion" (33, 7)
7. April 1930: S. Rychliński's paper on "The Influence of the Ideals of Organic Work on the Development of Science in the Kingdom of Poland after 1863" (17, 6).

In the Circle's first assembly of this period (a smaller meeting), Dr Ossowski spoke on "The Problematics of the Science of Science". He presented a schematic table of the

---

[1] In parentheses the number of participants and the number of individuals taking part in the discussion.

topics of the science of science and the relationships between them. The topics listed in this table can be categorized as follows: 1) questions such as what is science, what are its criteria, its methods, its classification (epistemological and methodological problems), 2) how is scientific work generated (psychological and partially sociological issues); 3) what is the impact of science on social life and vice versa (sociological issues). Thinking about the topics of the science of science, we should logically begin by establishing basic concepts, such as the criteria for "scientificity", "scientific work" and so on. Concerning the second category of questions, our point of departure should be either research on scientific works that are considered to be outstanding, or the typology of scholars. Here a number of interesting questions emerges: a) what are the motives for choosing one scientific specialization over another, b) what are scientific workers' inclinations beyond science, c) are the similarities among different scholarly types stronger than their similarities with creators and men of action of other types, d) what could be the motives for scientific work (both motives that inspire a person to do scientific work in general, and motives for the individual stages of scientific work), e) what are "revelations" in scientific work and is there a kind of major scientific creativity that does not rely on them, and so on. Among the historical questions, a particularly interesting one would be to examine the evolution of concepts of science, scientific value and the scholar. The discussion focused on completing the catalogue of topics in the science of science. It was proposed that besides social factors and the independent creativity of individuals, the development of science is determined, to some extent, by science itself, just as technology determines its own development. It is possible to search for intrinsic laws of this development. With regard to the typology of scholars, it was emphasized that it is necessary to analyse the type of the outstanding scholar who could perform creative work in almost every domain of human creativity, as well as the type of the artist working successfully in different fields of art.

[…]

The Subject Matter and Tasks of the Science of Knowledge

Florian Znaniecki

FLORIAN ZNANIECKI (15 January 1882, Świątniki, Congress Poland – 23 March 1958, Champaign IL, USA), sociologist, considered a key figure in the emergence of Polish academic sociology and in the development of empirical methodology. Originally trained in philosophy (Warsaw, Geneva, Zurich, Paris) he partnered with Chicago sociologist William I. Thomas to study Polish (labour) migration in 1913. Together they published the multi-volume work *The Polish Peasant in Europe and America* (1918–1920). Since 1920, Z. established the first sociological chair at a Polish univ. in Poznań and subsequently helped institutionalise the discipline in the country. He also was a visiting prof. at Columbia Univ. and finally became a prof. at Urbana-Champaign after the outbreak of World War II.

Z. published the Polish original version of "The Subject Matter and Tasks of the Science of Knowledge" in vol. 5 of Nauka Polska (1925). Christopher Kasparek's English translation, which is reproduced below, was published in 1982. Where Z.'s loose bibliographical practice could be completed unequivocally, the missing information was added for this edition. Minor typographical errors were corrected without special indication.

ORIGINAL REFERENCES
Florian Znaniecki: "Przedmiot i zadania nauki o wiedzy," *Nauka Polska*, vol. 5 (1925), 1–78.
— "The Subject Matter and Tasks of the Science of Knowledge," transl. by Christopher Kasparek, in: *Polish Contributions to the Science of Science*, ed. by Bohdan Walentynowicz, Dordrecht et al. (1982), 1–81.

CONTENTS: Historical roots of the theory of knowledge. The subject matter of the theory of knowledge. The analysis, description, and classification of cognitive phenomena. Causal treatment of the relationship of cognitive results to the circumstances of cognition. Causal treatment of cognitive activity. Bibliography.

Historical roots of the theory of knowledge

Although theoretical reflection on knowledge – which arose as early as Heraclitus and the Eleatics, stretches in an unbroken line through the history of human thought to the present day – nevertheless the most recent times have introduced into these reflections so many new questions and viewpoints so divergent from the earlier ones that we may safely say that we are now witnessing the creation of a new *science of knowledge* whose relationship to the old inquiries may be compared with the relationship of modern physics and chemistry to the 'natural philosophy' that preceded them, or of contemporary sociology to the 'political philosophy' of antiquity and the Renaissance. To be sure, we are still dealing with an accumulation of miscellaneous observations rather than with a systematically and consciously developed scientific whole, but gradually an order is emerging from this chaos and there is beginning to take shape a concept of a single, general theory of knowledge as a separate branch of human culture, endowed with special empirical properties and permitting of empirical study. This theory is beginning to take its place beside such sciences as economics and linguistics as it assumes the traits of a positive, comparative, generalizing and elucidating science. Thereby, too, it is coming to be distinguished clearly from epistemology, from normative logic and from a strictly descriptive history of knowledge. The distinction, whose guidelines we will briefly set down here, is not the result of some arbitrary *a priori* designation of the boundaries between the respective fields of human thought, but has developed spontaneously through the emergence – within each of the earlier types of reflection upon knowledge – of problems that have resisted accommodation within its traditional sphere. These problems, gradually concentrating on a common ground outside the scope of purely epistemological, logical or historical thought, constitute one of the main sources of the new science of knowledge.

Epistemology, which may be characterized as a metaphysics of knowledge, seeks to plumb the absolute essence of cognition in general, its limits, and the general conditions of its validity. Thus, for the epistemologist, the essence of knowledge may be defined absolutely and once and for all; his view of cognition is, in his own eyes, independent of the historical evolution of human knowledge. A classic example here is Kant's *Critique of Pure Reason*[1], which likewise contains implicitly a view of the meaning that research into the actual development of knowledge possesses for the epistemologist; that research presents itself simply as a record of the gradual realization in human history of that absolute essence of cognition that the epistemologist defines independently of history. However, in the past several decades there have appeared epistemological trends in which the earlier theories on the essence of knowledge are represented as being conditioned by the actual state of the disciplines in the respective periods, and thus as historically relative. These trends, while themselves unable to do without general absolute propositions concerning the essence of knowledge (inasmuch as they are still epistemological), nevertheless strive to reduce those propositions to the smallest possible number of formal principles and envisage the main task of the theory of knowledge as the empirical study of the actual development of science based on concrete historical examples. That is the trend, for example, of Schiller's humanism, of the pragmatism of Dewey, Mead and A.W. Moore; and

---

[1] Immanuel Kant: *Critique of pure reason*, transl. by J.M.D. Meiklejohn, London, Toronto 1934, esp. 'Transcendental Doctrine of Method', part IV: The History of Pure Reason.

of Rauh's extreme empiricism; and it is approximated by Mach's empiriocriticism and Poincaré's partial relativism. In short, in these currents, the centre-of-gravity of cognitive theory shifts from metaphysics to an empirical science of knowledge; and precisely to that extent these currents cease to fit within the boundaries of general philosophical epistemology and contribute to the building of a 'science of science' as a special humanistic discipline.

In referring to this evolution within epistemology itself, we scarcely wish to suggest that the latter loses all and any meaning it may have in our overall thinking, and that the empirical 'science of knowledge' can supplant it entirely. The question of the nature, limits, and validity of cognition remains a momentous problem of human thought, and a positive science of knowledge is in no position to resolve it. Like every special discipline, it must leave ultimate problems of this kind to general philosophy; it simply accepts the existence of knowledge as a given collection of facts, being unable as it is to delve into the transcendental nature of the entire field. Its limitation results from – among other things – the fact that it is itself a discipline, just like the disciplines that it studies, and its own limits and validity remain, for the philosopher, as problematic as the limits and validity of any other discipline. The far-reaching significance of the newer epistemological trends in the history of reflection upon knowledge rests mainly on the fact that the epistemologists themselves have begun to realize the necessity of *complementing the metaphysics* of knowledge with an empirical science of knowledge – a necessity arising from, among other things, the fact that the very metaphysics of knowledge expresses itself in historically given particular epistemological theories whose sources, character, and influence lend themselves in some measure to scholarly explanation in accordance with the tasks and methods of scientific investigation prevalent at the time, the personality and social standing of their authors, and so on.

Logic, in company with methodology, strives to create an ideal of perfect knowledge and to point the ways to that ideal. If we have described epistemology as a metaphysics of cognition, we may characterize logic and methodology as the *axiology* of cognition. Considered abstractly, its chief task is independent of the results of studies of historically given scientific and philosophical theories, since what it attempts to do is precisely to impose its ideal on any and all theories and it appraises them accordingly as they approximate to that ideal. An empirical theory of knowledge, in its structure, must submit to some logical ideal. Yet, in the more recent evolution of logic and methodology, a question has also arisen concerning the historical relativity of the logical ideals themselves. Schiller, in his works (among others, in his *Formal Logic*[2]), pointedly emphasizes the dependence of tests of logical perfection on the questions actually posed by science in various fields, periods and environments. American pragmatism attempts to accentuate this same dependence, working from the principle of the dependence of the formal process of cognitive thinking on the

---

[2] F.C.S. Schiller: *Formal Logic. A Scientific and Social Problem*, London 1912.

practical everyday problems that are thrust upon that thinking for their solution. Even in denying this principle we cannot shut our eyes to the fact that the diversity of actual processes in which cognitive questions are posed and solved does not submit to confinement in any logical system. Similarly, in France, in lieu of the old deduction of methodological rules from general principles of formal logic, we find a tremendous preponderance of monographic inductive studies on methods actually employed in various disciplines and branches of disciplines – without regard to the question of the possibility of their subordination to common ideals of formal perfection. In periodicals such as *Revue de Métaphysique et de Morale, Journal of Philosophy* and *Psychology and Scientific Method* these trends have found appropriate expression. In short, logic, just as epistemology, is beginning to demand an ever more empirical theory of knowledge as its complement, and many logicians and methodologists have already made important contributions to this theory.

The descriptive history of knowledge quite some time ago had already produced questions which resisted a convenient fitting within its own framework, questions which might generally be described as being concerned with regularities in scientific development. In Comte's classification of the sciences, the logical order of which is supposed to correspond to the sciences' genetic order; in the attempts at a synthesis of the history of philosophy by Hegel's school, in the works of Ernest Naville, in the well known history of philosophy textbook by Janet and Séailles based on the concept of the development of philosophical questions, in numerous studies of the histories of individual disciplines[3], treating the history of a given discipline as the gradual production or emergence of its systematic structure – in these and many similar deliberations we find the beginnings of a type of study logically incommensurate with history proper, i.e. with individualizing history. The history of knowledge, considered descriptively, presents itself to us as a history of creative individuals and groups. Of course, when we reach back into the remote past the discovery of creative individuals becomes impossible for lack of documents; still, a given branch of knowledge is always historically individualized at least in regard to the social group, to the people that have fostered it. To be sure, in recent times, in the most developed sciences the role of individuals of great creativity seems to be weakening in the face of the sizeable number of persons working in a field and the spreading expanse of that field; however, this means only that a strictly historical reconstruction of the evolution of science in recent times must take into account a considerably greater number of creative individuals than in the past, and not that history can today pass over the creative individual.

And between the reconstruction of the history of knowledge as the handiwork of concrete individuals and groups, and the study of an abstractly considered, rational

---

[3] For example in Aleksander Heflich and Stanisław Michalski (eds): *Dzieje myśli. Zarys historii rozwoju nauk* (4 vols.) [The History of Thought. Outline of the History of the Development of Science] (= *Poradnik dla samouków* [A Guide for Autodidacts], vol. 6), Warszawa 1907–1911. ||| Znaniecki refers to Paul Janet and Gabriel Séailles: *A History of the Problems of Philosophy* (2 vols.), transl. by Ada Monahan, London 1902 above.

development of problems and solutions, the difference is fundamental and obvious. A particular creative act which, logically speaking, is a link in the systematic evolutionary process of science, is connected for the historian with the totality of a creative personality, which after all takes in not only scientific or scholarly functions but also social, political, economic, religious, aesthetic and other kinds. Through his creative work the scientist or scholar is in communion with all other creative individuals in the discipline, independently of place and time – and with them only; as a concrete personality he comes in contact almost exclusively with his immediate social surroundings in the given period: among other things, and perhaps predominantly; he is in contact with people and institutions having nothing in common with learning. A historian who wishes to describe creative persons in all their concreteness, against their historical background, cannot at the same time seek for an internal regularity of the development of learning, which becomes manifest only in abstraction from the historical background against which its particular links have arisen. The study of this regularity thus inevitably parts company with historical description in order to become a separate, comparative theory of knowledge. Perhaps this latter will succeed in discovering interdependencies between learning and other spheres of human activity; but that will require first our isolation of scholarly and scientific activity from the primitive concreteness of life and an understanding of the internal unity of learning.

Thus from epistemology, logic and the history of knowledge there are emerging complexes of empirical questions which are clearly beginning to constitute a separate field of study. Nor is that the end of it. This field is also starting to become enriched with empirical studies of knowledge that have come into being thanks to the initiative of two relatively new disciplines whose early generalizations extended over the entire sphere of cultural life. We have in mind here the so-called psychology and sociology of cognition.

It is difficult to draw precisely a historical boundary between the 'psychology of cognition' and logic on the one hand, and between the former and epistemology on the other; for, as we know, these two disciplines have frequently employed psychological assumptions and theories. There is, however, no doubt that over the last decades psychology – having taken shape as a discipline of quite distinct methods – has approached the question of cognitive thought from a new point of view, different from the old epistemological and logical psychologism. We shall merely point out a few of the newer works that illustrate this viewpoint. Classical examples already today are Hobhouse's work *Mind in Evolution*[4] and Baldwin's *Mental Development in the Child and the Race*[5], and especially the latter's three-volume *Thought and Things: A Study of the Development and Meaning of Thought, or Genetic Logic*, whose subject, in the author's expression, is the '*modus* of psychic function known as cognition'[6]. Here

---

[4] Leonard Trelawny Hobhouse: *Mind in Evolution*, London, New York 1901.
[5] James Mark Baldwin: *Mental Development in the Child and the Race. Method and Processes*, New York 1895.
[6] Id., *Thought and Things. A Study of the Development and Meaning of Thought, or Genetic Logic*,

belong likewise studies by American psychological schools, influenced by pragmatism, of the biological role of cognition as a tool in the loving individual's adjustment to conditions in the environment. Generally speaking, a characteristic feature of the contemporary psychology of cognition is treatment of the cognitive function in its connection with the total biopsychical life of the cognitive individual as an organism endowed with consciousness; that exactly is the new viewpoint which we do not find in the old 'psychologistic' epistemology and logic and which is the special contribution of psychology, as an already developed discipline, to the theory of knowledge.

It is, however, not only in the construction of psychological abstractions that this viewpoint occurs. Earlier still and independently it makes its appearance in the biopsychological analysis of the concrete personalities of creative scientists and scholars, tied in with the question of heredity. Galton[7] and de Candolle[8] initiated this kind of investigation; de Candolle's work is continued by Ostwald[9]. Lombroso and his school, as is known, linked up studies of genius (extended to include creative scholars and scientists) with psychopathology.

Sociological studies of knowledge began, strictly speaking, with Durkheim's school in France. Of course, as early as Comte the connection between cognitive functions on the one hand and the social system and the entire culture of society generally on the other, epitomized in his famous law of three stages, had already endowed knowledge with a sociological significance; still, the concern here is with the dependence of society's life on knowledge rather than vice versa, and consequently we do not find in Comte any attempt at a sociological explanation of knowledge itself. It was only Durkheim, in his paper "De quelques formes primitives de classification"[10], who explicitly opened up a new path down which he was followed by Lévy-Bruhl[11], Simmel and others. Apart from these strictly scientific studies, we find still earlier in Germany, mainly under the influence of the romantic concept of the *Volksgeist* or *Volksseele*, philosophical investigations into the national character of knowledge, which have developed further in Poland and of late have been revived from a more positive standpoint in connection with the question of building up Polish learning (see, for example, articles by Gawroński, Minkiewicz, Kochanowski, Ossowski, Bujak and others in preceding volumes of *Nauka Polska*).[12] Clearly, we

---

London, New York 1906. ||| It has not been possible to verify the exact quotation as it appears in Baldwin [C.K.]. However, Znaniecki possibly paraphrases p. 9: "It is the *mode of psychic function called knowledge*, together with its objects and meanings, that is explicitly the topic of Genetic Logic."

[7] Francis Galton: *Hereditary Genius: An Inquiry into its Laws and Consequences*, London 1869.

[8] Alphonse de Candolle: *Histoire des sciences et des savants depuis deux siècles*, Genève, Bale, Lyon 1873.

[9] Wilhelm Ostwald: *Große Männer. Studien zur Biologie des Genies* (vol. 1), Leipzig 1910.

[10] Émile Durkheim and Marcel Mauss: "De quelques formes primitives de classification. Contribution a l'étude des representations collectives," *L'Année Sociologique*, vol. 6 (1901–1902), 1–72. ||| Znaniecki does not mention Durkheim's co-author Mauss in the original.

[11] Lucien Lévy-Bruhl: *Les fonctions mentales dans les sociétés inférieures*, Paris 1910.

[12] Andrzej Gawroński: „Nauka narodowa czy międzynarodowa" [National or International

have here a question whose scientific solution requires sociological studies of nationality. Of a similar nature are the well-known generalizations of the school of 'historical materialism' (or more precisely, historical economism) concerning the dependence of learning on forms of technical production and on economic and class systems; the establishment of the latter dependence requires thorough sociological studies of social classes and of their significance for culture. Generally speaking, sociology has introduced into theory of knowledge a thesis of the conditioning of cognition by the life of society – a thesis which in Durkheim's school has been expressed in such extreme fashion that that school seems intent on ascribing to scientific truths exclusively the character of social beliefs that draw their objectivity solely from their universal acknowledgment in a society.

It is clear, however, that whatever may have been the contributions heretofore of psychology and sociology to the theory of knowledge, the problems raised by the latter must perforce extend beyond the boundaries of the former. In the process of the fixing of the scope and methods of these two sciences, the audacious claims of their youth are gradually moderating. A psycho-biological interpretation of conscious life can be applied scientifically only to the most elementary phenomena of consciousness – to those that lend themselves in fairly accurate approximation either to experimental studies or to statistical computations. The 'psychology of cognition' in its higher manifestations, i.e., above all in those which constitute science and philosophy, must be something completely different from the theory of the psycho-organic reactions of a living being to influences from the natural environment, inasmuch as here the conscious individual is dealing not with material forces but with ideal categories, and his cognitive activities cannot be conceived in isolation from the entire ideal system of knowledge with which they are associated. And so far as sociology is concerned, we have shown elsewhere[13] that it cannot study all of human culture but must confine itself to its own special field of social phenomena in the strict sense of the word, i.e., in the sense of functions having for their subject individuals and human groups as well as the values created by these functions. Although it was sociologists who initiated the study of collective cognitive beliefs, those studies must be continued outside the limits of sociology.

Thus the theory of knowledge which is being built up can and ought to encompass, among other things, questions in this area that have been raised by psychology and sociology, without however making itself client to those sciences. This hardly means

---

Science], *Nauka Polska*, vol. 4 (1923), 36–44, Romuald Minkiewicz: „O niezależność nauki polskiej" [On the Independence of Polish Science], *Nauka Polska*, vol. 2 (1919), 493–502, Jan K. Kochanowski: „Kilka słów w sprawie nauki narodowej" [Some Remarks on National Science], *Nauka Polska*, vol. 2 (1919), 444–448, Stanisław Ossowski: „Funkcja dziejowa nauki" [The Historical Function of Science], *Nauka Polska*, vol. 4 (1923), 8–35, Franciszek Bujak: „Nauka a społeczeństwo" [Science and Society], *Nauka Polska*, vol. 3 (1920), 64–74. ||| Znaniecki does not list any specific information on published materials, so the most relevant texts published by the given authors before 1925 are mentioned here.

[13] Florian Znaniecki: *Wstęp do socjologii* [Introduction to Sociology], Poznań 1922.

that psychological and sociological facts and theories cannot at times shed light on particular questions in the theory of knowledge; cooperation among various disciplines is always possible and desirable. The point, however, is for the theory of knowledge, like every other discipline, to acquire its own point of view and its own methods, instead of making use of those of other disciplines. This is essential if only because, as we have seen, questions of the theory of knowledge have been approached from at least five different points of view – the epistemological, logical, historical, psychological and sociological. Thus if the study of knowledge remained client to these disciplines, it could not attain the homogeneous character of a systematic discipline but would persist as a chaotic collection of diverse, often mutually incommensurable, questions and views.

The need to isolate and emancipate an empirical theory of knowledge as a separate discipline with its own tasks and method springs, however, not only from theoretical but from practical considerations as well. The theory of knowledge should provide a basis for all practical efforts at inspiring, broadening, organizing and refining scientific work – and the need for it is more and more to be felt. Particularly in Poland the efforts of many years' duration to maintain learning at an appreciable level in spite of political obstacles and, subsequently, following the recovery of independence, the drive for a rapid expansion of learning on a scale far broader than hitherto, have imposed numerous questions whose solution necessitates a much more thorough, strictly scientific investigation of this, perhaps the most momentous, field of cultural life than has been possible up to the present time. The long-time activity of the Academy of Sciences, of the Mianowski Fund, of learned societies and of many other institutions supporting scientific work supplies plentiful evidence of the practical need for a theory of knowledge, and a number of specific publications of the past twenty-odd years, begun by *Poradnik dla samouków* [A Guide for Autodidacts] and continued at present by Nauka Polska, have been giving expression to that need. Of course, the question has hardly been ignored elsewhere; every learned association, every institution for the support of productivity in the field of knowledge, has of necessity had to ponder it. It is possible, however, that nowhere has it been the object of so comprehensive, persevering and concentrated investigations as in our country, and we owe this not only to the exceptional circumstances which we have experienced but also, and above all, to the fact that in these circumstances there have been people who have devoted their whole lives to the question and who have become centres about which the movement in this field could organize. We are the first country to have a separate Department of Science [*Wydział Nauki*] in its government; perhaps we will also succeed in becoming the first nation to apprehend the theory of learning in its totality and discover an adequate foundation for it.

While a new science is not created by way of methodological deliberations but by way of positive research still in certain crucial periods reflection upon the subject matter, tasks and method of such a science is essential to its further development. It is in just such a period that the theory of knowledge finds itself today. The numerous scattered contributions, of which we have merely mentioned a small part as examples, must be apprehended and organized from a uniform standpoint, if its subsequent development is to proceed consciously, systematically, with a greater than heretofore economy of effort and a greater fertility of results. Such organization must be preceded by abstract programmatic deliberations in order that the theory of knowledge may the more quickly achieve a planned and self-conscious character. And so let us try, if only tentatively and hypothetically, to draw at full length the boundaries of the new science and to indicate guidelines for its conduct. Certainly we realize full well that this kind of introductory outline cannot be in any case a binding program but at best a sketch suggesting certain ideas which may or may not influence the subsequent development of the science: but in any event they ought to provoke discussion and induce scholars to deepen their reflections on their own scientific conduct.

The subject matter of the theory of knowledge

From the empirico-scientific viewpoint 'knowledge' as a subject of research is of course a complex of cultural phenomena simply given to the scholar, much as are speech, art, law or any other field of culture. The theoretician of knowledge, much like the linguist, the theoretical economist or the sociologist, and as opposed, e. g., to the logician or the moralist, must take various phenomena entering into the scope of 'knowledge' in various ages and human communities as objectively existing in equal measure. In his capacity as a scholar he has no right to wonder whether the phenomena given to him as the 'knowledge' of a certain age and society are in fact knowledge from the standpoint of the cognitive ideal that he himself recognizes as definitive, whether and in what measure they are not only phenomena but also truths in his view, logically justified by the tests he himself at present accepts. Such elimination, from studies of knowledge, of value judgments upon the phenomena studied may be difficult considering the fact that, in creating or adopting scientific theories, we have grown accustomed to criticizing other persons' statements as well as our own. Equally difficult is it, at first, for a sociologist to abandon his claims to moral or utilitarian judgment of social phenomena, considering that as an active member of social groups he has grown used to applying moral and utilitarian tests to the social life in which he participates. Yet there surely is no need to demonstrate that a theoretician must maintain complete objectivity in relation to his data (which of course is no impediment to taking an axiological position concerning theories referring to the data) if he wishes to achieve results consistent with present-day standards of scholarship. A linguist studying linguistic phenomena does not distinguish between 'correct' and 'incorrect', between 'good' and 'bad' speech, but leaves any judgments in that way to grammatians and to writers, i. e., to the practical organizers and co-authors of language.

The concern now is with what phenomena the theoretician of knowledge must include in the field of culture that he is studying. Since he cannot apply his own tests of truth to people's views that are given him historically, since in other words he has no right to regard as 'knowledge' in the historico-cultural sense only those cognitive phenomena which in his conviction constitute true knowledge in the logical sense – it is clear that in the choice of phenomena to be studied he must apply the tests of truth recognized by the people who have accepted the given cognitive phenomena as true, that he must regard as 'knowledge' everything that in the periods and communities studied by him has been deemed knowledge. Only in this way will he avoid the danger of substituting for phenomena a conception entertained in advance concerning the nature of the phenomena, of bending the data of historico-cultural experience to his own subjective conviction as to what those data ought to be like. Such conduct on the part of the theoretician of knowledge will accord with the general guidelines of all the humanities, whose basic feature, as distinguished for example from the natural sciences, is the examination of the phenomena under study in the form in which they have been or are actually experienced by people involved practically with them, that is by individuals and groups living in the respective historic periods and places. Similarly the linguist studying a language considers its words and sentences in the sound and meaning that they possess for the individuals and groups speaking that language.

On this general basis we must find a purely formal criterion that, without presupposing anything in advance about the essence of the phenomena comprising 'knowledge' in a historico-cultural, humanistic construction, would still allow for distinguishing these phenomena from among any other cultural (economic, legal, social, linguistic, aesthetic, religious, etc.) phenomena, that would allow a judgment in every particular case concerning whether the given phenomenon constitutes a proper subject of study for the theory of knowledge or whether it belongs in the domain of a different humanistic discipline. The first, in a way an external, test by virtue of which some phenomena may from a humanistic viewpoint be classed in the domain of 'knowledge', is that for at least certain human subjects (individuals or collectives) these phenomena possess the characteristic of *'trueness'* and in that respect are opposed in their conviction to other phenomena regarded as *'false'*. This conviction manifests itself empirically in the intentional utterances of these human subjects or in the use which they make of the phenomena in question. Any phenomenon to which anyone at any time ascribes the quality of trueness as opposed to the falseness of another phenomenon, we term a *cognitive value*. This trueness or falseness (in the humanistic sense stipulated above) is attributable to cognitive values with respect to something else which occurs in the role of an object of knowledge, that is of an object to which the meaning of the cognitive values refers.

In defining cognitive value in the above fashion, we purposely leave its own content undefined. For it turns out, in examination of materials which history and ethnography furnish to the theory of knowledge, that in various times and human communities various categories of phenomena have been regarded as being capable of possessing the characteristic of trueness or falseness. And since in studies of knowledge we must abide by the guideline of the humanities and take phenomena in the way in which they present themselves to people in the epochs and environments studied, then we too must regard as real cognitive values the categories of phenomena that those people have treated as cognitive values. And so at the dawn of knowledge various words or other symbols have sometimes been regarded as true or false; trueness in the belief of peoples and individuals at lower levels of culture often consists in applying to an object of knowledge the symbol (word or sign) that has traditionally been applied to such objects, falseness – in applying a different symbol. Hence, for example, the innumerable disputes over words in the history of learning. Furthermore, as true or false may be taken not only single symbols but also combinations of symbols: sentences, arrangements of signs. This understanding manifests itself, e.g., in the efforts aimed at grammatical precision of statements and verbal definitions that so conspicuously characterize medieval knowledge. Finally, truth or falsity is sometimes ascribed not to sense phenomena but to ideal objects and to their combinations – to concepts, logical judgments, conclusions, theories. These objects are real from a humanistic point of view exactly to the extent to which people actually operate with them as being real, as manifested in the words or sensory signs being treated as pure symbols denoting those ideal objects, and those being regarded as true or false with reference to the objects which by their aid we come to know. We find this category of cognitive values, e.g., in modern physico-mathematical knowledge. The theoretician of knowledge has, of course, no right to regard any one of these categories of cognitive values as the sole justified one and to pass over others as not belonging in the domain of knowledge; nor may he substitute one category for another. His research will be fruitful only if in every case he is fully conscious of what kind of cognitive values he is dealing with. His task is not criticism but a thorough plumbing of the content, significance and relationships of the values that he finds in the history of intellectual culture, comparing, generalizing, classifying and explaining them as positively given and actually existing phenomena.

But the object of study in the theory of knowledge is not restricted to cognitive values. These, like all other values, owe their genesis, their historical evolution, their relationships and the role that they play in cultural life to appropriate human operations. These *cognitive operations*, as we may call them, constitute a special category of cultural activities of man's, much as do, for example, social, economic, legal or artistic activities. And although, in contradistinction for example to technical operations, they do not always manifest themselves in material movements, that is, they are above all 'mental' operations, still there is surely no need to prove their existence. For we experience those operations through their very execution; our every act of observation, of definition, of reasoning brings home to us their empirical character. And scholarly investigation can reconstitute and define every operation of another person from its manifestations in the sphere of cognitive values. For cognitive values are objects of cognitive operations in the capacity of their materials, tools or products; every change in cognitive values has its source in cognitive operations.

Appropriately to these objective manifestations we may divide cognitive operations into three classes. In the first class we will number operations that refer existing cognitive values to the objects to which they apply. Here belongs, e.g., the search for examples to illustrate the meaning of a scientific term, the carrying out of experiments to demonstrate theories, the application of the results of scientific research to everyday practical affairs. We might term operations of this class, operations of *cognitive experience*. They result in a cognitive orientation in the chaos of our data, in the introduction into our empirical world of a certain rational order consistent with the meaning and arrangement of our cognitive values (words, sentences, concepts, judgments, etc.) and conspicuous for the most part in the manner in which phenomena present themselves to us henceforth, and sometimes (in a practical application of knowledge) in the manner in which we attempt to materially transform these phenomena. In operations of this kind, cognitive values play the role of tools with which the cognitive arrangement of phenomena is brought about.

A second class of cognitive operations comprises those which form new cognitive values or modify old ones with the aid of material supplied by observation of cognitive objects. These operations are the best known ones, since it is chiefly they that have up to now been the focus of interest for various portions of the theory of knowledge that has been taking shape. As examples may be cited: the creation of a term to designate a newly isolated class of objects; the suggestion and proof of a new hypothesis on the basis of observed facts; the critique of an old hypothesis by new experiments whose purpose is to test its validity. We would term these operations, operations of *cognitive idealization*, for they create from materials isolated from the chaotic becoming of our empirical world, objects of a distinct type elevated above this becoming, playing, as we have already seen above, the role of regulators by whose help the chaos of empirical data may in some measure be ordered, rationalized. This idealization of course is not absolute since in fact, as the history of knowledge proves, cognitive values in their capacity as components of human culture themselves enter into the general current of becoming, are historically conditioned and mutable; in any event, however, the concern is only with their relative ideality as compared with the phenomena to which they refer.

The cognitive operations of the third class, which we shall term operations of *cognitive systematization*, work with cognitive values in abstraction from the objects to which the latter refer, associate them with each other and systematize them, in the process making use of preexisting cognitive values as material for the creation or modification of others. Here belongs, e.g., the linking up of sentences into the continuous course of an exposition, the deduction of a conclusion from given premises, the elaboration of a general theory to which are subordinated a series of preexisting particular theories, the logical criticism of a theory on the basis of its agreement or disagreement with other theories or on account of its internal content. In saying that these operations work with cognitive values in abstraction from the objects of the latter, we understand by this that they are not immediately involved in the representation, production or modification of the connection between cognitive values and their objects; but this connection is, of course, potentially implied as established in every act of associating cognitive values, or otherwise those values would be devoid of meaning. Much as in speaking we need not refer every expression to an object symbolized by it but employ the established sense of the expressions as something that is empirically associated with their sound and potentially indicates the designated object, so also in systematizing concepts and theories we know what they designate without having every time to illustrate them or to verify them by reference to experience.

Still, it is understood that the foregoing three kinds of operations in the actual course of cognition seldom occur in isolation. Thus, for example, the practical application of a theory often leads to its critical modification, the creation of new concepts is linked to the contemplation of old ones, the systematization of scientific values is sometimes accompanied by their verification and, conversely, the founding of a new theory on experience often takes place parallel to its systematic association with other theories.

In studying cognitive operations, just as with cognitive values, we must maintain the humanistic viewpoint, i. e., regard as cognitive those operations which are entitled to that characterization in the conviction of the acting or 'thinking' subjects themselves. From this it follows, among other things, that the known distinction among logical operations is to the theoretician of knowledge completely unfounded. All cognitive operations are logical in the sense that they possess, in the belief of the cognitive subject, an objective validity, in the sense that they refer to cognitive values as to their objects, actualize them in application to experience, create them, transform them and associate them. To be sure, their validity, considered comparatively, turns out to be only relative; no cognitive operation actually carried out is an absolute cognition; each may be complemented or supplanted by another operation. The theoretician of knowledge cannot, however, on that ground deny them a logical character since, standing as he is on a humanistic position, he is studying not absolute cognition in itself but historically given cognition as it presents itself to historically living people.

On the other hand, all cognitive operations may be called psychological, precisely in the sense that they are carried out by empirical individuals or by human communities and so are subject to psychological study just like any other kinds of operations. This psychological character of theirs does not, of course, mean that they are biopsychical processes; they possess a subjective-ideal character, as opposed to the objective-real character of the processes studied by the biological psychologist. The same may, however, be said generally of any conscious operation; the 'psychology' that must be practiced by the humanistic investigator, the psychology of art, religion or societal life, in spite of the common term actually has few viewpoints in common with the experimental psychology of the biological schools.

The most important source for the distinction between the logical and psychological processes in the domain of cognition actually has been the evaluation of cognitive operations from the standpoint of the standard of scientific or scholarly knowledge that at a certain stage in the evolution of knowledge has been regarded as supreme. Operations conforming to accepted criteria of scientific excellence have been opposed, as being logical, to operations that failed to satisfy those criteria, as being psychological. Thus, for example, syllogistic reasoning in accordance with norms has been distinguished, as being logical, from that which was not so in accordance and which, being 'illogical', has been treated not as a cognitive process but as psychological association. It is obvious, however, that this kind of distinction cannot be maintained in science, which deals with an unlimited diversity of historically manifested cognitive operations and cannot recognize any one of the tests of scholarly adequacy as final and absolute. A cognitively imperfect operation, one qualified as 'illogical' from the standpoint of one test, may present itself as logically valid from the standpoint of other tests; the opposites, known to history, of apriorism and empiricism, of rationalism and mysticism, supply good examples of such a difference in stance.

One might at most consider whether a given operation is cognitively adequate or not from the standpoint of its own criteria to which it seeks to conform, whether it is, in other words, 'logical' or 'illogical' subjectively. But even then, on close examination, it transpires that such a classification is pointless since in fact there are no perfect cognitive operations, even from the standpoint of their own criteria. For these criteria constitute ideal models which cognitive operations attempt with lesser or greater success to realize in the values they produce, but which they never realize completely in the face of the difficulties that cognitive thought encounters from the reality that is the object of cognitive values. An operation working with cognitive values which from the standpoint of certain criteria are appraised as false by other operations, is still a cognitive operation just as is one whose values under the provisions of the same criteria are recognized as true, since the operation would not be consummated did it not lay claim to the truth of its materials, tools or results; and the differences in validity of the claims of particular operations are not absolute but relative, conditional not only on the nature of the criteria but also on the degree of approximation to their realization. Likewise, e. g., technical operations producing objects defective or useless by their own technical criteria, are still technical operations just as are those whose products most nearly approach the acknowledged ideal of excellence.

The analysis, description, and classification of cognitive phenomena

It is obvious that the theory of knowledge cannot satisfy itself with a simple description of individual cognitive values and operations in their concrete relationships – that is the business of the history of knowledge. The task of the theory of cognitive phenomena, as of any theory of a scholarly character, is firstly the analysis and classification of those phenomena, and secondly the causal elucidation of facts occurring in the field.

In scientific analysis and classification the concern is above all with the isolation, from the concrete complexity of the phenomenal systems given to us in observation, of relatively simple elements and arrangements which can be discovered in various combinations and repeated in variable circumstances; and subsequently, with the systematic designation of basic similarities and difference among these elements and arrangements that might permit us to divide them into classes hierarchically ordered as to their generality.

Relatively the furthest advanced in the literature so far is the scientific analysis of cognitive phenomena. Of course, one might register numerous reservations as to the methodological premises of many works in the field; one might wish for a greater standardization of methods, a more conscious cooperation, above all a broadening of analytical studies to various, as yet almost untouched, sets of cognitive phenomena and a strict separation of exclusively theoretical studies of knowledge from attempts at a normative critique, evaluation and reform of knowledge. These are, however, questions whose consideration would require far more space than we presently have at our disposal; and other questions need more urgent attention.

Specifically, for example, the present state of the theory of knowledge is much more unsatisfactory in the classification of cognitive phenomena than it is in the domain of analysis. True, we do find on the one hand various classifications of cognitive operations in logic, epistemology and the psychology of cognition, and on the other hand the so-called 'classification of the sciences' to which Comte first strove to give a positive-scientific character: still, all these attempts have one flaw in common. Of the entire enormous wealth of cognitive phenomena that we find in the world of culture, they consider only a small part, most often selecting and systematizing it on the basis of criteria adopted *a priori* and associated with an ideal of cognition approved of by the given author. A truly scientific classification, however, ought to be completely objective and impartial, based on criteria deduced *a posteriori* from the empirical properties of the phenomena studied and embracing all the known phenomena in the field; moreover it ought to be not only a framework for the systematization of results attained but a tool for discovering new, as yet unobserved phenomena and new properties of phenomena already observed. The only kind of classification that can satisfy all these requirements is a *genetic* classification. The question is, however, whether such a classification is possible in the theory of knowledge and on what premises it should rest. To be sure, Comte already held that this classification of the sciences was at once logical and genetic; for reasons mentioned above, however, we cannot accept that assertion as definitive.

In considering the above matter we will have to limit ourselves here to the question of the classification of cognitive operations as being less complicated than the classification of values. For every cognitive value is first and foremost the product of a certain operation; consequently it is determined genetically above all by the nature of the latter, and beyond that depends on the kind of materials and the tools which the operation employed. We can express this still another way in traditional terms. Every cognitive value, every 'truth' presents itself as the solution to a certain theoretical problem. The posing of the problem depends solely on the cognitive operation; its solution, by contrast, is conditioned additionally by the data and investigative tools which the operation has at its disposal. This is expressed at its most elementary when the problem posed requires a quantitative solution; thus, for example, the problem of the size of the sun or of the earth's distance from the sun has been subject to many different solutions depending not only on the logical differentiation of the actual mode of thinking but also on the data on which astronomy has relied and on the instruments that it has employed. The number of possible solutions reduces to two, affirmation and negation, if the problem is of the kind such that its solution depends entirely on the existence or nonexistence within precisely defined limits of a certain hypothetical, just as precisely defined, datum; this is an extreme case, possible only in the realm of certain theoretical conventions.

Thus the diversity of cognitive values as products is greater than the diversity of the operations producing them. Moreover, though, a value already produced can serve further as material or continues to differentiate. Let us recall, for instance, the various changes of meaning that have affected the atomic theory, concepts of psychological association, and the assertion that the invention of the steam engine was the cause of contemporary capitalism – all depending on the questions for whose solution the above cognitive values have been used. The classification of cognitive values cannot, therefore, take them in all their historical concreteness but must define them in relation to the specific determinations to which they are subjected by certain kinds of cognitive operations. The latter are, to be sure, accessible to historical investigation only intermediately through the changes that they bring about in the domain of cognitive values; however, it is easy to obtain an exact definition of every operation by comparing its various manifestations under various circumstances.

If following these preliminary remarks we consider the historical evolution of human knowledge, it will present itself to us as the creative growth of cognitive values and the emergence of new cognitive operations. Clearly, the empirical theory of knowledge need not and cannot ponder a question belonging to the metaphysics of knowledge – namely, whether and to what extent scientific creativity is in an absolute sense the creation of objectively new truths or merely the discovery of subjectively new truths which have existed outside the realm of human consciousness always, or rather timelessly. For in any case the act of introducing, into the empirical domain of human cognitive values, a truth that had not previously been in that domain constitutes, empirically, an original and creative act, just as does the creation of a new work of art unlike any existing up to that time, or of a new technical invention.

This creative development, precisely to the extent that it is creative, is impossible to *elucidate scientifically*. For scientific elucidation as applied to phenomena is basically causal elucidation, i.e., the indication of a necessary *real* relationship among *facts*; thereby does it differ from *explanation*, which expresses an *ideal* relationship among the elements of a logical system, and from *philosophical* elucidation, which seeks to deduce secondary properties of phenomena from their basic *essence*. Creativity eludes the principle of causality and must in some way be excluded from scientific elucidation, as we shall see later. The science of knowledge may, however, systematically describe the latter in its creative development, basing on the genetic association of cognitive operations their essential definitions and division into classes.

For every new cognitive operation is associated with some preceding operation, is a result of the latter's differentiation, is a new variant of it, or arises as a new intellectual function in a logical synthesis with other operations. Thus, to take an example, the question of the relationship of the culture of human societies to their racial composition arose as a modification of the old problem of the relationship of the 'soul' to the 'body', as applied to a community instead of to the individual; the new problems in Newton's physics sprang from problems of Galileo's, Kepler's and Huygens' in a synthetic union; the new components of Kant's philosophy arose genetically partly as formal modifications, partly as functional differentiations of certain philosophical concepts of Hume's, Wolff's, Rousseau's and others' and of certain mathematical physical and ethical questions. Every new cognitive operation is thus logically and genetically a continuation of cognitive operations previously carried out. We may call this assumption, indispensable to the scientific treatment of knowledge in its historic becoming, the *principle of continuity in the evolution of knowledge*.

In accepting this principle, the theory of knowledge must nevertheless beware of two pitfalls: historical empiricism on the one hand, and teleological rationalism on the other. Although every new cognitive operation harks back genetically to previous ones, this hardly means that continuity in the evolution of knowledge rests on the historic continuity and cultural homogeneity of the society nurturing the field of knowledge. New cognitive problems do not necessarily arise in connection with questions that have concerned scholars working in the society in the immediately preceding period of its history. They can constitute a continuation of questions posed and solved a thousand years earlier and in entirely different social groups characterized by fundamentally different cultural complexes. The genetic connection between Renaissance thought and ancient Greek thought, and the taking up by Japanese scholars of scientific research in the mode of European, and American scientists, here supply obvious examples. We shall see later in what sense and in what measure one may speak of the relationship of learning to factors of a societal nature; that relationship, however, is in no event one of such a kind that the creative evolution of learning might take place solely within the historic continuity of human communities and constitute merely a part of the general evolution of the cultures of those communities.

Similarly false would be the assumption, typical if different examples of which we find in Comte and Hegel, that there exists in the evolution of knowledge an immanent rational order, definable once and for all from the standpoint of the requirements of a scientific final cause; that this evolution is essentially a gradual, objectively necessary realization of some perfect type or system of cognition and that its every step constitutes a logically indispensable link in some universal regularity. On the contrary. The actual evolution of knowledge at every step displays features of irrationality and unpredictability in the generation of new evolutionary lines. Every cognitive operation is, to be sure, logically a continuation of some preceding operation, but there is no objective necessity, real or ideal, for just this new operation and no other to be consummated in connection with the preceding operation, or even for just these operations and not others already carried out to find continuation at all in new operations. The principle of continuity in the evolution of knowledge has to be complemented by a principle of the *freedom and unpredictability* of that evolution. Having some cognitive operations already effected, we cannot aver that any particular designated new operation had actually or logically to arise from them. Precisely to the extent that the new operation is original and creative, its accomplishment cannot be subject either to actual or to logical determination.

In studying the history of knowledge, we repeatedly find that some cognitive operations, in defiance of historical probability and the requirements of final cause, do not find a logical continuation, that an opened evolutionary line breaks off for many years or even forever. Thus, for example, the heliocentric hypothesis of Aristarchus of Samos languished for centuries and had to be recreated anew by Copernicus; the Cartesian theory of vortices and the *characteristica universalis* of Leibniz have not found a logical continuation and, despite recent attempts at their resurrection, it is uncertain whether the respective evolutionary lines ever will have any continuation. And conversely, new cognitive operations sometimes turn out to recur not to some established evolutionary line which, rationally considered, 'ought to' lead further, but either break with the immediate tradition and, to initiate a new line of evolution, reach back to a remote, forgotten past, or search for new points of departure outside the field to which we would like logically to restrict them, e.g., in some other discipline or even in a cognitive activity attendant to practical life and hitherto unconfined by scientific method. And so, e.g., certain trends in contemporary philosophy have made recourse to medieval problems which modern thought is accustomed to regard as sterile, incapable of further development; numerous currents that have arisen in the past century in the humanities have had their sources not in the past histories of those disciplines but in new theories and methods of the natural sciences; we find the genesis of American pragmatism first and foremost in cognitive reflection on practical questions – of technology, economics, religion.

Furthermore, even when the point of departure for subsequent evolution is an operation or set of operations whose prolongation presents itself to us as in fact probable and requisite to the rational progress of learning – this subsequent evolution can still go in a completely arbitrary direction, even one totally inconsistent with our anticipations. Looking into the future, we should realize that there may be various ways of logically extending a certain operation, that various new operations can recur to the latter and that there is no determination, binding in advance, on whose strength a future operation that will recur logically to the operation presently under consideration, would have to be just so and not otherwise. The genetico-logical association between operations comes into being only at the moment of execution of the respective operations; it is not limned in advance prior to their appearance in the sphere of cognition. A genetic relationship is negative, not positive; this means that a new operation could not appear if there were not already another operation of which it is logically a continuation, hence that the new operation must stem from a designated operation, rest on it for its actualization; but an operation already existing in the realm of knowledge does not require the actualization of designated new operations. There could have been no Newtonian physics without Galileo, Kepler, Huygens; but Galileo, Kepler and Huygens did not have to be followed by Newton and his problems; their theories might have gone undeveloped, or a savant might have happened along who would have rested other problems on them, today perhaps altogether impossible to conceive.[14]

Undeniably the scale of possibilities for the future evolution of a science differentiates into various levels of probability. That differentiation has its source in the real and ideal limitations that human creativity sets for itself. The real limitations stem from the fact that cognitive thought, like all human activity, is in lesser or greater degree dependent on the circumstances with which it deals and as a result is in fact capable of realizing only some of its theoretical possibilities; in just the measure to which the dependence has come into effect (and we see an example of this in common mental habits), the possibilities realized will be precisely those to whose realization the given circumstances are most conducive. And the ideal limitations of future possibilities stem from the voluntary conformity of scientific thought to certain guidelines marked out by its previous development. For, while the growth of knowledge is not originally and spontaneously purposeful, still at higher levels of scientific culture attempts at introducing a teleological element into it multiply. We find in history numerous attempts at a planned organization of future scientific work attempts based for the most part on results and criteria in each case already achieved. The programmes of scientific and philosophical schools, and normative works in logic, methodology and the classification of sciences supply guidelines of this kind to which within more or less broad limits, cognitive creativity consciously conforms. In spite of that, as we know, sometimes in lieu of the cognitive operations for which the most conducive cultural conditions are ready, there unexpectedly appear other operations for which reality up to that time provides no support, and in face of all external limitations they create for themselves, despite difficulty, the indispensable scientific situation; occasionally too the best-laid plans for the organization of scientific work are disregarded or are altered by some new and unforeseen individual initiative.

This relatively unpredictable character to the development of science, however, hardly impedes the descriptive systematization of cognitive operations, provided

---

[14] It is only within the closed confines of elaborated scientific systems that certain operations positively require others.

only that the systematization does not attempt immediately, working from aprioristic categories, to be exhaustive and to possess a perfect logical form. However original a certain cognitive operation may be, it is never absolutely unique in its kind, if only because it is never an absolute beginning. Cognitive operations can be similar to one another in various degrees, and that makes their classification possible. And the history of knowledge demonstrates that the number and diversity of cognitive operations has grown enormously, even in the relatively short period of civilized life that we know from historic documents; evidence of this is, among other things, the emergence of numerous new disciplines and branches of disciplines. On the strength of the principle of continuity in evolution formulated above it may be concluded that, notwithstanding many cases of the breaking off and inhibition of evolutionary lines, the history of knowledge may be represented as the gradual emergence of a whole immense wealth of cognitive operations from one or several original operations through the progressive differentiation of forms and functions originally undifferentiated.

We thus find here a basis for a genetic classification of the elements of cognitive thought. Much like the biologist in the genetic classification of living organisms, the theoretician of knowledge must assume that operations arising by differentiation from a common source resemble each other more closely in a formal (or a functional) regard than do operations derived from diverse sources, and that this similarity among them is the greater the closer is their common genesis, the fewer the intermediate links between them and the more primitive operation in which the historic lines leading to the operations under study have had a common beginning. It is easy to realize how by this method, all the operations that we shall discover in the history of cognitive thought, can gradually be defined and brought into a classification system based now, not on arbitrary criteria drawn from one or another logical, epistemological or psychological theory, but on the actual bonds of genetic kinship. Naturally, such a system will be far more complicated than is, say, a Kantian classification; and further complications will arise when we pass over from elementary cognitive operations to the characterization of systems of operations – and, finally, of those great complexes of cognitive operations and values that we call 'disciplines'; in fact the main charge that must be lodged against all the attempts heretofore made at a general theoretical treatment of cognitive phenomena is that of illegitimate, we might ever say naive, simplification of the object studied, an insufficient awareness of the gigantic wealth and complexity of this field of culture that we call knowledge.

It should further be stressed here that a genetico-descriptive systematics of cognitive phenomena is by no means merely a scholarly re-creation of the cognitive past. With the emergence of new, genetically later cognitive operations, the old, more primitive ones do not at all atrophy. They persist and continue to be carried out with only slight modifications of their original form. Likewise old types of cognitive values continue to recur in communities of the highest cultural standing and in the intellectually most highly developed individuals. There is no discipline in which there do not in fact coexist cognitive operations and values corresponding to various historical stages in the evolution of knowledge. Stages whose genesis lies in prehistory are represented in all their vitality to this day in the thinking of so-called savage peoples and among the uneducated classes of civilized societies. They manifest themselves even in individuals familiar with the newest criteria of contemporary learning, in respect to questions of everyday life, both in so-called common sense and in numerous uncritical so-called superstitions. Finally, the most primitive, the most elementary cognitive operations and values are probably to be found in the thinking of small children and perhaps also in that of mentally subnormal individuals. The genetico-descriptive treatment of cognitive phenomena should, certainly, rest mainly on the history of knowledge, since it is there that cognitive phenomena stand out most particularly and clearly in documented forms and symbols and it is there that the emergence of truly, objectively new phenomena is best traced; nevertheless, the domain of those studies spreads likewise over the entire field of contemporary intellectual life, not only to its peaks. For here we can observe the content, meaning and changes in the cognitive values of others and even evoke cognitive operations experimentally. Comparison of currently experienced contemporary cognitive phenomena with those that we recreate on the basis of historically available materials should become equally fruitful a method in the theory of knowledge as is the comparison of currently observed and artificially produced biological processes with the data of paleontology, comparative anatomy, animal and plant geography, etc., in genetic studies in biology.[15]

There are, of course, basic differences between the creative emergence of a new cognitive operation for the first time in the history of knowledge and its subsequent reproduction by an individual or social group that in its intellectual evolution has attained the appropriate cognitive level. In the second case the development of the operation is very seldom spontaneous; usually it occurs under the influence of unreflective or of reflective instruction which shortens the evolutionary series, considering only superficially or skipping certain links; and often the instruction strives to alter its direction, leading the individual or group to an idea with which they are unfamiliar along a totally different route from that by which mankind first arrived at it. Thus for instance the order of systematic exposition of a discipline in schools often diverges radically from the evolutionary order of the discipline. Nevertheless, in spite of this we find numerous data permitting at least a hypothetical conclusion that, appearances to the contrary notwithstanding, the same evolutionary operation possesses essentially the same genesis, the same indispensable antecedents in the intellectual development of the nation, class or child as it does in the intellectual development of mankind.

---

[15] In the cited sections of Baldwin's *Mental Development* (footnote 5) and Hobhouse's *Mind in Evolution* (footnote 4) we find a partial application of this method, but as yet without adequate factual basis.

This hardly means that the nation or child must pass through all the cognitive experiences that mankind has gone through, that cognitive 'ontogeny' must be an exact, if abbreviated, recapitulation of the cognitive 'phylogeny'; to the genesis of a certain cognitive operation, the countless past deviations from the course that has led to that operation, the disrupted collaterals, have of course no logico-genetic significance, although they may in fact have been indispensable in one or another set of historical circumstances. The point is only that – as we see clearly in considering the great differences in intellectual levels required for certain operations – a genetically late operation cannot be carried out for the first time by an individual or group assimilating past scientific achievements prior to a genetically earlier operation from which the first operation in its historical evolution has logically derived. One cannot learn differential calculus prior to addition and subtraction nor understand the heliocentric theory prior to at least a schematic appreciation of the geocentric theory from which we begin our reflections on astronomical phenomena, nor grasp Kant's philosophy without at the least an approximate acquaintance with the problems of the sophists and of Plato. If artificial instruction passes over these earlier levels, the nation or person in reality does not reconstruct ideas for whose reconstruction he is unprepared but only repeats words or symbols. Unless – as is generally the case – he makes a spontaneous effort and independently reconstructs the logical evolutionary sequence in his own thinking.

Observations of this kind have, as we know, led to new trends of thought in psychology and pedagogy, according to which the old teaching methods distort the 'normal' evolutionary course and must be supplanted by new ones, in which for example instead of the teaching of geography beginning with a description of the earth, it begins with a description of the school and of the village or town in which the school is located. Similarly, applying an analogous general premise to the intellectual development of backward nations, we ought to conclude that we should above all, emulating the evolutionary order of the past, support the highest types of intellectual creativity and realize the greatest possible cognitive ideals, which will radiate over the generality of society, rather than, as has usually been done up to now, place the main emphasis on the diffusion among the broad masses of superficial presentations of the results of other people's work.

If this principle of *irreversibility in the evolutionary order of knowledge*, which we are here positing, does indeed prove correct following deeper and more detailed study, it would open up an extremely important field of practical applications for the theory of knowledge. Here, however, we are approaching a second great category of questions that the new science faces – questions concerning not the systematic description of cognitive phenomena but a causal treatment of the relationships obtaining among them. This treatment must seek, in the creative evolution of new operations and the growth of new values, an element of repeatability on which could be based the formulation of laws, the forecasting of the future and the planned steering of the further evolution of knowledge.

Causal treatment of the relationship of cognitive results to the circumstances of cognition

Every discipline aspires to the greatest possible broadening of the field of its causal explanations, this both from theoretical considerations – inasmuch as the discovery of causal laws permits the subordination of the unlimited multiplicity of repeatable processes to a few general concepts – and from practical reasons, inasmuch as only a causal treatment makes it possible to foresee future phenomena and to direct them. The theory of knowledge must therefore also be concerned, without in any way ignoring the creative character of cognitive evolution, to nevertheless find such access to cognitive phenomena as to explain causally with all possible approximation every particular moment in the enormous system of processes that go to make up the history of human knowledge.

To this end it is essential first of all to isolate, from the cognitive life of mankind, relatively simple processes which might be divided into classes; only then, and considering the processes in each class as being approximately repeatable, will it be possible to apply to them the principle of causality. What we have said before about a genetic classification of cognitive operations, can serve here as a point of departure. If the concern is above all with operations that are merely the reconstitution of preceding ones, than the element of repeatability is already intentionally contained in them. Every cognitive operation – even the most original – at its next repetitions (e.g., in school exercises) is of course subject to a certain regularity which renders the cognitive operation more or less exactly accessible to causal treatment. It is in this very repetition, however, that the theory of knowledge is least interested, important as the repetition is to pedagogy. The more creative, the newer, the more uncommon a certain cognitive operation is, the more it draws the attention of historians and theoreticians of knowledge. Clearly, though, these same traits make its causal treatment difficult. Consequently a compromise of sorts imposes itself.

Between the most original operation and the most passive repetition of another operation exist countless intermediate degrees of originality or repetitiveness. If the precision with which we can treat cognitive processes causally varies directly as their uniformity, then the theory of knowledge must reconcile itself to the fact that the more original and creative a cognitive process, the less precise will be its causal treatment, in other words, the less significance will be possessed in the whole by that element of it which may be subject to the principles of causality. That element, as we shall see, is detectable in every cognitive process; the only question is that of its relative importance. That which, with a suitable method of causal explanation, remains unexplained in the operation of a pupil doing an assignment, possesses little significance in establishing the laws of intellectual life and in foreseeing the future; on the other hand, even the best scientific methods, applied to the creative work of a Copernicus, Newton or Hegel, will have to leave unexplained aspects of scientific activity that possess a first-rate importance for the coming into being of knowledge and for the direction of its future development.

Combining the foregoing deliberations with the question of the genetic classification of cognitive phenomena, we shall say that in every class of these phenomena laws of causal dependence will no doubt be detectable; the more diverse, however, are the phenomena contained in that class, hence, other things being equal, the broader is the scope of applicability of the given law, the smaller a role the law will play in elucidating any particular phenomenon to which we apply it.

Of course, no complete, actually occurring, concrete process of creation of a cognitive value will yield to reduction to any causal law. For its causal treatment would have to consider both the mental act itself and the overall conditions in which the act operates and which co-determine its results. And this actual, concrete combination of a certain operation with certain conditions presents itself from a causal standpoint as incomprehensible and unpredictable, unique in its kind; for any attempts at its explanation would require an inexhaustible causal regression, the consideration of all preceding facts that account for just this act happening upon just these circumstances at just this time and place. Therefore it is necessary to isolate, from this whole complex process, certain basic moments to which causation would be applicable; it is necessary to break it down into relatively simple processes each of which can separately be a link in causal chains.

Such analysis of a process of cognitive creativity is possible and leads to two different and mutually complementary types of questions. In questions of the first kind we will be concerned solely with the relationship of the results of cognitive creativity to its given circumstances, and we will not elucidate the act of creativity itself; while in questions of the second kind we will consider only the causal elucidation of the actualization of a designated cognitive operation, leaving aside its given circumstances and results.

In posing the first kind of questions we must regard a certain category of cognitive operations as being constant. In other words we posit that, in all cases studied and compared, there have been and will be carried out cognitive operations of a certain class, differing among themselves only within the limits set by the definition of the class, and that we need not take into consideration perturbations that may have resulted from not carrying out appropriate operations or from carrying out operations of a different class or such original ones that no generalization is suitable to them. We find further that given the afore-mentioned constancy of a certain category of cognitive operations, there obtains a constant relationship between a certain set of circumstances with which such an operation works and the value produced by the operation. There exists a necessary and sufficient set of circumstances for the realization of a certain cognitive value by a certain cognitive operation. These circumstances are necessary in the sense that until they have been realized, the operation in question cannot create precisely the value that it intends to create, but certain other operations must first be carried out in order that these circumstances may be realized; and they are sufficient in the sense that if they are given and an appropriate operation makes use of them, it will produce the required value. Thus if we find that cognitive operations of a certain class (insofar, of course, as they are carried out at all) produce values of a certain class only when, and whenever, they deal with a certain set of circumstances, then, assuming the operations to be constant, we may conclude that between this set of circumstances and values of the corresponding class there obtains a constant relationship.

The principle formulated here is simply an expression of a conviction of which we make use at every step, in popular reflections on knowledge and in practical activity seeking to stimulate or to restrain scientific creativity in a given direction. Thus for example we are convinced that a chemist who is given a certain stock of previous scientific achievements and a chemical laboratory supplied with materials and instruments, insofar as he wishes and knows how to work scientifically at all, is going to discover new chemical truths, and not astronomical or linguistic ones; what is more, having a precise knowledge of the branch of previous chemical achievements that presently interests him, the kind of materials and instruments that he has at his disposal, and his working technique, we can place predictions about his discoveries within fairly narrow bounds, especially if he is not an exceptional genius but simply a gifted scientist, which is to say, if his cognitive faculties do not extend beyond a certain set range of differentiation.

This relationship between a certain set of circumstances and a cognitive value of a certain class produced under these circumstances, given a constant category of operations, is still not, of course, a causal relationship in the modern sense of the word; for a causal relationship, as it is construed by recent science, is a relationship among processes, or changes, and not a relationship among substances or objects. Nevertheless, on the basis of that relationship there may be realized relationships possessing all the features of a causal relationship. Let us suppose that there occurs a change in cognitive circumstances; inevitably the cognitive value produced will likewise change (always, of course, assuming the constancy of the given category of operations). Let the chemist learn that some previous chemical assertion on which he has been relying has been proved erroneous; let the material or instruments that he employs change: the results of his work will be different than they would have been without the change. Insofar, indeed, as there exists a strict relationship between a given set of cognitive circumstances and a given class of cognitive values produced by their help, we may postulate that a specified kind of change in circumstances will always and everywhere evoke a specified kind of change in the cognitive values produced. And a relationship of this kind among changes corresponds to the concept of a causal relationship.

The task of the theory of knowledge in this realm of questions would therefore be the discovery of laws governing that relationship, i.e., of assertions formulated more or less as follows: wherever and whenever, given a constant relationship obtaining between a set A of circumstances and a class B of cognitive values, there occurs in cognitive circumstances a change of type m, then there must occur a change in cognitive value of type n. The studies in question must of course be conducted by way of comparative observation and (within certain limits) experiment, removing individual elements from systems of cognitive circumstances and studying the results of the changes produced in cognitive values by the changes in those elements, given the demonstrated or postulated immutability of the remaining circumstances. Obviously we cannot expect investigations of this kind soon to attain a complete precision; nevertheless, what kind of significance, theoretical and practical, even their tentative and approximate results can have, will become clear when we have enumerated a few kinds of special questions belonging to the category under discussion here.

(a) *The question of the relationship of the development of particular fields of knowledge to factors introducing previously unavailable materials into the research* (or, conversely, removing certain materials from the research). This is an old and familiar question; much has been said and written on the influence that, e.g., the telescope's broadening of the field of observation has exerted on astronomy, that the consideration of microscopic data has had on biology, that the opening up of new territories – thanks to scientific expeditions – has had on geology or anthropology, that the accumulation or destruction of records has had on history, etc. In spite of that, however, attempts at a scientific treatment of the problem are rare, albeit the very formulation of the problem in terms of an 'influence' (by which is usually understood a causal action) has seemed to call for a more exact posing of the problem. It will not suffice to describe historically the new truths that have been worked out by scholars using the microscopic data or the new discoveries that have been made by this or that geological expedition and the conclusions that have been drawn therefrom; in order to base on these facts any generalizations as to the relationship of biological truths to the microscopic data or of geological truths to the results of the scientific expeditions, it is necessary to discover in them some regularity that warrants the drawing of conclusions about unknown facts on the basis of known ones. As long as we put the question in terms of the creativity of this or that scientist who has discovered these or those truths with the aid of microscopic observation or thanks to materials supplied by a scientific expedition, we have no right to speak of the 'influence' of microscopic observations on scientific theories but only of the 'use' made of these observations by scientists; in other words, we have to formulate the relationship between the microscopic data and scientific theories not in terms of cause but in terms of final cause. Yet popular thinking about science actually and rightly does assume the existence of a regularity of a causal nature in facts of this kind, although it does not know how to isolate the regularity.

Isolation of this regularity will become possible as soon as we reduce each of the facts involved to a formula containing: (1) a statement of a specified field of scientific work in whose course materials of a certain kind have served scholars of a certain range of faculties and interests in discovering truths of a certain type; (2) affirmation of the appearance, in the sphere of the research, of a given type of new materials (sometimes the removal, from the sphere of the research, of specified materials that had previously been accessible, e. g. as in the destruction, by some cataclysm, of paleontological relics or historic documents); (3) the determination of a change brought about in the type of truths discovered by the research, hence in the type of cognitive values produced through the research, due to introduction of the materials.

If it turns out that in each case when new materials of the same kind have been made available to researchers, the results of the research undergo changes of the same kind, and supposing that other circumstances have remained basically unaltered, then we shall have a right to regard the change in results as an effect of introducing those materials into the sphere of the given science. So, for instance, let us suppose that we have compared the results of biological research in which microscopic data have been used with the results of otherwise similar research prior to the use of those data; we will probably be able to determine precisely, though of course making allowances for a fairly broad range of possible differences, what kind of specific consequences are brought about by the introduction of microscopic materials into biological research, and foresee with complete certainty, though only again within specified limits of possible differentiations, what will be the result of introducing microscopic observations into biological studies in which they have not yet been taken into account. In the same manner we will probably succeed in generalizing and predicting modifications in the type of scientific results that are produced, e. g., by introducing judicial records into research on a social system, psychological materials into economic questions or technological theory, etc.

The examples cited here are familiar and topical, and we have formulated them rather cursorily; however, it is clear that countless further questions can be posed here and that the answers to those questions will not only have meaning for the elucidation of many facts from the past but will also provide valuable practical hints for the future organization of scientific material collection. One must only realize that we face here questions of an experimental nature requiring numerous comparisons of facts from the early as well as the most recent history of knowledge – of facts accurately described and precisely defined. Deductive explanation or prediction that materials of one kind or another *logically* had to, or logically will have to, lead researchers to one or another kind of results, will not suffice; the actual relationship between the results of research and their material has never been and probably never will be completely in accord with the predictions of logicians.

(b) *The question of the relationship between scientific results and technical instruments.* This question is contiguous with the preceding one since some technical instruments, such as the telescope and microscope, serve mainly to supply observation with materials which otherwise would be unavailable. We shall, however, pass over that aspect of them; at the moment we are interested in something else. A technical instrument not only broadens the range of data (of objects or of their properties) but likewise influences the kind of scientific problems posed and solved in respect to the data. Typical in that respect are physical instruments. The materiality and precision of any technical tool suggests and facilitates problems of a certain kind, while on the other hand diverting the researcher's attention from other problems to whose solution the material tool affords no assistance. Through introduction of the instrument into the cognitive circumstances, the value produced is clearly altered, and the changes seem to be so specific and so extremely characteristic of all facts of this kind that the attainment of positive results in comparative studies of this question ought to be relatively easy for the theoretician of knowledge. We will point, by way of example, to the differences between Aristotle's physics and modern physics, between introspective and social psychology on one hand and laboratory psychology on the other, between the position of the surgeon and the internist in diagnosis and therapy, etc. We shall venture to hypothesize that the result of introducing technical instruments into cognitive circumstances is the introduction of a quantitative element into the cognitive values produced, so that wherever a material tool enters a field of scientific research, there will appear (barring counteracting causalities) a tendency to impart to discovered truths a measurable character – in extreme cases, mathematically calculable. More thorough research will confirm or overthrow this hypothesis, and in any event will uncover further relationships of this kind.

(c) *The relationship of the results of research in particular fields of knowledge to cognitive values serving as tools*, most prominently to so-called heuristic concepts and methodological principles. Various aspects of this question have, as we know, often been discussed. We find at every step, in methodological studies and critiques, deliberations on the scope of applicability, e. g., of causation, of the concepts of matter and energy, etc., and on the results of their application. The objective, however, is to introduce a new viewpoint – experimental-theoretical instead of methodological-normative – into these deliberations. Methodology is generally content to point out the logical consequences stemming from use of this or that principle or concept, drawing therefrom normative conclusions concerning what principles or concepts ought to be used in a given area of research in order to achieve a certain ideal type of truth. But the logical consequences to which the use of certain cognitive tools leads in methodological reasoning, are hardly identical with the actual results of their use in an empirical, historically given course of scientific research. Logical analysis of heuristic principles and concepts can only indicate the direction in which research employing these principles or concepts would develop if it made exactly the use, and only that use, of them which is consistent with their meaning as given by the methodologist. In reality, though, that use may be entirely different; the influence exerted by heuristic principles and concepts on cognitive values produced with their help can only be stated *a posteriori*, on the strength of their actual application, since it depends not only on their logical essence as independent cognitive values but also on their historic function as tools of cognitive activity, for which they are merely elements in cognitive situations. Therefore, in order to discover in this influence an actual regularity other than the normative regularity required by methodology, it is necessary to compare changes arising in cognitive values due to the introduction, into the research in question, of one or another heuristic principle or concept – e.g., the principles of the conservation of energy or of entropy into the realm of sociological research, the concept of 'human nature' into economic research, etc. Of course, the actual regularity, as we have already emphasized, likewise does not exhaust that entire aspect of cognitive creativity, but in any case it will throw a new light on it.

(d) *The relationship of the results of cognitive activity to symbols serving to designate cognitive values or their objects.* This question is tied to the preceding one, to the extent that at a certain stage in the evolution of cognition symbols themselves play the role of cognitive values and may serve as tools in the same sense as do heuristic principles and concepts. Nevertheless, in more recent science there prevails the sense of a symbol as a special kind of medium which is not itself a cognitive value but only a distinct type of tool for operating with cognitive values. Appropriate studies in the theory of knowledge would therefore cover the following particular questions: the cognitive influence of specified changes spontaneously occurring in language (e.g., the influence of the atrophy of certain flexions or of the rise of auxiliary words); the consequences of translating from one language into another, in connection with the question of the influence of differences among various languages on scientific creativity; the results of efforts to create an international scientific language; the consequences of introducing artificial terms; further, the influence of the invention, and of various phases in the evolution, of writing and, subsequently, of printing on cognitive values; the influence of artificial written symbols on the evolution of mathematics, logic and other disciplines. We have here an enormous field for research, one which has, to be sure, often been an object of attention, but one which ought to be addressed finally with methods more exact than heretofore. Precise scientific induction, abundantly supported by a skilful selection of data from the entire scientific material available to us, ought to supplant the erstwhile semi-popular reflections based on a small number of arbitrarily selected and superficially observed facts.

One might name a few more possible questions from this general area of concrete relationships between the results of scientific work and circumstances; the examples given up to now, however, indicate sufficiently what an interesting and momentous task here awaits the theory of knowledge. It is obvious that the point here is hardly to remove the creative element from the evolution of knowledge, or to replace entirely the ideal regularity of norms, which logic and methodology would like to see realized in the process of scientific work, with the real regularity of causal relationships. The empirical function of the 'science of knowledge' comes down to considering, studying and designating the actual aspects of cognitive activity which demonstrate a partial limitation of free and spontaneous creativity – a limitation stemming not from conformity to norms in advance imposed by logic and methodology, but from dependence on real circumstances. No doubt these circumstances will hamper in only very modest measure creative geniuses in moments of supreme inspiration; but there are few geniuses, and even in a genius's life the supreme levels in the scale of creativity are seldom attained. For the most part the development of knowledge takes place through the accumulation of more or less important contributions, and in this enormous mass of average facts, neither exceptionally original ones nor simple undifferentiated repetitions, the theory of knowledge can and ought to discover and precisely determine the component of a constant dependence of results upon circumstances whose existence popular reflection and everyday practice permit us to surmise. No doubt, too, scientific work at a high critical level is partly susceptible to interpretation as being subject to the regularity of normative logic and methodology; yet everyone who has devoted himself to it knows how divergent actual scientific procedure often is from ideal norms, which in the fullness of their meaning appear only in the retrospective critique and final formulation of the obtained results. And if we take the entire history of knowledge under consideration, it will turn out that even after removal of the unpredictable element of uncommon originality, logical and methodological regulation can contribute only in small measure to elucidation of the tremendous majority of cognitive processes; causation will probably prove considerably more fruitful than the assumption that the human mind always follows in cognition the route indicated by logic. Naturally, however, one should always bear in mind the limits of applicability of causation.

Causal treatment of cognitive activity

Finally we pass to what is by all odds the most important type of question in the science of knowledge, the concern of which, as we have indicated generally, is the causal elucidation of the actualization of a designated cognitive operation.

Here we must first of all exactly delimit this area of questions from others. In speaking of elucidating the *actualization* of a cognitive operation, we have in mind not the question of its absolute genesis in the history of human thought, not where this kind of operation arose in the first place, in what previous operations it has its source – for that is a question of the creative evolution of cognitive thought elusive to causal elucidation. Our task is a more modest one: it touches only, so to speak, the time and place of the operation's emergence. More precisely, supposing that a certain operation was in fact to emerge in the creative evolution of cognition, then we should seek to explain what relationship obtains between its emergence and phenomena of one kind or another that occurred contemporaneously, previously or in consequence in the consciousness of the same man and the same community. It is not its logical connection with the past and future of knowledge, not its theoretical function, not its ideal role in learning, but simply the actual historico-biographical fact of its accomplishment that is here the point of departure for our interests.

And furthermore, this fact of its accomplishment can only be grasped as a link in a causal relationship if in principle we can regard it as being typical and repeatable in its type, discovered or at least discoverable in other times and places. For only then will its realization in the given time and place present itself not as a chance consequence of an exceptional coincidence of which human creativity has spontaneously and unpredictably taken advantage, but as a fact connected in a necessary fashion with other ascertained facts. On the other hand, however, since, any cognitive operation interests us as theoreticians of knowledge only when it is relatively original and its next repetitions, unless they are again fairly original modifications of it, possess comparatively little significance for us, as we have already stipulated in the previous section, we are interested not in the one special operation but in the entire class of operations, more or less original on a certain scale of potential originality. Therefore we will attempt to explain not merely a certain individual fact of the accomplishment by a certain individual in a certain society of a certain operation A^1, but any and every fact generally of the accomplishment by anyone, anywhere, of an operation of class A, be it A^1 or A^2 or A^3.

But that is not the end of it. We have mentioned previously that the complete, concrete process of creating a new cognitive value does not lend itself to formulation in a causal law and must be divided up. Thus we pondered first the relationship of the value produced by a cognitive operation to the given circumstances with which the operation has been working, and we took the actual operation indispensable to the production of the value in the given circumstances as a constant component of the process, not considering its possible changes outside the limits of differentiation possible for the given class of operations and not attempting to elucidate the fact of its accomplishment; for it is only on condition of the hypothetical immutability of the operation's class and of its assured accomplishment that the influence of circumstances on the result can be stated exactly. And at present, when the above assumption itself is to be formulated as problematical, whom we are concerned with elucidating just this fact of the accomplishment of a certain operation and consequently when we recognize the possibility of its not being accomplished or of an operation of another class being accomplished in its stead, we must eliminate the problem of circumstances influencing the result independently of the nature of the operation. Therefore we consider here both the circumstances and the results of the operation only insofar as they are determined by the operation itself. This means that we prescind completely from the result which the operation actually attains; we are concerned only with the result that it *wants* to attain, with its cognitive intent, in the presumption that this intent has been defined by the operation and that it possesses everything necessary to the realization of that intent. And we take the circumstances into consideration only to the extent that they are already adapted to the attainment of the intent, are already determined cognitively with regard to the result that the operation wishes to attain; consequently we assume that the operation has already selected its object, its materials and tools and that it has posed the question which it intends to solve, so that the concern now is only with solving that question.

We term the cognitive operation thus manifesting itself in its intent, disregarding the question of the agreement of intent with execution, the *cognitive purpose*. Hence the purpose is that element of the operation that characterizes it in its subjective, psychological aspect, apart from the objective role that it carries out with respect to objects given to it with which it works. And we term the system of objective elements determined by the operation, within whose limits and on whose basis the given question has been posed, that special combination of object, materials and tools of which the operation intends to make use in carrying out its assignments, the *cognitive situation*.

Clearly the purpose constitutes the most essential element in the cognitive process; it is preeminently on the purpose that the process depends, hence the purpose is the chief basis for the process's definition. Although the same purpose may express itself in different processes and lead to different results, still these differences are secondary compared to differences in creative processes stemming from dissimilar purposes. For example, the purpose of quantitatively determining the ratios of elements in chemical combinations and the purpose of physico-chemically elucidating, biological phenomena have expressed themselves in different methods and results, but both characterize a certain specific class of cognitive processes. Individual manifestations of each of these purposes differ less among themselves than any of them does from any manifestations of the purpose of qualitatively deducing the 'essence' of a chemical combination from the 'essences' of its components or of the purpose of vitalistic explanation in biology, and the more so from manifestations of any purposes of discovering truths of a certain kind about astronomical or psychological phenomena. Similarly, too, in a cognitive situation we find that which is most essential to the characterization of the circumstances of a given cognitive process, since that is what ultimately conditions the solution of a given question; differences in circumstances originally given to cognitive operations are comparatively secondary so long as an operation of a certain kind is able to produce similar cognitive situations from them, and their importance is the greater, the more diverse are the situations that they suggest to cognitive thought.

The questions that we can put concerning the causes of the emergence of certain cognitive purposes at certain moments in the lives of individuals belonging to certain social groups, must rest on a proper understanding of the nature of psychical causality. Every purpose occurring in the consciousness of a subject (of an individual or a group) requires for its causal elucidation the cooperation of two factors: of another, preexisting purpose in the same subject; and a change in the objective set of phenomena given to the subject that presents itself to him as being pertinent to his preexisting purpose. Neither the psychical process by itself nor the action of 'external influences' by itself can be the cause of a psychical fact. For every psychical process refers to objects of some kind, is intentional, possesses the character of a purpose. It becomes a factor in the emergence of a new process only when there occur, in the objects to which it refers, changes not covered by its intent, hence depriving it of its original conscious role. And external influences are only true influences, they only act on the psychical life of the individual (or the group) if they fall upon prepared ground, if they happen upon some actual or potential intent of the subject's, if the latter gives them some meaning for the sake of his own purposes – in short, if irrespective of the subject's intent they modify some ready *situation* of his. The nature of the imprecisely termed 'reaction' which the given external influence will evoke, depends precisely on the nature of the purpose which corresponds to the situation modified by that influence.

Thus we may say one of two things: either that an emergent new purpose b is a result of the cooperation of a previous purpose a and of an external purpose on a situation A associated with that purpose, or again more simply, that the exchange of purpose a for purpose b (or for a combination $a+b$) is a result of a change in situation A associated with purpose a. It is understood in this connection that a presupposition for this causal relationship is an association between situation A and purpose a, that is to say, membership by purpose a and situation A in the same active process. Thus a new social purpose (e.g., a desire to humiliate an acquaintance) will emerge in an individual when the individual's social environment 'reacts', in a manner not intended by the individual, to some other social purpose, disclosed externally or not, but expressing itself in some social situation of which he is aware (e.g., if the acquaintance reacts by scoffing at an attempt to win his admiration). Independently of the individual's intent, changes that have taken place in the tools or material which the individual intends to use in a certain technical activity, evoke in him a new technical purpose (e.g., a desire to learn a different way of operating with given material) or economic purpose (e.g., a desire to acquire a better tool), etc.

In the same way the cognitive purposes of the individual (or group) are explained causally by the influence of objective changes connected with certain preexisting purposes in the individual. No external factors, natural or cultural, will produce cognitive purposes in the psyche out of nothing; they can only exert a modifying influence on already, existing purposes, replace them or complement them with new ones. On the other hand, so long as we stand on a causative ground and ignore the creative element in spiritual development, we have to say that within the limits of causal elucidation no cognitive purpose will emerge in the individual (or group) simply as a result of previous purposes without some objective coefficient, some change in the preexisting cognitive situation.[16] And wherever and whenever a certain cognitive purpose of an individual (or a group) encounters a certain change in the cognitive situation independent of the intent of the individual or group, a necessary result will be the emergence of a different purpose.

Let us now consider some categories of questions that the theory of knowledge can pose and solve with the help of the above premises.

(a) *The influence exerted on knowledge by other fields of human activity.* This question has long been familiar but has acquired a greater significance and broader appli-

---

[16] We emphasize once again that this is not a postulate that might in fact be extended to all of conscious life; since free and creative development is a fact, a causeless actualization of new purposes cannot be philosophically ruled out. But the bounds of science end where causal explanation ceases to be sufficient.

cation in connection with the biological theory of consciousness in general and of cognition in particular, and further in connection with pragmatism. From the standpoint of biological evolution, conscious activity in general has been treated as a broadening and extension of the material or energetic processes in which the adjustment of a living being to its environment manifests itself. This broadening assumes in the species the form of instinct, in the individual – the form of habit. Cognition is a specific function of consciousness that comes into being when, as a result of a change that has come about in the environment, an instinctive or habitual activity encounters obstacles that make difficult or impossible the effective conclusion of its established course; the consequence of such an impediment is reflection seeking to adjust the manner of action to the new circumstances, and a fundamental moment in that reflection is the purpose of identifying the circumstances from the standpoint of the needs of the activity. This biological theory of cognition has been linked in pragmatism with the results of observing more complicated manifestations of cognitive thought accompanying various forms of cultural activity – technical, economic, religious, etc. – which, in contradistinction to theoretical thought, have been given the common term of 'practice' or of practical activity. From the mentioned observations pragmatism has drawn the conclusion that all new, original cognitive thinking is evoked by some practical difficulty requiring solution.

We need not argue how momentous this matter is to the theory of knowledge. If in fact the final and complete causal explanation of every cognitive purpose lay in some other field of human activity, knowledge could be explained in its entirety as being a result of the gradual broadening and accumulation of our entire culture, which in the course of its broadening and accumulation continually encounters difficulties that provoke theoretical reflection. Knowledge would not then be a separate field of culture with its own order but a set of phenomena of the most diverse origin, growing up on the soil of various other fields of activity and organically associated with them.

Precise analysis of facts adduced in support of this thesis, however, readily shows its one-sidedness. New cognitive purposes do indeed emerge, among other things, in connection with questions of 'practice' (in the broad sense of that word as used by pragmatism), but only as changes in previously existing cognitive purposes. For instance, the failure of some practical intent due to altered circumstances compels us to ponder those circumstances, to reflect theoretically, but only when our intent has been linked up to a theory of reality to which that intent has referred, and our activity has been accompanied by the expectation that new experiences will confirm the theory. It is exactly to the extent, and only to the extent, that a practical activity carried out is explicitly or implicitly treated by theoretical thought as an experiment which is to confirm a hypothesis, that is, to the extent that the practical activity is accompanied by a theoretical purpose for which the intent, circumstances and process of practice constitute a *theoretical* situation – that unexpected difficulties encountered by practice strike us as having theoretical significance, as modifying the theoretical situation. It is only through this that they exert an influence on our reflections, that they lead to the emergence of a new cognitive purpose in lieu of, or as complement to, that which appeared in our original hypothesis and was impeded by an unforeseen fact discovered or produced in the course of the practical activity.

The far-reaching significance of practice for the evolution of knowledge does not, therefore, rest in practice being a wellspring of theoretical thought but in its imposing, on the theoretical thought associated with practice, unforeseen changes in theoretical situations, in its not permitting theoretical thought to shut itself up in old problems and solutions but, so to speak, forcing it to notice the inadequacies of its previous theories and to seek new ones. Obviously, however, practical activity only carries out this role if, firstly, it makes use of theoretical reflection in order to attain its aims, and secondly, if it is itself bold, original and creative, if it not only does not steer clear of difficulties raised by altered circumstances on the road to the realization of old goals but moreover seeks out new goals and new difficulties to overcome. On the other hand, theoretical thought hardly need wait for changes that practice may inject into its ready situations. For not only is theoretical thought capable of independent creative development, but, even when it holds to a plotted direction, when its creative spontaneity falters in any field and consequently it becomes dependent on the causal action of external influences, those influences giving rise to new purposes can come not only from other fields of culture but also, and perhaps first and foremost, from other branches of knowledge itself.

Bearing these reservations in mind, let us nonetheless recall several well-known and empirically determined relationships of science to practical questions, in order to illustrate the matter. We know how materially engineering has contributed in recent times to the development of physics and chemistry, and if we were to reach further back into the past – to Archimedes or even to the ancient Egyptians – it would turn out that this influence was no smaller in the first stages of those sciences. The dependence of the biological sciences on medicine is manifest throughout the course of their evolution: the protracted initial period of magico-religious concepts of life phenomena is explained in large measure by the magical character of primitive medical technique, just as in the later gradual emancipation of biology from religious beliefs the progress of positive methods of treatment played an important role. Until recent times questions of politics and morality have been nearly all-powerful in conditioning the evolution of theoretical thought on societal life, and even the very emancipation of modern sociology from this direct dependence on practical life is explained not only by the trend of intellectualism proceeding from philosophy and from the biological sciences, but requires consideration as well of the influence of the needs of social practice, which in the face of the hitherto unprecedented complexity that we find, e.g., in South America, demands ever more thorough and detailed theoretical studies of social phenomena in order to be able to base its reforms on their results. History to this day has not freed itself completely from political and practical-social considerations; philosophical problems, as is generally known, for long periods display the action of religious or of antireligious motives; we know that even in mathematics and astronomy progress in the beginning occurred largely thanks to motives of a technical and magico-religious nature.

In all these examples, which could be multiplied further, the generally formulated course of action on knowledge by one or another field of culture breaks down, upon closer analysis, into a series of particular processes of the type that we have described above, each of which should basically be amenable to interpretation as a manifestation of a more or less general law or synthesis of laws. Thus, for example, engineering, medical and political practice, whenever they plan some relatively complex operation, need a preliminary theoretical orientation in the phenomena constituting the object, the material and the tools of the operation. The technical or social purpose is therefore accompanied, as a special cognitive situation, in a subsidiary capacity by the purpose of a cognitive treatment of the practical situation in question, together with an intended course of action; the cognitive problem that emerges here is that of applying the theory on which a plan for practical activity has come to rest, to future facts which the attempt to realize this plan will produce. This always happens in planned action; it is only in an unplanned operation that cognitive purposes do not appear, except perhaps when the unplanned operation encounters unforeseen obstacles to the realization of its intent and must form a plan of alternative procedure.

So long as the practical plan lies within the same limits of differentiation as a former activity of the given type; so long, e.g., as an engineer wishes to build a machine of the same kind as have formerly been built, a physician to treat his patient for a known illness by known methods, a politician to win triumph for his party in the usual way at the elections, so long there also suffice, in the creation of the plan, cognitive operations of some old type, working with ready-made values. The engineer employs familiar physical formulas in a familiar way, the physician makes his diagnosis in accordance with scientific traditions, the politician informs himself by a common method about the people's position on his party's program and candidates and employs traditional premises concerning the influence of popular slogans, arguments, promises, threats, calumnies aimed at opponents, etc., on the disposition of the people. A theoretical problem always exists, but it is so similar to others already solved that science is not interested in it.

Let us suppose, however, that the practical question at hand is a new one, either because the activity involved sets itself new tasks (e.g., the engineer wishes to build a novel kind of machine, the physician to apply a novel method of treatment, the politician to push through a novel social reform) or because there occur in experience new circumstances making impossible the application of the previous procedure (e.g., the engineer, due to war, cannot obtain some needed material, the physician encounters a hitherto unfamiliar clinical case, the politician cannot agitate for his party on account of repression by the governing party). In that case the framing of a plan is always likewise the posing of a certain, at least relatively vital, problem: whether cognitive values which have up to now repeatedly proven true in relation to facts of a known type brought into play by the corresponding practical activity, will likewise prove true in relation to the new facts that the planned new activity will discover or produce. Thus a cognitive purpose comes into play, to be sure one of a familiar category (since the corresponding cognitive values have already on various occasions been tested), but in any case one rousing a certain scientific interest.

If now the plan proves operative, the cognitive values are considered to be confirmed and no new cognitive purpose arises. If, however, on the contrary the plan proves inappropriate in action, does not lead to the intended results (or if, on a higher level of technical reflection, it turns out in advance that the previous cognitive operations and values do not suffice to build a plan with any chance of success), then that failure, modifying the cognitive situation given to theoretical reflection, will take on significance as a new cognitive problem requiring a new cognitive purpose for its adequate posing and solution. Failure of the engineer's plan produces the purpose of a new interpretation or of a deeper investigation of certain physical phenomena, the physician's failure – the purpose of discovering unknown pathological processes or a new interpretation of known ones, the politician's setback – that of deepening knowledge of certain social phenomena or of changing the premises on which rest methods of social action on the masses. And from there on the new cognitive impulses can continue to develop independently of practical considerations, in association with purely cognitive motives springing from strictly scientific interests.

The concern of the theory of knowledge will therefore be to discover, in processes of this kind involving the emergence of new cognitive purposes from old ones under the influence of modifications produced in cognitive situations by an activity other than a theoretical one, regularities in the form of causal relationships of the kind: wherever and whenever a cognitive purpose of class A encounters a change B produced by activity of a certain category (technical, social, economic, religious, etc.) in a situation corresponding to it, the result will be the appearance of a new cognitive purpose of class C. The future will show in what measure the theory of knowledge will succeed in carrying out this task. In any event, only after the attainment of laws of this type will we be able in fact to state precisely the actual results of the influence of particular fields of culture on knowledge, instead of, as today, contenting ourselves with shallow generalizations; only then will we know what can be ascribed in one or another branch of knowledge to the action of practice, and what to the mutual influence of scientific theories. This is also a matter of first-rate importance with regard to the future of knowledge: whether, when, in what departments and within what limits are we to support the association of learning with practice, or rather seek its isolation? What forms most advantageous to knowledge should this association take, insofar as we regard it as desirable at all in the given domain? How to harmonize the needs of science with the requirements of practice? In the present day state of the theory of knowledge we cannot give a conscientious answer to these questions.

(b) *The education of individuals in theoretical thinking*. The theoretical education that is the main task of our whole educational system takes in two basically diverse matters: the imparting, to youth, of a certain stock of information, i.e., of certain systems of theoretical values, and the development of their capacity for theoretical thinking, i.e., for carrying out cognitive operations. Here we may pass over the first matter, for it affects only the diffusion in society and the fixing in the individual of a ready stock of knowledge, which exerts no direct influence on knowledge itself; the indirect influence of those processes, we shall consider later. On the other hand, the second matter possesses an extremely important significance for the theory of knowledge, since it is obvious that the entire evolution of knowledge depends directly on the education of scientific workers in theoretical thinking.

In discussing this matter, we will pass over the nonsensical, if (or rather because) today popular, opposition of two tasks of the educational system: one general, consisting in preparing youth for everyday practical activities, and one special, limited to a few almost exceptional individuals, aimed at making savants of those individuals. Naturally, we are interested here in the matter of intellectual education, not of education in general; for no one doubts that with respect to the development of character there can be no other differences in methods of education save those linked to individual and group differentiation of youth on the one hand and to differences in their future professional functions on the other, and no one proposes to make any special exceptions for future scientists and scholars as opposed to all the rest. However, a conviction has taken firm root that in the intellectual sphere extreme differences obtain between methods of successfully educating scholars and methods of successfully preparing people for any other vocation. And since future scientists constitute a very small percentage of the alumni of schools of all kinds, not excluding universities, a view has grown up that the educational system should take account above all, almost exclusively, of the requirements of practical life; often in this connection it is assumed implicitly that an individual possessed of the abilities and taste for scholarly work will get along in any educational system. Interestingly, such sacrificing of the needs of intellectually outstanding individuals to the supposed needs of the generality is often considered to be a symptom of and a prerequisite for the democratization of education.

It is not difficult to show that the intellectual needs of people intending to act in any practical field are in their essence completely identical with the intellectual needs of future scientists and differ from them only in degree. There is only one effective method of educating people intellectually, and that is the one that in its ultimate outcome leads to the development of scholars and scientists. The so-called man of action: the official, the legislator, the civic leader, the teacher, the physician, the technician, the businessman, indeed as the priest, the artist and the poet, needs in addition to training in theoretical thinking still another kind – of background, in some field of activity basically different from cognitive reflection – a background which, parenthetically, today's professional schools do not give him to an adequate degree. Insofar, however, as he ought to possess a certain measure of strictly intellectual ability, that ability must be basically the same and be developed in the same way as the ability of an individual specializing in an appropriate field of scholarship, only the degree of its development need not be equally high, since the objective here is not further scientific creativity but merely the application of learning already generated.

The trend discussed here, of adapting methods of intellectual education to the needs of practical people, considered to be different from the needs of scientists, has behind it one weighty motive namely, a reaction against the overloading of youth with detailed erudition, which in life for the most part turns out to be an unproductive, if not a harmful, ballast. But as we shall presently find, that reaction is equally desirable in the interests of knowledge: overburdening with erudition is just as incompatible, perhaps even more so, with the development of scientific creativity as it is with the ability to act practically. The future savant, like the future practitioner, needs above all a thorough mastery of basic information in a given field of knowledge, information that is indispensable both to any further, detailed research in this field and to any practical applications of it. Only on this common basis, one in fact sufficiently broad to fill a minimal program of higher studies, can there arise a further specialization of knowledge, from the standpoint of practical or theoretical objectives, which should be left for the most part to the independent effort of each individual. However, as we have indicated, the main objective of education meant to produce scholars is training in more or less original intellectual operations. The point is that the practitioner needs this training as much as the scientist does. Although the range of originality required of him is considerably smaller, still he must be capable at least of the minimum of creative thought indispensable to the elaboration of plans for new practical questions which, as we have previously seen, includes at least hypothesizing about the applicability of existing theories to new facts. Without that ability he will always remain either a dilettante feeling his way along or a routinist slavishly keeping to beaten paths. And that ability is but a lower rung of the ability to create new theories.

Therefore, in considering the question of intellectual education, the theory of knowledge should at once champion the highest requirements in this realm, which *eo ipso* already subsume all the lower requirements, and should reflect upon methods of developing independent scientific researchers and upon the premises on which those methods rest. We postulate that the supreme task of our school system in the educational domain is the formation of scholars and scientists and that all other tasks in this domain should be subordinate to this simply because their realization is likewise completely contained within the realization of that supreme task. From the same considerations we will here take under consideration only the highest stage of schooling, that which directly leads to the formation of scientists and scholars – the university stage. For it is here that are manifested most fully and in the greatest detail the general principles of strictly intellectual education, which at lower levels is still linked in diverse ways with moral and physical education and with the development of faculties possessing more of an auxiliary and preparatory character (e.g., perfection of the means of social expression of thoughts through the study of language, the development of the formal faculty of observation and reasoning in general, etc.).

The education of an individual to be a scientist reduces basically to two functions: to giving him a certain technique, that is, the ability to prepare circumstances appropriate to his cognitive intents, and to rousing in him more or less original cognitive purposes. The first matter belongs, strictly speaking, to the category of questions discussed in the previous chapter. Exactly to the extent that we know the relationship between the circumstances and the results of research, we will know how to indicate what kind of materials should be gathered together in certain kinds of questions, how to prepare them for use, what material and cognitive tools to use, what symbols to use, etc. The attainment of certain results in certain circumstances depends, however, as we have seen, on the accomplishment of appropriate cognitive operations; the most perfect technique is utterly fruitless without creative thought. Consequently the main task of university education is to influence the development of cognitive operations. And since it is impossible to impose a complete creative cognitive operation on anyone, the task reduces to arousing in the individual relatively new cognitive purposes whose actual realization, assuming that he possesses the appropriate technical information, depends now only on him.

We can describe this function of intellectual education more exactly in the following way. We are dealing with a person who brings with him certain cognitive purposes but not yet the ones, or in any case not all the ones, that are needed in order that he may become creative scientifically in a given field of knowledge, in order that he may be a more or less original chemist, biologist, sociologist or philosopher. Thus we are gradually to bring him, by acquainting him with the previous achievements of the given science, to the evolutionary stage at which such cognitive purposes will arise in him whose realization under certain circumstances, with the help of certain materials and tools, will lead him to elaborating cognitive values that truly augment the scientific attainments in that field. Leaving aside influences of a social character about which we shall speak later, and which develop in the individual a desire for creative scientific specialization and a drive for self-education complementing and frequently taking the place of the pedagogical activity of a professor, the latter has at his disposal, for the speeding up and guidance of the student's intellectual development, the most important resources in the form of intellectual influences.

Intellectual influences consist simply in application of the general principle of psychological causality, that new purposes arise from preexisting ones as a result of modifications in situations to which those preceding ones refer. In the intellectual sphere a new purpose arises if there are introduced into a cognitive situation corresponding to some old purpose new elements making it impossible for the old purpose to produce the intended cognitive value – in other words if, into a problem whose solution seems to us assured, there enter new cognitive values or new objects of cognition requiring new intellectual efforts and a new posing of the problem. A cognitive purpose emerging due to such an intellectual influence can either completely replace the previous purpose – e. g., as when a theoretical critique demonstrates the total insolubility of the problem posed – or it can merely complement the previous purpose, e.g., when it turns out that solution of the problem posed necessitates posing and solving some other problems. In any case, though, the change is a result of strictly intellectual factors, not of ones stemming indirectly from practical changes as in questions previously considered (Section (a)). An essential feature both of intellectual education and self-education is that they rest on a completely different causality governing the evolution of cognitive purposes, that they are possible solely thanks to the fact that new cognitive purposes arise from previous ones as a result of changes in cognitive situations whose source lies exclusively in a theoretical activity introducing new elements into those situations. In education to situations with which a pupil operates, the new elements are introduced by the theoretical (critical or complementary) activity of the teacher. In self-education, which following the completion of one's higher education remains the sole (if we leave aside the mutual interactions of scholars) means of reflectively, consciously guiding the individual's intellectual development, the person himself introduces new elements into his cognitive situations, since his sphere of theoretical activity encompasses numerous and diverse operations, in part mutually independent, and each of these operations can influence another one, modifying its situation.

It is clear that a truly efficient reflective influencing of intellectual development is possible only given an adequate knowledge of the laws governing changes in cognitive purposes due to intellectual factors. This knowledge is more important in education than in self-education; for in the latter a spontaneous, causally indeterminate, creative development of the individual often plays a relatively greater role than does conscious, planned guidance of one's own evolution based on more or less explicit causal premises; whereas in educating others we rely exclusively on the conviction that certain influences will achieve certain results. As is generally known, it is easier to excel in an intellectual field without any guidance rather than by relying entirely on incompetent guidance. To be sure, numerous centuries of intellectual education have produced a certain approximate, empirical knowledge of some general relationships in this sphere of phenomena; how insufficient that knowledge is, however, is shown even by superficial observation of our university system. Although we personally make no claims to a thorough knowledge in this field, nevertheless some errors in that system reveal such glaring contradictions to the most elementary generalizations, striking the eye of anyone who has studied the psychology of intellectual creativity, that we cannot forgo a few critical remarks.

Today's approach to university education, aside from the imparting of technical instruction, rests mainly on two expedients: on conveying, in the form of lectures, the most important results in a given field of knowledge, i.e., a systematic collection of cognitive values in the field, and on scholarly exercises, i.e., on inducing students to work up, in relative independence, certain typical questions from the given field and occasionally to participate in a subordinate capacity in the professor's scientific works. It is surely not difficult to realize that the first expedient in its present-day form is hardly a practical application of the results of theoretical studies on the evolution of cognitive purposes in students but constitutes a relic from times when due to a lack of books the lecture was the chief, and sometimes the sole, means of passing on the accumulated acquisitions of knowledge to the young generation.

Undeniably an acquaintance with previously produced cognitive values is an indispensable *condition* for subsequent creativity, nevertheless that condition can be satisfied far better through properly selected literature than by listening to a lecture. And it is clear that assimilation of scientific results cannot in itself be a factor in the development of scientific creativity. A familiar observation surely states this adequately: those who best assimilate ready scientific values in a given field, are most often themselves scientifically unproductive; perfect erudition not only does not constitute a stimulus to creativity but, on the contrary, if left to itself it hinders the emergence of creative purposes. We can easily understand this by considering its mechanism of action. The emergence of a new cognitive purpose requires a preexisting purpose and a change in the situation corresponding to it, which change the subject perceives as making impossible the present satisfaction of the purpose and thus forcing its supplanting or supplementing with another. But in giving the student a ready system of scientific results we thereby seek to satisfy in advance not only the cognitive purposes he already possesses but also those that may arise within him, we attempt to give him a ready answer both to the questions that he has already put to himself and to those that we are now suggesting to him ourselves. Not only do we not attempt to arouse in him purposes which in the ready-made system of knowledge will not find satisfaction, but often we intentionally impede the spontaneous emergence of such purposes by overloading him with erudition, by making him passively repeat the problems and solutions given him, leaving no time or energy for independent thought.[17]

The only justified role for lectures, thanks to which they would constitute a true complement to information drawn from books and would become a factor in the development of cognitive purposes, would be the reconstruction of the operations that have led to the production of cognitive values, instead of passing on those values in a cut and dried form. Every lecture should give an exact description of the genesis

---

[17] I know of nothing more deadly in this respect than the attempts currently prevalent in Poland at a detailed regulation of the entire university curriculum, especially in law and medical schools and in courses preparatory for the teachers' examinations. No future lawyer, medic or teacher who wants to conscientiously satisfy the requirements set him in regard to erudition, and does not want to extend the period of his studies by 2-3 years, has time or energy to think anything through independently during his entire stay at the university. What will come of a generation whose minds in their best age, just when intellectual spontaneity should develop the most, are systematically and relentlessly pressed into a framework of ready-made formulas? Of course, there are individuals whose sense of conscientiousness in preparing for examinations yields pride of place to instinctually felt needs to think independently, as there are others too – and they are the majority – in whom the same sense of conscientiousness is overshadowed by indolence or by a desire for amusement. Still, what is one to think of a system which makes creative intellectual development possible for some only at the cost of unconscientiousness, and in others cannot develop that minimum of intellectual interests which is an indispensable feature of an intellectually mature man?

of certain truths in a given field of knowledge, in order thereby to show students, based on examples, what is the essence of creative cognitive thinking. Retracing during the course of university studies the most important of the intellectual processes to which one or another branch of knowledge owes its genesis and its present development, the student would thereby learn really to understand knowledge as a lively and developing domain of the human spirit, rather than regarding it as a collection of inanimate formulas or absolute axioms. In the process he would acquire practice in the methodical and critical posing and solving not of fictional questions *ad usum delphini*, which popular pedagogy enjoys posing, but of substantial questions that human thought has in fact encountered and does encounter on its path.

Such lectures might be of two types. In the one, the student would gain a genetic acquaintance with the whole present state of a given science and finally would have pointed out to him the newest still unsolved problems and fields open to future creativity. In the second, the professor, as a scientist or scholar, in discussing a question on which he himself has worked or is working, would give his students an example of his own intellectual creativity, not in the form of ready results but in an exact description of the process that has led or is leading him to certain problems and solutions. To be sure, we know that many scholars are loath to initiate anyone into the course of their work, that they do not like to speak of their hesitations and errors. As far as work not yet completed is concerned, this reluctance may be justified by the personal temperament of the researcher; however, in relation to research already concluded it is all too often caused by fear of losing prestige, a fear somewhat reminiscent of the stance of the old sorcerers and the present-day prestidigitators. Without a doubt, as we shall see later, a certain esoteric atmosphere about knowledge is often useful in the scientist's relationship to society at large; but where a teacher's task is to develop his pupils, such an atmosphere conflicts with that task.

However, even lectures acquainting the student with creative scientific work, though they lead to a true orientation in the tasks of science, are not yet sufficient to develop spontaneous cognitive purposes. For that the student must be faced with questions which are important and vital to him, and which at the same time lie in the direction of development of the science and require new cognitive operation for their solution. The very *raison d'être* of scientific exercises is to gradually develop in the student not only technical skills but the capacity for creativity. However, a flaw of the present system of seminars and laboratory exercises is the assumption that all students possess the same kinds of predispositions for a given science and that for all there is one and the same route, the same line of development leading to independent thought. On this assumption rests the uniformity of the exercises, the commonness of the topics, be it in the reading and discussion of the same work or in the assigning of the same laboratory experiments or in the working up of the same scientific materials. At best, in the process the student himself is expected to show an ability to select that part of the topic, the work or chapter, that aspect of the materials or questions that especially interest him or of whose elaboration he feels himself more capable; at worst, the same point of departure and the same course of research schematically set by the professor is imposed as compulsory upon all.

Now, it is obvious that despite our absurdly leveling system of secondary education, every student brings with him to the university a different cognitive disposition, and this diversity does not coincide at all with the differentiation of scientific specialties. Although in everyone there may potentially exist a certain minimum of cognitive purposes which he shares with others, the actual significance of particular purposes, of particular interests in his overall intellectual life and the relative degree of his development in particular intellectual directions are usually completely incommensurable in different individualities. Thus the only effective method would be to discover in everyone his dominant interest, his most current purposes, and gradually to develop, differentiate and complement those purposes in order that, proceeding from them, he may gradually take in the most important departments of the field of knowledge, thus binding them into an organic whole under the action of his most vital intellectual current; arriving at the stage when every new cognitive purpose that may emerge within him will manifest itself in objectively original creative work. And with that intent he should be given problems which, while lying in the direction of his purposes, nevertheless compel him to constantly expand the sphere of his previous interests and knowledge and demand an ever more independent performance of ever more operations, until at last he attains that level at which he will no longer find models in the past and will have to continue on his own.

In the intellectual development of the individual, however, there are acting not only cognitive factors that are either spontaneous or associated with practical questions. An individual who plans a practical activity or seeks to master and develop a certain field of knowledge is not the isolated subject of an operation or thought but a concrete member of social groups which impose on him certain requirements involving every aspect of his conscious life, and hence also of his intellectual activity. Thus, for example, the directions and methods of intellectual education – such as those which today predominate in our school system and of which we have here given a critique – are in every age and in every community the result of certain social traditions and purposes, and today's university student and professor, much like the pupil and director of a philosophical school in Greece, much like the ancient Egyptian priest and his disciple, cannot be understood as social individuals otherwise than against the background of their social environment. That brings us to a new extensive field in the theory of knowledge.

(c) *The question of the social determination of a scientifically active individual.* Cognitive activity, fairly early in the history of culture becomes a *social institution*, that is, certain individuals specialize in it temporarily or permanently, and this specialization is subject to sanction by the group to which the individuals belong; the cognitive activity of the individuals is treated by the group as if it were carried out in the group's name, by its consent and for its benefit, much as any professional work or special function. Here above all we must distinguish the permanent profession – in which the individual devotes his whole life partly or completely to knowledge, as does the shaman (priest, sorcerer and physician in one person) in lower tribes, later the priest, then the philosopher in the Greek sense, and finally the savant-specialist – from the temporary specialization, in which the cultivation of knowledge is linked to a certain age class, as during preparation for other social functions (the pupil), or on the basis of experience acquired in the execution of other social functions (the old man as social adviser). A further important difference occurs between involvement in esoteric knowledge – i.e., that limited exclusively to circles of professional specialists – and involvement in exoteric knowledge, intended for dissemination among society generally; and finally, both with esoteric and with exoteric knowledge, the social function of cognitive activity may consist chiefly in the preservation and transmittal of existing knowledge (as in ancient Egypt and medieval Europe) or in the creation of new knowledge (as in Greece and modern Europe).

The social determination of individual cognitive activity manifests itself preeminently in the social group – by its positive opinion, by the glamour investing the position of the professional theoretician, by a guaranteed living for the individual who devotes himself full-time to knowledge or by attachment to other social positions that secure a living, a precondition for acquiring a certain knowledge, by diplomas and special awards and honorary or material distinctions – associating certain values generally regarded as being positive with manifestations of cognitive purposes which it treats as being desirable. Conversely, condemnation by group opinion, deprivation of professional position, the creation of financial difficulties, and occasionally the imposition of a material penalty (from detention of a lazy pupil after school, to the death penalty for a dangerous thinker) are methods associating values generally regarded as being negative, as the case may be, with lack of signs of cognitive purposes in an individual of whom the group requires those signs, or with signs of cognitive purposes which the group happens to condemn for any reason. These methods cannot, of course, directly produce or remove the cognitive purposes that they have in view, since cognitive purposes, as we know, change only under the influence of cognitive situations, and neither positive nor negative social pressure produces cognitive situations. Nevertheless, they do have an indirect influence, and one of two kinds.

Acting on other purposes of the individual's – social, financial, hedonistic – they cause the individual, in order that he may attain positive values in these realms or avoid negative ones, to make use of certain cognitive values or to shun certain cognitive values: in order to pass an examination he absorbs certain scientific or scholarly results, in order to obtain a scientific or scholarly position he displays his erudition in the required field, in order to win popularity he occupies himself with popular questions and does not touch truths socially dangerous for him, etc. In this way, to be sure, the character of his purposes is not (but the sphere of his cognitive values is) shaped to conform with the requirements of the milieu, being the more explicitly delimited the better are the requirements defined, as to content and significance, of the intellectual values desired by the group. And if an individual of this kind occupies himself with knowledge professionally, operating constantly with a certain domain of values, sooner or later he undergoes an evolution in this field, and some more or less new cognitive purposes will arise within him. Under the influence of these same motives, giving the broadest possible expression to purposes of his own whose manifestations the milieu sanctions positively, and suppressing the active manifestation of purposes negatively assessed in his society, such an individual can become a fairly productive professional, exactly adapted to the cognitive needs that his group currently feels.

Understandably, though, the range of his creativity will not be great. For the emergence in him of a new purpose depends on whether, in the domain of values imposed on him by society, there occurs a new cognitive situation that will in fact stir some preexisting purpose of his and effect a change in it. Whereas it is well known that commonly in sets of cognitive values imposed on the individual by society there are no frankly and conspicuously new situations, that they are most often ready-made and solved problems and that only either unexpected perturbations in a group's traditional cognitive scheme unwanted by the group (e. g., the demonstrated uselessness of acknowledged truths in the planning of new practical questions, or the influence of values brought in from other societies) or the creative initiative of individuals independent of tradition can introduce genuinely new problems into the domain of socially acknowledged knowledge. And even if a new purpose does emerge in such an individual, it can never become revolutionary, it can never develop to such a degree as to transform fundamentally some basic part of the traditional sphere of cognitive values, since the social group always fears revolutionary transformations in any field. There are, to be sure, periods and societies in which, nominally, originality and even revolutionary views are positively appraised by the community; closer study, however, shows that only those scientific revolutions (as, indeed, sociopolitical revolutions) enjoy immediate acceptance by society at large, in which there appears not a fundamentally new direction of thought but either merely a new formulation of familiar truths or a return to some earlier direction, apparently forgotten but in reality still strong among the masses. All of us are familiar with the type of person that might be termed the 'cognitive philistine', openly conservative in societies that set obstacles to any and all innovations, superficially original in groups which in the intellectual realm are subject to fashion rather than tradition. This type flourishes mainly on the soil of exoteric knowledge, that which is imparted in the schools to youth (not specializing in scientific or scholarly work) and propagated in popular lectures and popular science writing, etc.; for only exoteric knowledge can be subject to the kind of social control of which we are speaking here.

The influence of social circumstances on the individual's cognitive creativity can, however, be different, deeper and more fruitful, if again only indirect. Namely, the group does not always nor over the full range of knowledge set specific requirements as to the nature of the cognitive purposes desirable and permissible in the individual. Its demands are very clear in regard, e.g., to the range and kind of knowledge that is to be assimilated by the generality of youth, and generally enter imperatively into the intellectual life of those members of the group whose cognitive operations do not constitute their total and exclusive task in life. On the other hand, they usually become less strict, more general, in respect to individuals whose cognition is supposed by the generality to rise above the level of popular comprehension. The cognitive professional, the priest, the philosopher, the scientist, insofar as he has not occupied himself with popularization, has always been surrounded by a certain aura of sanctity, mystery, inaccessibility; every society has not only permitted but in some measure even viewed positively a certain esoterism in cognition. This esoterism, at first based chiefly on an external sequestration of the cognitive elect from the crowd, and later stemming simply from the internal inaccessibility of the deeper cognitive values and operations for the unprepared, has always left, at least to some individuals, a relative freedom of scientific or scholarly action within more or less broad limits.

This freedom may again sometimes have been hampered by esoteric associations, by special groups formed within the broader society for the cultivation of esoteric knowledge; we shall have more to say of this later. In any case, though, the associations themselves have usually defended their members from excessively strict demands and sanctions by society at large, have guarded the independence of knowledge from popular opinion; while on the other hand the individual has sometimes been able to appeal to the broader society against pressure by the association, when that pressure has been excessively tyrannical, when the association has been too ossified in tradition whereas broader social spheres had undergone a certain, if inadvertent, evolution. In this, as in any field of cultural life, simultaneous membership by the individual in two or more groups has been conducive to the individual's liberation from overly exclusive dependence on one group, has enabled him to attain a certain sphere of private personal freedom. This sphere has further grown markedly in more recent times thanks, on the one hand, to the plurality and differentiation of national societies possessed of a more or less intensive scientific life, and on the other to the contact in which the scientific associations of the various countries find themselves.

This freedom in the realm of cognition has never been a license free of all regulation. It is true that often individuals and even small groups have taken advantage of some cognitive esoterism, or even artificially manufactured such an esoterism, for the sake of scientific charlatanism; still, the 'charlatan' in the realm of learning, like the hypocrite in the religious and moral realms, is only an aberration from the norm that in itself points to the existence of the norm. By the very fact that professional cognitive activity was institutionalized, that society had, as it were, officially recognized the priest, the philosopher, the scientist as an individual in a way exalted above others, this recognition was accompanied by certain demands of a general nature which such an individual had to satisfy in order to preserve his right to esoterism. The broader social group, nation or state, as well as the esoteric association, have placed before the cognitive elect a certain personal *ideal* whose realization in life has been the price of his exaltation above particular observances and constraints to which exoteric knowledge has been subject; the generality's conviction that the cognitive professional submits to that ideal, has been the basis for their confidence in him, even when they have not understood his activity.

The personal ideal of the 'thinker' (if we may use this term in common for all the afore-mentioned classes of cognitive specialists) has been different in various societies and ages. Often, also, an ideal prevalent in an esoteric association has been somewhat different from the ideal acknowledged by the society that has sustained the association, and an individual who, in his struggle with the generality, has leaned on the association, or who, in his struggle with the association, has appealed to the generality, has championed one of these ideals against the other. Nevertheless, despite all these differences, the community of the traits is marked enough for us to be able to consider these temporally and locally differentiated ideals as variants of one basic ideal. Negatively it is always defined by the requirement that the thinker be independent of those values that are the chief motives for the activity of the generality – first and foremost the economic and hedonistic ones, and in some measure even the social ones which manifest themselves in the search for power or popularity. It is characteristic that this is exactly the criterion differentiating the esoteric thinker, whom the generality is prepared to trust and vouchsafe a certain freedom, from the exoteric-cognitive philistine, whom it does not trust and whom it controls at every step, although on the other hand the philistine is often a favourite of the masses and recipient of the most rewards, whereas the thinker tends to arouse in the masses a cool respect and a desire to isolate.

Positively the ideal of the thinker unites within it two requirements: excellence in carrying out his special function, and public service, or action for the good of the group. The criterion of excellence, obviously, will be different when the main task of knowledge is the preservation of old truths, and different again when the objective is the discovery or creation of new truths; usually, however, the two elements are conjoined in various proportions. In any event, the actual criterion of excellence is not imposed by the generality, but is developed by the thinkers themselves and fixed by the professional associations. The requirement of public service was satisfied to the highest degree, in the conviction of the generality, by priestly knowledge, inasmuch as it was on the latter that the capacity to obtain the beneficent influences of religious and magical powers for the group was deemed to rest. Secular knowledge became public service when the generality began to realize that it was such knowledge that was the source of the intellectual power indispensable to any improvement in the sphere of material and social practice. Hence the demand, universal today, that esoteric knowledge sooner or later be applicable in practice, and the growth, alongside the class of pure theoreticians, of a class of technicians mediating between knowledge, incomprehensible to the generality, and the needs of the collective life.

These two requirements – of excellence in the domain of cognitive values and operations, and of the practical applicability of cognitive results for the good of society – not always mutually harmonious, and mutable in their content, nonetheless constitute guidelines for the individual's development as a thinker, and that is their significance for knowledge. The purpose of realizing the social ideal of the thinker in one's own personality arises as a social purpose, as a complex result of numerous influences by the social milieu to which the individual educating himself to be a thinker is subject from the first moment when there begins to take shape in his consciousness the intention of specializing in a cognitive activity. However, that social purpose gradually leads the individual outside the realm of social life in the strict sense of the word, into the domain of objective theoretical criteria. Desiring to conform to this ideal from social considerations, he develops a conscious measure of his own intellectual progress, and with time that progress in itself becomes his ultimate task. And despite the fact that, as we have said in the previous section, techniques of educating scientists and scholars and techniques of self-education are still of a comparatively low standard, yet in view of the fact that the existing standard of knowledge is also, with few exceptions, still low, the individual, given intensive and persistent pursuit of the ideal of cognitive excellence, by way of numerous wasted efforts and vain attempts, nearly always reaches the standard at which his further activity is necessarily the creative addition of something to the existing store of knowledge. And then the social influences that were the first cause of his purposes may to such a degree be neutralized by the influence of exclusively intellectual factors that the individual sometimes consciously opposes both the broader society and the professional association in the name of the very ideal of cognitive excellence that those groups had initially fostered in him.

As we see, here again there is rich ground for the theoretician of knowledge. Research into the regulation of exoteric knowledge and into ideals of esoteric knowledge, in connection with comparison of the various social types of cognitive philistines and of thinkers that we find in history, will permit precise statement of the actual influence of social conditions on the scientific worker, determination of the indirect consequences of this influence for particular fields and varieties of cognition and the drawing therefrom of practical conclusions for the future. Of course, here too, as everywhere else, exceptionally creative and original individuals resist total confinement within any framework of theoretical generalizations and practical guidelines; in fact, we find creative scholars and scientists outside the circle of professional and socially recognized workers. But every individual owes something, positively or negatively, to social influences; in some measure his evolution is explained by the action of the social environment, if only because he has rebelled against that environment. Hence it is only a matter of degree, how far elucidation through social influences permits understanding his spiritual process.

(d) *The question of the intellectual life of social groups.* The relationship of the individual's intellectual life to his social environment, considered above, leads us on to further questions deriving from the borderline between the theory of knowledge and sociology. The concern now is the creation and the adoption by the social group of those common cognitive purposes which find expression in the collective recognition of a set of values and in the imposition on individual members of certain rules of thinking and a course of intellectual evolution.

Here again we have to assume the existence, in members of the group, of cognitive purposes preceding the development of a collective intellectual life and of any cognitive evolution generally of the group as a whole. These may be purposes innate to members of a genetic group, i.e., of one based on common origin and taking in, from birth, all descendants of individuals belonging to the group at the moment (as, e.g., a family, tribe, nation); they may be purposes partly innate, partly acquired in members of groups formed via co-opting older individuals (e.g., a school, a scientific society). It is understood that these cognitive purposes preceding the collective intellectual life of the group are never uniform in all members of the group; for not only may there exist, in various individuals, purposes qualitatively more or less at variance with those that other members possess, but even the purposes that are found in all, may manifest themselves differently among them. In some they are more lasting, actualize more frequently and consequently play a greater role in the totality of their conscious life, whereas in others they are transitory, seldom actualize and consequently have a smaller personal importance. Later we shall put the question as to whether and to what extent we may nevertheless assume, despite all these individual differences, a relative uniformity of primitive cognitive dispositions in certain groups. Of course, in any case such a primitive uniformity can only be partial; no innate similarity among members of the group, by itself, would ever suffice to explain the community of the convictions and cognitive purposes that sometimes reign omnipotent in the group.

The community is hardly explained by the fact that all the members of the given group are subject to more or less like external influences, be it from the natural environment or by the action of other groups. For like influences produce like results only insofar as they are linked to like purposes; given a diversity of individual purposes, the results of like actions will of necessity be diverse. If we transfer various individuals separately into the same natural or social environment; then their differences, instead of blurring, will become still more conspicuous. Only if such diverse individuals form a unified social group will common new circumstances produce, in addition to divergent results in every person, also certain uniform results in all. This means that the sources of social uniformity lie in something that is to be found only in the social association of many individuals into a single group. Especially in the cognitive realm sheer identity of external actions obviously does not suffice to produce social identity. It seems even a less improbable assumption that individuals of various dispositions, finding themselves in like natural circumstances, will be led to the discovery of like technical tools, than that there will develop in them purposes of elucidating like phenomena in a like way.

There is a well-known theory, originated by Tarde, according to which uniformity of 'beliefs', similarly to uniformity of 'desires', to the extent that such uniformity is not innate, constitutes a result of imitation. This theory, however, treats social processes too superficially. Above all, why do just these beliefs spread, and not others? If there is no special reason inclining all the members of the group, regardless of the individual differences in their purposes, to accept a certain belief while rejecting others, in that case the choice of a belief imitated by each individual will be conditioned on the one hand by objective factors (by the nature of the belief or by the personality of him who initiated the belief) and on the other hand by the subjective inclinations of the individual: various individuals will imitate various beliefs, and no exclusive uniformity of certain beliefs will ever come about. And, even when a certain number of individuals choose the same belief, the latter will take a different shape in each of them and, if there is no reason to consider all these variants as subjective interpretations of the same objective belief, we shall once again have diversity instead of uniformity. In short, imitation would lead to uniform results only if the imitating individuals were already uniform; the theory of imitation thus assumes implicitly that which it intends to explain.

However, if we pass from this theory to its antithesis, represented most strikingly in Durkheim's school, according to which it is the social objectivity of collective 'presentations' imposed on every individual that is the reason for their acceptance by all individuals, we notice that we again stand before an unresolved problem – how and why just this 'presentation' and not another has taken on that objectivity. Certainly this theory considers that the sole positive, scientific method of solving the problem is to seek the source of the objectivity of a collective presentation in previous collective presentations that have possessed that objectivity. That, however, again confronts us with the problem of the cause of the replacement or complementation of those previous collective presentations by the present ones, in which ultimately there will again come into play the question of the uniformity of individual reactions to particular influences. Furthermore, at every step we find the emergence of new 'collective presentations' not at all amenable to reduction to previous ones, since their emergence is linked with the organization of new social groups that are only in the process of evolving common convictions.

This seemingly insoluble problem, of how collective uniformity arises from individual diversity, is, however, solved with ease if we realize (as, incidentally, Durkheim also emphasizes) that the essential, fundamental thing in the collective purpose of a social group is not that it is uniform in all members, but that it is *common* to them all, and if we inquire about the sources of that community. Community may be associated with uniformity, e.g., as when all the members of a given group have the same cognitive purposes in connection with a religion or with a political life, but it can also coexist with considerable differentiation, as, e.g., in a scientific society which is jointly interested in the development of the same discipline, though each member is actively occupied with a different special branch of it while conveying his results to others and in turn having their results made available to him. The only thing that is uniform to all the members here is recognition of the development of the discipline as a common task for all members of the group, whereas modes of carrying out that task are different with each.

Among cognitive purposes present in members of a given community, besides those which each member shows on his own in relation to situations that interest himself above all, we find as well those which refer to situations indirectly or directly *touching the group as a whole*, and it is these that are common *par excellence*. For example, every member of a semi-savage tribe is interested in the magical explanation of events befalling him personally; beyond that, however, both he and all the other members are also interested in the magical causes of events that befall the entire group and are momentous for all. Likewise, at a higher level of development, priestly knowledge is regarded as being a collective value of the entire society, and the cognitive purposes of the priests are regarded as representing the purposes of the generality, since the latter's existence depends on a proper understanding by the priests of the nature of religious powers and of their action. When at last knowledge becomes a secular matter, once again a group – the community, city, state, nation – treats as common, as 'public', those cognitive purposes which present themselves to its members as useful for the whole group. Furthermore, numerous cognitive purposes referring directly to the private situations of particular members or to objects not pertaining directly to the group's existence, assume a public significance through the association that they acquire in the general consciousness with truths pertaining to general matters. For instance, a group takes an interest in theories of economic phenomena if these theories, in the group opinion, can exert an influence on certain social institutions, and recognizes physical and biological hypotheses as a common matter if they are linked in its eyes with religious dogmas which have long since had significance as common values.

Thus the source of the community of cognitive purposes in a social group, and consequently of any similarity in convictions and modes of thinking that comes into being despite the diversity of cognitive dispositions introduced into the collective life by each member, is the community of societal interests of the individuals comprising the group. Recognition of certain cognitive purposes as being common is bound up with recognition of the situations corresponding to those purposes as being important to the group as a whole; and that does not in the least require *uniformity* of individual preexisting purposes, but their *synthesis* for the collective existence. Naturally, as regards those situations the attainment of unanimity is essential in the eyes of the group; that uniformity, however, need not be an actual uniformity of cognitive purposes, only a uniformity of *convictions*, i.e., of a general recognition of certain purposes as being positive in connection with certain situations.

Decisive to the collective conviction is the position taken in regard to a common question by individuals who are, in a sense, authorized to think cognitively in behalf of the group, whose social function consists in posing and solving such questions. Any other member is demanded merely to accept that position as the position of the group, much as the military decision of the commander must be recognized as binding upon all by each of the warriors subordinate to him – which obviously does not mean at all that they must always imitate his acts. 'Imitation' in the cognitive realm, insofar as it occurs at all, springs from a conviction based on recognition of the opinion of the intellectual leader. An individual who has accepted as true the professional leader's manner of posing and resolving a question involving the group – to the extent that he is going to think about that question at all – is going to be subject to the influence of the collective conviction and, in posing and solving his problem, is going to strive for a result compatible with the recognized one. However, this will not be imitation in the sense of repeating another's operation, but only in the sense of imitating its results. Hence we may speak of a common cognitive purpose expressing itself in the treatment of a common situation, although actually that common purpose may be the result of a synthesis of diverse individual purposes striving to conform to the purpose of the intellectual leader.

In the same way that a common purpose manifesting itself in the posing and solution of a common problem binding upon all, has once been formed and fixed, it can later change. Some individual may pose and solve the problem differently or replace it altogether with another; his new rendering of the problem either will be rejected or will again be accepted as binding. Both alternatives depend not on whether this rendering of the problem encounters uniform cognitive purposes in all, but on whether in the overall conscious life of the group a more important role is actually played by cognitive or social purposes which find their satisfaction in the older rendering of the problem or by purposes that, for one reason or another, are satisfied by the new rendering. The sign of victory for the second set of purposes will be the imposition, upon all members of the group, of an appropriate new conviction under whose influence there will again take form, from the diverse purposes of the individuals, a common collective purpose. This schematic representation of the matter should not, however, suggest the conclusion that every collective purpose is originally the creation of some one individual. Most often we find gradual, imperceptible modifications and complementations of existing common purposes by small individual transformations and additions, so that it is impossible to distinguish the part of particular intellectual leaders in the posing and solution of a collective question; neither they themselves nor the group are sometimes aware of the changes taking place in the collective intellectual life.

The matter of the emergence and evolution of collective cognitive purposes assumes a different character when the concern is not with common recognition of some particular question and solution, but with the common conscious creation of new truths, as in associations organized for the collective development of a discipline or group of disciplines: in academies, scientific and scholarly societies, partly in universities, etc. Although here basically every member is required to create new truths on his own, and hence to develop new cognitive purposes for himself, nevertheless in these associations too we usually find, if not community of problems and solutions, then at least a unanimity as to the direction of evolution of the discipline and as to the meaning of the results obtained. To be sure, it is often implicit in the founding of the association that the proposition that the only common cause of all its members is the very development of the discipline and not the conferring upon that development of a specific framework or results, and that every new cognitive purpose of any member working in the common field is to win recognition from all. In reality, however, the results of cognitive purposes of particular members may clash, and in resultant conflicts the group adopts a position. In this way, through a series of such acts of collective recognition, there comes into being a certain common line of development and there emerges a framework, at times a very rigid one, beyond which the cognitive activity of the association does not extend.

Up to now we have considered a social group in abstraction from other groups, as a separate, self-contained entity. However, the fact that the intellectual life of a stated group, whether it be a nation or a scientific society, stands in an objective relationship to the intellectual life of other nations or societies, and at the same time differs from them, holds far-reaching significance both for the theory of knowledge and for sociology. In this relationship and opposition to other groups, the sphere of intellectual culture of a given group becomes as it were the collective property of the group, something that appertains to the group and sets it apart; and consequently the group strives to bind that sphere to the domain of its purely social interests, introducing into its intellectual life motives of internal solidarity, collective egoism, antagonism to other groups, collective pride, group idealism, etc.

The commonest manifestation of this association of cognitive with social purposes is the practical emphasis that the group places on those features of its intellectual life that distinguish it from other groups. Every group is prone to an exclusively positive appraisal of its own culture and a negative appraisal of an alien culture, solely because the latter is alien and different from its own. This is linked with a desire to isolate itself from alien influences, to preserve its own individuality. In the cognitive domain these desires are perhaps less conspicuous than, e.g., in the domain of religion or manners and morals since, in the intellectual sphere, a critical attitude and objective tests of values are more easily developed. Nevertheless here too, at every step, we encounter attempts to emphasize and enhance the group's individuality, either on the basis of an actual sense of the superiority of its own concerns, its own modes of thinking, its own results over those of others, or at least by artificially fostering in the group's members a faith in their own superiority. We need not reach to the remote examples of China or of ancient Egypt for evidence: each of the modern civilized nations, every learned organization, will supply numerous facts of this type. This subordination of the intellectual life to social considerations reaches its apex when a nation's knowledge is used as a tool in a struggle against other nations and when scientific truths are consciously twisted to the purposes of social propaganda.

Narrow exclusivity and antagonism with respect to other groups is transformed into rivalry in the scientific field when the intellectual spheres of various nations or various cognitive associations approach one another in a fundamental recognition of all knowledge as being the common property of *all civilized mankind* and when the sense of distinctness of the nation or association, instead of manifesting itself in a recognition of its own cognitive purposes and values as being absolutely positive as opposed to those of others being negative, expresses itself in a desire for one's own nation, for one's own association to add more than do others to the common treasury of knowledge. That to this end it is oftentimes desirable to cultivate certain special interests, methods and procedures of research, a certain collective specialization, is beyond doubt; acknowledgment of that fact, however, need hardly be linked to a hostile self-isolation from other groups but, on the contrary, can bring in its train an intensification of efforts to develop the best possible acquaintance with, and as objective as possible an appreciation of, the intellectual life of others.

(e) *The question of innate cognitive dispositions.* In the deliberations of the previous sections we have sought to outline a general framework for study of the influence of practice on theory, of theoretical education, of the social determination of individual cognitive types, and of the social conditioning of the collective properties of groups in the realm of cognitive life. There still remains, however, the most difficult and just possibly the most momentous question: *what, if any, innate cognitive dispositions should be posited in individuals and groups*? These dispositions being independent as well of the influence of practical questions, as of methods of intellectual education, and of social circumstances, but constituting, as it were, the basis for action of all cultural agencies. There are chiefly three problems involved here: whether mankind possesses any stock, common to everyone, of cognitive dispositions, identical in all ages and in all varieties of the human species; whether in particular races and nationalities we find hereditary and scientifically determinable cognitive faculties and dispositions distinguishing each race and nationality from others; and finally whether and to what extent the individual differences that we find in the faculties and dispositions of various people can be divided into acquired and innate, and whether on the basis of the latter it is possible to divide individuals into cognitive classes or types paralleling or cutting across the division into races and nationalities.

The chief difficulties in this domain derive from the fact that inborn cognitive dispositions, indeed like most other inborn dispositions brought into the world by conscious beings, do not manifest themselves all at once but show up fully only by degrees, in the course of active contact with appropriate fields of experience. And since every human being, from the earliest times accessible to us, has always been subject from birth to certain cultural influences modifying his inborn purposes, at first glance it is impossible to distinguish what is inborn in him and what is developed by culture. Likewise where a group is concerned. The intellectual life of any social group, like any field of its activity, is certainly shaped in each generation on the basis, to be sure, of inherited dispositions, but under the predominating influence of all the intellectual values accumulated by preceding generations, within a framework of traditionally transmitted logical forms and scholarly or scientific methods; nor ought we to pass over the influence of other groups, continually impinging from outside. How do we separate, in the overall outcome, that which is passed down from generation to generation by way of biological heredity, from that which is a result of cultural heredity, of the continuity of traditions, and lastly from that which has been brought in from outside? Nor are we in a better position when we wish to define the cognitive faculties and dispositions of mankind in general: indeed, as far back into the past as we care to reach, we always find a stock of culture common to all peoples, if more or less differentiated, and we cannot ignore the fact that in every period the intellectual life of mankind accessible to us is a result of the interaction of that stock of culture with the biological predispositions of the species.

The numerous essays at a scientific treatment of these questions, whether as deduction of the human species' cognitive functions from biological instincts, or as characterization of the intellectual inclinations of particular races and peoples, or as the well-known classification of individuals by intellectual types, or lastly in still other forms which we need not detail here, largely ignore or fail to appreciate the difficulties mentioned. Therefore their results up to now should be considered as indications of certain general possibilities which future methodical research in time may state precisely and confirm or discard, rather than as actual solutions or even as exact statements of the problems. It is clear that we should begin not from the definition of innate cognitive purposes but, on the contrary, from the study of *acquired* cognitive purposes, which, precisely because they are acquired, are easier to designate exactly and the genesis of which can be empirically traced. Only through the gradual exclusion of that which turns out to be acquired will we arrive at the primal sources of the intellectual life of the individual, the nation, the race or of mankind.

This process of exclusion should make use of two mutually, complementary methods: the genetic and the comparative. Studying genetically the causes of the emergence of each cognitive purpose from previous ones under the influence of changes in cognitive situations, we may at length arrive at purposes which will have to be recognized as existing primally, as not being deducible causally from others; in relation to these, external influences would not be coefficients of their emergence but only causes of their passage from the potential state in which they persist prior to contact with experience, into an actual state. However, inasmuch as the innate dispositions of an individual or group may be expressed not only in the existence of genetically primal purposes but also in certain characteristic variants and combinations to which genetically later purposes are subject in the individual or group, then comparison of various purposes in the same individual or group with the purposes of other individuals or groups should work together with the genetic method in determining what we might term the *cognitive character* of the man, the nation, the race or the species.

*

The guidelines for the theory of knowledge sketched above cannot be, of course, either exhaustive or final. Innumerable new, today not even foreseeable, questions will no doubt suggest themselves even as research progresses, and the methodological canon that today, at the inception of the discipline's evolution, seems indispensable and obvious, is also bound in the course of events to undergo unforeseen refinements and transformations, or even to be completely discarded. For the present, the main objective is to frame at least a theory of knowledge as a single positive science based on experimental material and directed toward the most exact possible description, classification and causal elucidation of cognitive phenomena as cultural phenomena given to us on a par with linguistic, economic, social and religious phenomena. Another objective is for those interested in human knowledge as an object for cognitive reflection to realize that this subject may not only be approached from the standpoint of epistemology, logic or concrete and individualizing history, but that there exists another, as yet unexploited vantage point from which there open up to the inquiring mind new, enormous vistas.

*

Bibliography

In addition to the works mentioned in the text we shall also cite, among works that shed light on various questions touched upon here: John Dewey: *How We Think*, Boston 1910, and *Democracy and Education: An Introduction to the Philosophy of Education*, New York 1916, Émile Durkheim: *Les règles de la méthode sociologique*, Paris 1895 and "Representations individuelles et collectives," *Revue de la Métaphysique et de Morale*, vol. 6 (1898), 273–302, Karl Jaspers: *Psychologie der Weltanschauungen*, Berlin 1922, Lucien Lévy-Bruhl: *La mentalité primitive*, Paris 1922, Herbert Spencer: *The Principles of Sociology* (several vols.), New York 1874–1896[a], William Stern: *Werden und Wesen der Persönlichkeit*, Leipzig 1913, Gabriel Tarde: *Les lois de l'imitation: Étude sociologique*, Paris 1890, Edward L. Thorndike: *Individuality*, Boston 1911, and *Educational Psychology*, New York 1913, Ferdinand Tönnies: *Kritik der öffentlichen Meinung*, Berlin 1922, Florian Znaniecki: *Cultural Reality*, Chicago 1919, and *Wstęp do socjologii* [Introduction to Sociology], Poznań 1922. See also for the psychological aspect of cognitive activity, the bibliography in William Stern: *Die Differentielle Psychologie in ihren methodischen Grundlagen*, Leipzig 1911, 425–438, and for the social aspect, the bibliography in Robert E. Park and Ernest W. Burgess: *Introduction to the Science of Sociology*, Chicago 1921 as well as Polish bibliographies in *Ruch Filozoficzny* [Philosophical Movement], *Poradnik dla Samouków* [A Guide for Autodidacts], *Dzieje myśli* [History of Thought] and *Nauka Polska*.

---

[a] In 1925, Znaniecki quoted the Polish edition: *Zasady socjologii*, Warszawa 1889–1898.

The Science of Science

Maria Ossowska and Stanisław Ossowski

MARIA OSSOWSKA (16 January 1896 as M. Niedźwiedzka, Warsaw, Congress Poland – 13 August 1974, Warsaw, Polish People's Republic), philosopher of ethics and morality. After philosophical studies in Warsaw and at Collége de France in Paris she attained a doctorate from Warsaw Univ. In the following years M.O. worked as an assistant at her *alma mater* and for the Polish Teacher's Union. In 1945 she became a prof. at Lodz Univ. and from 1948 to 1966 she taught at her *alma mater*. Politically very engaged, for example against antisemitism and censorship.

Her husband STANISŁAW OSSOWSKI (22 May 1897, Lipno, Congress Poland – 7 November 1963, Warsaw, Polish People's Republic) finished philosophy at Warsaw Univ. as well before turning to sociology. Apart from science studies, St.O. conducted extensive research on social classes and social bonds, e.g. a critical sociology of race concepts. A univ. assistant in Warsaw before 1939, St.O. later became a prof. of sociology in Lodz (1945–1947) and Warsaw (1947–1963).

Both M. and St.O. were connected to the Mianowski Fund since student times and frequently attended the meetings of the Science Studies Circle. While both actively contributed to Nauka Polska and Organon with essays and reviews, M.O. also was an active member of the Foundation's scientific section and part of the redactions of both journals.

The text *Nauka o nauce* was published in Nauka Polska in 1935 (vol. 20) and translated into English for vol. 1 of Organon one year later. St.O. presented a draft at the Science Studies Circle on 25 November 1929 (▶ **IN THIS VOL.** Sprawozdanie drugie 1930), when both M. and St.O. were in consideration for organising a Polish research institute for science studies in Rapperswil, Switzerland. The text was republished several times (Ossowska/Ossowski 1964, 1965, 1966, 1982) and prominently acclaimed in international science studies after 1945. This edition follows the English version from 1936. It omits a first footnote to the authors' names introducing them as "Ph.D. Lecturers at the University of Warsaw". Square brackets mark linguistic clarifications, minor typographic changes were not indicated.

ORIGINAL REFERENCES
Maria Ossowska and Stanisław Ossowski: "Nauka o nauce," *Nauka Polska*, vol. 20 (1935), 1–12.
— "The Science of Science," *Organon*, vol. 1 (1936), 1–12.

FURTHER TEXTS MENTIONED
Maria Ossowska, Stanisław Ossowski: "The Science of Science," *Minerva*, vol. 3/1 (1964), 72–82.

— "The Science of Science," in: *Science and Society*, ed. by Norman Kaplan, Chicago (1965), 19–29.
— "Die Wissenschaft von der Wissenschaft," in: *Forschungsplanung. Eine Studie über Ziele und Strukturen amerikanischer Forschungsinstitute*, ed. by Helmut Krauch, Werner Kunz and Horst Rittel, München, Wien (1966), 11–21.
— "The Science of Science," in: *Polish Contributions to the Science of Science*, ed. by Bohdan Walentynowicz, Dordrecht et al. (1982), 82–95.
▶ "Sprawozdanie drugie z działalności Koła Naukoznawczego," *Nauka Polska*, vol. 13 (1930), 166–169.

CONTENTS: Two Points of View in the Investigations which Take Science as their Subject. Problems of the Science of Science. Question of the Autonomy of the Science of Science. Actuality of the Science of Science. Practical Application of the Science of Science.

Two points of view in the investigations which take science as their subject

For a number of years Science[1] itself has become the subject of scientific investigation, both Science in its functional sense (the whole of research activities), and in its static sense (the whole of the products of these activities, i.e., the whole of the scientific truths).[2] How vivid the interest taken in this research is, is eloquently proved by the bibliography published in vols. 18 and 19 of *Nauka Polska* (*Science and Letters in Poland*).[a]

In the first, as well as in the second sense, Science may become a subject of research in different ways. It seems, however, that two chief points of view may be established to which all investigations into the science of Science can be subordinated: the investigator is interested in Science either as a way to the knowledge of the world, or as a

---

[1] The meaning in which we use the word "Science" with a capital has been pointed out in the Foreword. ||| Cf. ▶ Preface 1936, 105: "The term »Science« is used here in its broadest sense, including both Natural Science and Humanities, covering practically the whole field of organized knowledge."

[2] For this distinction see Tadeusz Kotarbiński: *Elementy teorii poznania, logiki formalnej i metodologii nauk* [The Elements of the Theory of Knowledge, of Formal Logic, and of the Methodology of the Sciences], Lwów 1929, 367. ||| In 1936, the authors gave the title of Kotarbiński's monograph in English and finished the footnote with the remark "(in Polish)".

[a] Here the edition follows the order of the journal's Polish and English titles as given in the 1936 version to illustrate the representative function of the Organon journal. The authors refer to the overview "Treść wydanych tomów Nauki Polskiej" [Contents of the Published Volumes of Nauka Polska], *Nauka Polska*, vol. 13 (1930), 271–279 and an extended 13-page version published as an annex to *Nauka Polska*, vol. 19 (1934) with an individual page count. For a systematic instead of chronological overview see "Bibliografia artykułów zawartych w tomach I–XX 'Nauki Polskiej'" [Bibliography of the Articles in volumes 1–20 of 'Science and Letters in Poland'], which was annexed to Nauka Polska, vol. 20 (1935).

field of human culture. Let us call this first attitude the *epistemological point of view*, remembering, however, that we are concerned here with a wider range of problems than the one embraced by traditional epistemology. And let us call the second attitude *the anthropological point of view*, bearing in mind anthropology in that wide sense in which it is understood in the English-speaking countries (the science of man and his culture).

The distinction of these two points of view is not a division of the problems: besides problems specially connected with the cognitive functions of Science, and besides problems specially connected with that second attitude, there exist problems which may be approached from both sides. For example, the psychology of scientists may interest both those who study the connection of Science with man's creativeness and the development of culture, as well as those who study the question of a scientific cognition of the world (what qualifications he who is to discover or prove new scientific truths has to possess?).

The interest taken in cognitive functions is very old. The history of investigations of this kind goes as far back as the times of Xenophanes. The interest taken in Science as a field of human culture is something new. It was partly derived from historical research, partly called forth by the development of modern sociology, and partly by practical needs (the question of the encouragement and organization of Science). Research in this field is much younger than the science of religion, than the science of economic production, than the science of art.

Those interested in Science from the epistemological point of view treat the existing scientific works and research methods applied by scientists only as examples and as material for conceptual analyses, for classifications, for considerations of the cognitive value of various possible activities and theses. The degree of importance, attributed from this point of view to the separate cognitive activities and fields of Science, does not depend on its proper rôle in the development of human culture: the criteria of importance are established on the basis of the general postulates of the theory of knowledge and of methodology.

The theory of scientific cognition, treating its subject *sub specie aeterni*, may be in fact cultivated without a closer contact with the life of Science.

In investigations about Science, resulting from the interest taken in man and his culture, we have before us a concrete reality: the life of Science with its triumphs and errors. And though we look forward to general formulations we must study this life of Science in the fullness of historical circumstances.

Problems of the Science of Science

The problems concerning Science in its manifold forms – both those already found in studies devoted to the science of Science and those which have not yet been voiced – could be grouped according to various principles. They could, for example, be divided in three groups: the one taking into consideration all problems connected with the *personality* of the creative worker in Science, the second pursuing all problems connected with the *activities* leading to the formation of Science, the third concerned with the problems relating to Science as a completed human *product*. This is one of the numerous possibilities. According to another, the same problems could be put under three other headings, analysing under the first all problems which answer the question what Science is, under the second how Science originates, and under the third – what effects it has.

While the principles of classifying the problems quoted above only as examples – must lead to the separation of problems which hitherto were usually considered jointly[3], the grouping given below takes into account the hitherto prevailing division of labour among the scientists. According to this proposition the problems concerning Science would be grouped under five different headings, three of which would form the backbone of our new branch of science.

a) Among the first of these three fundamental groups we would propose to count all those problems which hitherto have been usually treated by philosophers and which could be called the *philosophy of Science*. Here would be placed such problems as that of the conception of Science and the numerous controversies connected with this question, as to what is still, and what is no longer Science. Here the classification of the different branches of Science would find a place. Here would be included a large group of general methodological problems: the analysis of the way of proving various kinds of statements, the analysis of the rôle of "as if" in scientific cognition[4], analyses of such conceptions as that of scientific law, hypothesis and such like. In this group the epistemological point of view, spoken of at the beginning, would be dominant.

b) The second division of research about the science of Science could be called *the psychology of Science*. Here will belong such problems as, for example, the problem concerning the psychic development of a scientific worker, the distinguishing of certain types of investigators, the psychological analysis of various types of research activities and of various stages of scientific activity, the question of capacities necessary for the cultivation of certain branches of learning etc.

c) In the third group Science does not appear as in the first one, detached from any historical background, but against the background of social life and of the whole of cultural life in general. This division, in which the point of view, called by us anthropological, predominates, could be given the name of *the sociology of Science*.

---

[3] So, for example, according to the first proposition the problems which we include in psychology, would be found both in the group of problems connected with the personality of the scientist and in that connected with the activities leading to the formation of Science.

[4] We refer to the book of Hans Vaihinger: *The Philosophy of as If: A System of the Theoretical, Practical and Religious Fictions of Mankind*, transl. by C. K. Ogden, London 1925. ||| While the 1936 English version refers explicitly to Vaihiniger's work, a year earlier the Polish version solely mentions the "role of fiction" ("roli fikcyji") without any reference.

Here will be included such problems as that of the relations between Science and other products of culture, like art or religion. Here will be voiced the dependence of the development of Science on economic conditions, on the structure of a given society, on the organization of education. Here not only the factors by which Science is influenced, but also its effects on cultural life, its influence on the economic conditions, on law, morality etc. will be examined.

One group of these problems is at present being discussed by German sociologists in the so called "Wissenssoziologie"[5]. These sociologists are in vigorous opposition to the widely spread treatment of the evolution of Science as if in this evolution only immanent factors played a decisive rôle.[6] Science like all other human products is in their opinion "sozialbedingt," and "the problem of knowledge being socially conditioned in its various forms" is one of the central focuses of their interests[7]. In accordance with the meaning of the word "Wissenssoziologie" their investigations do not only embrace Science. They are interested in knowledge in general. They are interested in the formation of public opinion, the arising of certain popular conceptions of life, characterising a given epoch, the shaping of certain ideologies. Related to these studies was the article by F. Znaniecki, entitled "The Subject and the Tasks of the Science of Knowledge," published in vol. 4 of *Nauka Polska* ("*Science and Letters in Poland,*" 1925)[b]. Like the works mentioned before it had a rather programme-like character. As this is a new field of research it contains, of course, more plans than realisations.

d) In close contact with the group of sociological problems, singled out by us a moment ago, remains a large group of problems of a *practical and organising character*. These problems have been hitherto chiefly undertaken by institutions carrying on the promotion of Science and using for certain practical ends the theoretical results of the former groups. Here belong such questions as the general problems of the organization of Science, social and state policy in relation to Science, the organization of higher institutions of learning, of research institutes and of scientific expeditions, protection of scientific workers etc. etc. These problems deserve to be singled out from the sociological group on account of their practical character and on account of

---

[5] See Karl Dunkman: „Die soziologische Begründung der Wissenschaft," *Archiv für systematische Philosophie und Soziologie*, vol. 30 (1927), 145–164, Ernst Grünwald: *Das Problem der Soziologie des Wissens. Versuch einer kritischen Darstellung der wissenssoziologischen Theorien*, Wien, Leipzig 1934, Karl Mannheim: „Wissenssoziologie," in: *Handwörterbuch der Soziologie*, ed. by Alfred Vierkandt, Stuttgart 1931, 659–680, Max Scheler (ed.): *Versuche zu einer Soziologie des Wissens*, München, Leipzig 1924, Alexander von Schelting: „Zum Streit um die Wissenssoziologie I. Die Wissenssoziologie und die kultursoziologischen Kategorien Alfred Webers," *Archiv für Sozialwissenschaft und Sozialpolitik*, vol. 62 (1929), 1–66, Karl August Wittfogel: „Wissen und Gesellschaft. Neuere deutsche Literatur zur „Wissenssoziologie," *Unter den Bahnen des Marxismus*, vol. 5/1 (1931), 83–102. ||| The Grünwald reference was only added in the English version in 1936.

[6] Mannheim: Wissenssoziologie (footnote 5).

[7] Wittfogel: Wissen und Gesellschaft (footnote 5).

[b] ▶ Florian Znaniecki: "The Subject Matter and Tasks of the Science of Knowledge" (1925), i. e. "Przedmiot i zadania nauki o wiedzy," *Nauka Polska*, vol. 5 (1925), 1–78.

the fact that in the hitherto prevailing division of labour they were allotted to somebody else.

e) Finally, the fifth and the last of all the groups of problems enumerated by us, consists of *historical problems*. The history of the conception of Science, the history of the conception of the scientist, the history of the separate disciplines and of learning in general, these are materials which can be used by all the groups of problems, which have been mentioned above.

Undoubtedly these groups overlap in various ways, but perhaps there are in them fewer problems of a doubtful character than in other attempts at a classification of this diverse material. This attempt is of course only a provisional arrangement at which it would be difficult to stop. Looking at the question more closely, a further differentiation within the groups would prove indispensable, for the time being, however, we disregard this, sacrificing the accuracy of our presentation for its greater lucidity.

The set of problems proposed by us does not cover the same ground as those that have been previously classed by some under the name of science of Science. The history of German philosophy has for a long time made us familiar with the name "Wissenschaftslehre," but those who used it did so in a different sense from that in which it has been employed here.

Leaving aside Fichte who called his whole philosophical speculation by that name, this term was used in Germany chiefly to denote logic with general methodology, or logic with general methodology and the problems usually included in epistemology. In almost the same sense the term "Wissenschaftslehre" was used in the work of Bernard Bolzano, the subtitle of which explained "Versuch einer ausführlichen und grössentheils [sic] neuen Darstellung der Logik".[8] It was logic understood in a very wide sense, with which we were later made familiar by German textbooks of the end of the XIX century and the beginning of ours.

Limiting philosophy to problems of logic, methodology and epistemology, Prof. Kotarbiński in the quoted "Elements" proposes to call philosophy the science of Science. It is clear that the territory of the science of Science, as spoken about in our article, does not correspond with the area marked out in these propositions. It is too wide an area in relation to ours, and at the same time too narrow. Only a part of the problems of logic and epistemology will go into our first division of the science of Science, while all the remaining divisions singled out by us will be in relation to that "science of Science" something new. Closer to our problems is W. Schingnitz in his article entitled "Scientiologie"[9], but his remarks, though frequently valuable, are fragmentary and do not compose a distinct whole. Problems related to those singled

---

[8] Bernard Bolzano: *Wissenschaftslehre. Versuch einer ausführlichen und größtentheils neuen Darstellung der Logik mit steter Rücksicht auf deren bisherige Bearbeiter* (4 vols.), Sulzbach 1837.

[9] Werner Schingnitz: "Scientiologie," *Minerva-Zeitschrift*, vol. 7/5–6 (1931), 65–75 and vol. 7/7–8 (1931), 110–130. ||| The Polish version (1935) referred to Maria Ossowska's review of Schingnitz' text

out by us in our psychological and sociological group, are also enumerated by E. Rádl in an article entitled "Zur Philosophie der Wissenschaftsgeschichte".[10] They are allotted by him – with which it is hard to agree – to the historian of Science. This research, according to him, should be conducted from the point of view, which we have called anthropological in the present article.

Schingnitz in the article just quoted did not only enumerate various possible types of research in the science of Science, but also proposed a name for it: he wished to include them under the term of *scientiology*. Those who wish to replace the expression "science of Science" by a one-word term sounding international, in the belief that only after receiving such a name a given group of problems becomes officially dubbed an autonomous discipline, could be reminded of the name *mathesiology*, proposed long ago for similar purposes[11]. This term might satisfy those purists who in the term "scientiology" may feel offended by its combined Latin and Greek origin.

Question of the autonomy of the Science of Science

The conception of a new branch of science usually becomes finally crystallized only when this discipline has gained admission to the university curricula. Before this takes place it has to overcome the doubts as to whether there are sufficient reasons to recognize in it a separate, autonomous branch. These doubts were raised also in relation to the "science of Science".

From the theoretical point of view this is an insignificant matter. To a large extent it is a matter of convention whether we shall recognize a certain system of problems as a separate branch of learning, or whether we shall subordinate it to a more general science or assign it to several various branches. But as certain practical consequences are involved here, we shall for a moment consider whether in relation to the science of Science there are more reasons for such doubts than in other fields.

Against the autonomy of the science of Science the argument is raised that its problems already have their places in other formerly separated fields (psychology, sociology, logic with methodology, the theories of the separate sciences); the science of Science would be, therefore, a discipline without any such problems which would be peculiar to itself. Let us, however, consider that the areas of the separate sciences are not in general marked out like political or administrative territories; within their confines condominia cannot be avoided. These condominia are usually widest in the recently separated branches of learning, because usually, before a new discipline can

---

in addition: Maria Ossowska: "Review of: Werner Schingnitz. Scientiologie. Minerva-Zeitschrift 1931, 7. Jahrg. Hft. 5–6 i 7–8. S. 65–75 i 110–114," *Nauka Polska*, vol. 17 (1933), 362–364.

[10] Emanuel Rádl: "Zur Philosophie der Wissenschaftsgeschichte," *Scientia. Rivista di Scienza*, vol. 54 (1933), 309–315.

[11] This term was proposed by Ampère (see the article "Wissenschaftslehre," in: *Wörterbuch der philosophischen Begriffe* (2nd, revised ed.), ed. by Rudolf Eisler, Berlin 1904, 804–805).

emerge, the problems which are to form its foundation already exist, but up till now have been assigned to other divisions. Therefore it seems quite natural that in the first stages of the history of a new science the whole repertory of its problems can be taken from other fields. It is only in the course of time, when we have already become sufficiently familiar with the new frames, and when from the borrowed problems new ones arise, that we begin to treat the problems of the new branch as peculiar to itself. What matters, is whether these various problems possess a sufficient internal unity.

This very internal unity of the science of Science was also frequently questioned. The scope of research in "science of Science" comprises investigations concerning widely different subjects; in some cases these will be into psychic phenomena, from which scientific works arise, or into the physiological conditions of scientific creativeness; then again certain social institutions are the subject of research, and still again the relations between scientific assertions, or between scientific assertions and reality etc. These various problems are solved by various methods. But also in this respect the science of Science is in a similar position to many officially recognized disciplines. A complete uniformity of subject and a uniformity of methods can probably be found only in the aprioristic sciences. The link between the problems of one branch of science does not necessarily consist of the unity of methods and points of view. Such a link can be formed by the subject which is the centre of interest; and this certainly does not mean that this subject should be the only subject of research of the given branch; we are also right in speaking of the unity of a subject when heterogeneous objects are being investigated, so long as they are objects of research on account of their connections with that common centre of interest. Let us take general linguistics as an example. The student of general linguistics is concerned with psychic as well as with acoustic and physiological phenomena and also with the field of meanings. He is concerned both with the processes which take place in the mind and body of the speaker, hearer and reader, and with morphological and semasiological changes in the history of language, with the mutual interdependence of different languages and the peculiarities of various dialects of the same language. He applies extrospection as well as introspection, historical methods as well as those used by natural science. But the autonomy of general linguistics is never questioned, because all heterogeneous subjects which come into the scope of linguistic research remain in definite relations to *human speech* and therefore interest the student of linguistics.

The case is the same with the science of Science. Here Science is the centre of interest (it doesn't matter, whether we have in mind Science as a product, or as certain activities). All other [subjects] entering the scope of research in science of Science remain in this or that relation to Science, and it is for this very reason that they are here the subject of research.

Another argument advanced against the autonomy of the science of Science, is the lack, not so much of objective, as of personal unity. We have here in mind the doubt whether one man may successfully carry on research so manifold and requiring such a many sided preparation; can there exist a specialist of the science of Science, a specialist who would have to be experienced in general methodology, to be familiar with the achievements of psychology, theory and history of culture etc?

If we wish to decide whether these doubts hit at the autonomy of the science of Science, we can again refer to the above mentioned science of linguistics. The qualifications required to investigate thoroughly a given language are extremely manifold: phonetics, semantics or dialectology of a language are fields of research requiring a very diverse preparation. And what shall we say of one who wishes to become a specialist in general linguistics, where, besides the preparation in logic and psychology, besides historical and ethnological knowledge, it is indispensable to possess a knowledge of various languages which would make comparative studies possible.

The decision whether a certain group of problems can form a separate discipline, is not in fact dependent on whether one person is able to embrace all these problems. The scientific unities singled out in the traditional division of labour continue to live as unities, though, due to their growth, they have become impossible for a single individual to master. Such a principle of grouping the problems would take into account extremely casual factors: what one individual is able to master today, may be impossible for one to-morrow.

The contemporary life of Science makes us more and more familiar with group research, which, though it is done collectively, does not violate the unity of the discipline to which it is allotted. Such research is forced upon us especially where the same subject is to be investigated from various points of view i. e. in situations in which our science of Science is found. Its future will have to be entrusted to groups of people rather than to individuals.

Apart from this whole discussion we think that the legitimization of a new science should be brought about not by aprioristic considerations but by empirical ones. Only the pragmatic point of view should be decisive: is the given group of problems sufficiently vital for its importance to be emphasised by giving it a separate name? are there any concrete didactical considerations in its favour? have we any reason to expect fruitful consequences by grouping these problems in a common separate frame?

For if the question of the singling out of a certain group of problems into a separate discipline be insignificant from the theoretical point of view, it is not so from the practical one. A new grouping of problems lends additional importance to the original problems and gives rise to new ones and new ideas. The new grouping marks out the direction of new investigations; moreover it may exercise an influence on university studies, the foundation of chairs, periodicals and societies.

Actuality of the Science of Science

Studies in culture, classed under the name of social or cultural anthropology are in general not concerned with Science in the present sense of this term, because Science in this sense is not a component of all the heterogeneous cultures with which the student of anthropology or ethnology is concerned. Almost every culture is characterized by some religion, some magic, art, technology, morals, but Science in the perspective of time and space is a rather exceptional phenomenon. It plays, however, an extremely important rôle in one culture which is of special importance for us: in modern European culture.

The majority of contemporary sociologists think that it is owing to Science that modern culture is a completely new type in history, quite incomparable with all pre-scientific cultures. It has no more than three centuries behind it, and the gulf which divides it from all former cultures grows with every year as the influence of Science on all spheres of life becomes more powerful.

The history of Science goes back, of course, to earlier times. Certain characteristic features of our "scientific culture" had already had their precedents in history, be it in the Hellenistic or the Chinese or even the medieval cultures. There is, however, something entirely new in this uninterrupted development of Science in the course of the last three centuries.

The realisation of this fact will facilitate the proper appreciation of the science of Science. Therefore without entering upon an analysis of the conception "scientific culture" or "culture based on Science" it may be worthwhile to glance at some of its features which are most significant and which have been much written about in publications of recent years.

1. Pre-scientific cultures had to be stable in their foundations. Whatever had claims to complete stabilization met in them with highest approval. When, for example, religion was changed, it was either done in such a way that nobody distinctly realized it (slow transformation of religious tradition) or by means of a revolution when, instead of one "unchangeable" religion there was introduced another religion with claims to the same unchangeability. On the contrary, modern scientific culture is not only in a stage of constant changes, but this dynamic quality is accepted as a postulate by those who create it.

2. An equally new feature of the scientific culture is its universality. There is only one scientific culture, absorbing all scientific achievements, wherever and by whomever they are attained. There are no competing scientific cultures, there are no competing sciences as there are competing religions or codes of law. All incongruity between various scientific theories is considered a provisional stage which has to be overcome in this or that direction. Our culture becomes also universal in the geographical sense: it has encircled the globe from pole to pole, on the waves of the wireless it penetrates the widest thickets of the jungle, it reaches the most isolated islands of the Pacific.

3. Modern culture enables man in an ever growing degree to transform his environment according to his own aims instead of adjusting himself to it. In relation to former cultures this is a rather quantitative difference, but in such a powerful degree, that it is impossible not to see in it a qualitative difference as well. Not limiting itself to ruling over an extra-human environment, Science endeavours to assume control

over human instincts, over social and economic forces. In the last few decades scientific plans have been forming for the organization of human life on a world scale. This was unknown to any pre-scientific culture.

4. Changes in the conception of life, caused by modern Science, are incomparably more far-reaching than the changes occurring at the transition from one pre-scientific culture into another. We have in mind here not only changes, directly called forth by certain new scientific discoveries, as, for example, the widening of the limits of time and space due to astronomical and geological discoveries, or the change of man's position in the world of living creatures, brought about by the discoveries of biology. Here are also involved certain general tendencies of thought. On the one hand the dynamic conception of human culture, which has been already mentioned, and on the other, rationalism. On the basis of scientific culture a conflict has arisen between Science and the spheres of culture which have survived from the pre-scientific era. This conflict is presented in various ways: intellect and subconsciousness, rationalism and the traditional mental habits, Science and Religion. The future will show whether this conflict is only something transitional, or whether it will be a permanent characteristic of our new type of culture.

Scientific culture is only in a stage of formation. The speed of its development is growing, but its most characteristic features have not yet been wholly attained. At any rate the changes, brought about by Science in social life and in human minds, suffice to lead us to the belief that the science of Science will have to occupy a special place in the rank of the sciences of man and his products. It was brought into existence not only by new interests, but above all by the *new reality*. As the discovery of electricity brought with it the creation of new divisions of physics, so the development of Science in modern society will bring about the rise of the science of Science.

Practical application of the Science of Science

Besides theoretical interests in Science, and besides its more and more powerful influence on the whole system of human life, *practical needs* also come into play. The growth of Science requires an extremely wide and many-sided supplementary apparatus and the building of this apparatus requires theoretical studies. Practical needs, as we have mentioned, played an important rôle in the genesis of the science of Science.[12]

The practical applications of science of Science do not require an explanation; the organization of scientific work, individual and collective, the organization of institu-

---

[12] Cf. preface to the present volume and the tables of contents of the 20 volumes of Nauka Polska. ||| In 1936, the authors referred to ▶ Preface 1936, V–VI. A year earlier, in the Polish version they instead referred to ▶ Editorial Introduction 1923 for a description of the formation of a knowledge about science and also recommended a look at the tables of contents in the 20 vol.s of Nauka Polska for illustration.

tions, the protection of Science by the state and by social organization, the education of the scientist, all this – if it is to be fruitful – cannot to-day do without studies just as specialised and complicated as those which are required for the construction of large industrial establishments.

But the investigations of which we speak play yet another rôle, though not such a conspicuous and direct one: teaching what Science is, contributing to form in the minds of scientific workers this or that conception of Science, they at the same time influence their further creativeness; for Science, like all other fields of culture, is a part of that particular sphere of reality whose history depends on what we think of it.

The Problem of Scientific Creativity

Stefan Błachowski

STEFAN BŁACHOWSKI (19 May 1889, Opawa, German Empire – 31 January 1962, Warsaw, Polish People's Republic), psychologist, studied philosophy, psychology and philology and later also physics and biology at Univ.s in Lviv, Vienna and Göttingen (1907–1913). He worked at Lviv Univ. from 1914 till 1919 when appointed prof. at Poznań Univ. in 1919, where he taught mainly experimental psychology and the history of his discipline until he retired in 1960.

B. got involved with the Science Studies Circle due to his work in memory and mathematical talent. He published an extensive article in vol. 9 of Nauka Polska, in which he explicitly refers to Antoni Bolesław Dobrowolski's project on *biographies of creative thought* (► **IN THIS VOL.** Dobrowolski 1918, 1927) and closely analyses some of the materials from the series "Researching the Genesis and Development of Scientific Creativity" (*W sprawie badania i genezy rozwoju twórczości naukowej*) published in Nauka Polska, vol. 6: Czesław Białobrzeski's autobiographical sketch (► Białobrzeski 1927) and Franciszek Bujak's report "Ways of my Intellectual Development", which were only published under initials. In the passage, which was excluded from this edition (23–31 in the original), Błachowski discusses his discrimination of "the synthetic and the analytic way of thinking" against the background of recent literature (he refers to Alfred Binet, Alfred Hock, Felix Klein, Ernst Kretschmer, Ernst Meumann, Józef Nusbaum-Hilarowicz, Wilhelm Ostwald, Henri Poincaré, Eugenio Rignano, Wilhelm Stern, Jan Tur and Władysław Witwicki). Translation by Tul'si (Tuesday) Bhambry.

ORIGINAL REFERENCE
Stefan Błachowski: "Zagadnienie twórczości naukowej," *Nauka Polska*, vol. 9 (1928), 1–67.

FURTHER TEXTS MENTIONED
► Czesław Białobrzeski ["C.B."]: "Szkic autobiograficzny i uwagi o twórczości naukowej," *Nauka Polska*, vol. 6 (1927), 49–76.
Franciszek Bujak ["F.B."]: "Drogi mojego rozwoju umysłowego," *Nauka Polska*, vol. 6 (1927), 77–136.
► Antoni B. Dobrowolski: "O pilnej potrzebie wychowania umysłowego w Polsce: o konieczności zasadniczej reformy nauczania w szkołach średnich oraz stworzenia w związku z ową reformą nowych placówek pracy naukowej," *Nauka Polska*, vol. 1 (1918), 489–502.
► —"Archiwum materiałów do badania twórczości," *Nauka Polska*, vol. 6 (1927), 140.

184 *Part II: Sources*

CONTENTS: I. The notion of science and scientific creativity. II. The degrees of scientific creativity. III. [Categories of Scientific Creativity and] types of scientific creators. IV. The imagination, memory, erudition, the subconscious and their roles in scientific creativity. V. Problems and ideas.

I. The Notion of Science and Scientific Creativity

Science, its goals and tasks, has been described in a multitude of ways. Every scholar who is more than a specialist in some very narrow field of scientific research comes to form an opinion about the nature of science. But scholars' judgments about science remain astonishingly divergent. For instance, some view science as "a changeable system of theories and hypotheses that explain reality and in some way modify it,"[1] others see it as "a joint operation aiming methodically to explore a range of subjects and to make that exploration accessible in a responsible and systematic manner".[2] "A joint operation" and "a changeable system of theories and hypotheses" are unquestionably very different notions, which suggests that science is a formation with an extremely complex expression in which we can discern diverse characteristic traits.

For our purposes it will suffice to remark that science is the constantly changing product of the efforts of the human mind, efforts that aim to explore and to master reality. These sustained efforts of the human mind in the field of science express themselves in opinions that represent real progress in relation to the state of research at a given moment. Zaleski rightly claims that scientific creativity is a permanent and vivid effort to broaden and improve the system of theories and hypotheses, as well as an expansion of humanity's victorious relationship with the world, in a cognitive sense as well as in a moral and practical one.[3]

Thus scientific creativity aims to create new opinions, ones that do not yet exist in science. But not all new opinions are signs of scientific creativity. A hundred thousand singular, new and true opinions pronounced about as many cobblestones will not represent actual progress with regard to the scientific system of mineralogy. Any researcher would categorically reject the idea of such a research project, would deem it to be futile and to have nothing to do with science. And yet how close to such research is some travel writing, or accounts in zoography, ethnography, museography etc., inventorying and describing all the objects found in a given sphere and pronouncing truthful statements about each of them. These statements do not exhibit the traits of creativity as defined here. They might enrich the catalogue of statements

---

[1] Zygmunt Lubicz-Zaleski: "Rola wyobraźni w twórczości naukowej" [The Role of Imagination in Scientific Creativity], *Nauka Polska*, vol. 6 (1927), 1–23: 5.

[2] Władysław Witwicki: *Psychologia. Dla użytku słuchaczów wyższych zakładów naukowych*, vol. 1 [Psychology. For Students of all Institutions of Higher Learning], Lwów 1925, 6.

[3] Lubicz-Zaleski: "Rola wyobraźni w twórczości naukowej" (footnote 1), 5.

that one day may provide material for creative scientific research, but they represent no real progress in relation to the state of research.

Scientific creativity is not expressed in such certain (truthful) statements that merely enunciate a fact. Its role is about something else.

In his work "O nauce," Jan Łukasiewicz stated very clearly and emphatically that "not all true statements are scientific truths. There are truths that are too trivial for science, because there are facts that are too trivial"[4]. Łukasiewicz takes a pragmatic approach on the question of what truths become accepted parts of the scientific system. The supplementary value that "every statement ought to possess in addition to truthfulness in order to belong to science," he maintains, "can be described as the ability to evoke or to satisfy, directly or indirectly, universal intellectual needs, i.e. those that can be felt by every person who has attained a certain degree of intellectual development"[5].

Science is not limited to reproducing facts, but must produce theories that contain creative elements. Łukasiewicz aptly points out where creative elements of theory can be found. Let us try to summarize his arguments, as far as they are related to the problem of scientific creativity that interests us. Every theory is built on reasoning, and all reasoning rests upon the relationship of entailment, the relationship of cause and effect. We know that if the cause is true then the effect must also be true. If it is true that "every S is M and every M is P", then it must be true that "every S is P". But the relationship of entailment is asymmetrical, so that the truthfulness of the effect does not always entail the truthfulness of the cause. E.g. we cannot conclude that if "every S is P" then "every S is M" or that "every M is P".

Łukasiewicz called the progression from cause to effect the direction of entailment. It is important that the direction of reasoning, i.e. the progression from statements that are points of departure for reasoning to other statements that are the goal of reasoning, may or may not comply with the direction of entailment. This is why all reasoning falls into two great categories: deductive reasoning, i.e. deduction where the direction of reasoning complies with the direction of entailment, and reductive reasoning, i.e. reduction where the direction of reasoning goes against the direction of entailment. Deduction seeks effects for given causes; reduction seeks for causes for given effects.

Reasoning can take as a point of departure either a certain statement, one of whose truthfulness we are convinced, or an uncertain statement. Depending on whether that point of departure is certain or not, deduction can be divided into inference and verification, while reduction can be divided into explanation and justification.

---

[4] Jan Łukasiewicz: "O nauce," in: *Poradnik dla samouków* (vol. 1: Wskazówki metodyczne dla studiujących – Poszczególne Nauki) [A Guide for Autodidacts, vol. 1: Methodological guidelines for students – Individual Sciences], Warszawa 1915, XV–XXXIX: XVI.

[5] Ibid., XX.

In inference, which is a type of deductive reasoning, we seek certain effects for certain causes (example: every prime number of the form of $4n + 1$ is the sum of two squares; therefore 53 is the sum of two squares).

Verification is a type of deductive reasoning that seeks certain effects for a given uncertain statement (e.g. "(perhaps) I am experiencing a visual hallucination," "therefore (perhaps) someone else does not see what I see," and all the while I try to ascertain the truthfulness of that effect).

Explanation is a type of reductive reasoning that seeks the cause for certain statements (e.g. "I did not get a letter from my friend"; trying to find the cause, I reason: "my friend is ill and cannot write" or "the letter got lost"). The causes for that certain effect need not be certain.

Finally, justification is a type of reductive reasoning that seeks certain causes for given uncertain statements. E.g. assuming that Fermat's theorem about the equation $x^n + y^n = z^n$ (an uncertain statement) is an effect, I search for certain statements from which that assertion would follow.

The empirical sciences, which examine reality and are based on experiential facts, ascertain those facts in unitary and certain statements. "Ascertaining facts, however," Łukasiewicz writes[6], "cannot fully satisfy the desire for knowledge; facts have to be supplemented with *theories* created through reasoning".

In the empirical sciences the points of departure for reasoning are unitary and certain statements about facts considered as effects for which one seeks the causes. This is reasoning of the type called explanation.

Łukasiewicz further demonstrates that in the empirical sciences "the simplest and most common type of explanation [...] is *incomplete*, i.e. *proper induction*," a type which takes unitary and certain statements of the kind "S_1 is P," "S_2 is P," "S_3 is P,"... as its point of departure, which considers them to be effects and searches for their causes in the form of a general statement such as "every S is P," whereby S also includes unobserved individuals. "The general statement, i.e. the *rule*, that 'every S is P' explains then why 'S_1 is P,' 'S_2 is P,' etc."

Another way of providing explanations in the empirical sciences is to develop hypotheses, which Łukasiewicz understands to mean statements about facts "which we do not find directly in our experience, but which in connection with some general statement explain facts that our experience does supply. For instance, someone observes that some 'S is P,' but doesn't know why. Desiring to find an explanation, he assumes that 'S is M,' even though experience does not supply that fact. But he knows that 'every M is P'; so if he assumes that 'S is M,' then those two premises will allow him to assert that 'S is P'. The hypothesis here lies in the statement 'S is M,' which with the help of the rule 'every M is P' explains why 'S is P'."

Scientific theories in the empirical sciences are sets of rules and hypotheses gained through reductive reasoning. A particularly interesting and relevant quality of re-

---

[6] Ibid., XXIV.

ductive reasoning is that the cause comprises something more than experientially given effects, and that that excess comes not from experience but is created by the human mind. Each rule gained through induction, each hypothesis based in reductive reasoning, contains creative elements. Thus through creative effort we gain new information whose truthfulness we cannot ascertain, since, as we know, the truthfulness of the effect does not necessarily imply the truthfulness of the cause. The mind gains *new* information with the help of reduction, but at the cost of its *certainty*.

Thus developed rules and hypotheses are tested through deductive reasoning, by trying to find certain effects for uncertain statements. But testing does not lead to certainty, since the truthfulness of effects does not imply that the reason is true; that is because the truthfulness of the effect need not be related to the truthfulness of the cause. The truthfulness of the effect would only need to be related to the truthfulness of the cause only if all effects were known, which never occurs in research on the facts of reality.

Scientific creativity is not about deducing effects from given causes, about inference in the form: if "every *S* is *M*" and "every *M* is *P*," then "every *S* is *P*". According to Łukasiewicz, such deductive reasoning, which goes from cause to effect, is a *mechanical* operation of the intellect. Creative reasoning is reductive reasoning, where "the relationship of entailment does not progress in the direction from effect to cause; there is no logical node; the mind must perform a *leap* in order to find the cause for a given effect".

The role that deduction and induction play for scientific creativity has already been very aptly described by Whewell[7], who argues that while deduction proceeds methodically and step by step, induction takes place in leaps that lie beyond the scope of method. *Ex post*, an outcome of induction must therefore be verified with the help of deduction.

Deductive reasoning, proceeding from cause to effect, is a mechanical operation of the intellect, while reductive reasoning moves from the effect to the cause. Lacking a logical node, reductive reasoning forces the mind to make a leap in order to find the cause of a given effect. This is why we must agree with Mach's statement: "Logic does not offer new insight."[8] Syllogism and induction, according to Mach, only ensure that the mind does not become involved in contradictions, they afford protection from errors and serve to establish relationships between statements. The actual source of cognition lies elsewhere. The researcher's task is not only to classify but also to identify characteristics and relationships between them, which is much more difficult than classifying things that are already known.[9] This or that relationship be-

---

[7] William Whewell: *The Philosophy of the Inductive Sciences, Founded Upon Their History* (vol. 2), London 1840, 92, cited in Ernst Mach: *Erkenntnis und Irrtum: Skizzen zur Psychologie der Forschung*, Leipzig 1906, 318.

[8] Mach: *Erkenntnis und Irrtum* (footnote 7), 314. ||| Translation by T.B. The Polish original quotes Mach in German: "Die Logik liefert keine neuen Erkenntnisse."

[9] Ibid., 312.

tween elements catches our attention and becomes conceptually fixed. A result produced in this manner represents cognition if it can be upheld in the face of other results; if not, it is an error. The basic process to gain new cognition is also the (sometimes involuntary) process of abstraction, of ignoring negligible elements, by which the individual case takes on the character of a more general one, one that represents numerous singular cases. Finally, a fundamental role falls to the imagination, as Mach observed very clearly.

This is an extremely general overview of the question of creativity in the empirical sciences seen from the perspective of logical analysis. In the formal sciences, especially in logic and mathematics, the problem of creativity takes on a different form. Here, too, we can draw on Łukasiewicz's argument. The formal sciences are deductive. Their points of departure are not individual opinions about facts, but certainties, such as the principle of contradiction or the theorem of two quantities being equal to a third, which are general and obvious statements.

The most common form of reasoning in the formal sciences is inference, a form of deductive reasoning. Another kind that occurs frequently is justification, which is about seeking certain and already causes for given uncertain statements. "Formal reasoning," Łukasiewicz writes, "does not bring about creative elements. However, the entire field of those sciences is creative, as they are not based on experience and are not about reproducing facts".[10] The subjects of the formal sciences are *ideal* constructions. In the world of real subjects there exist neither figures of non-Euclidean geometry, nor dimensionless geometrical points, nor one-dimensional lines, integers, fractions, irrational or imaginary numbers. Creativity in the formal sciences is therefore about producing ideal constructions in the mind, and about connecting, through relationships of implication, statements that make some judgment about those ideal constructions.

It would be misguided to believe that there are no formal elements in empirical science. The fact alone that all science originates in and develops thanks to reasoning based on logical axioms, that is to say that it is based on ideal logical constructions, indicates that all scientific theories are imbued with some idealism and creativity.

At the end of his interesting disquisition Łukasiewicz reflects on the questions of truthfulness and fallacy in scientific statements. He concludes that in science there are two kinds of statements: 1) individual statements reproducing facts that can be experienced (reproductive statements), 2) individual and general statements that express truth and hypotheses. The former are true. About the latter we cannot say with certainty that they are false, but there are no grounds to deem them truthful either, since we usually do not know if reality corresponds to them.

For Łukasiewicz, this line of reasoning suggests that truth does not represent the sole goal of science. "The goal of science is to construct theories that satisfy universal intellectual needs."

---

[10] Łukasiewicz: "O nauce" (footnote 4), XXXIV.

The scientific creator who formulates laws and hypotheses usually has no way of gauging the truthfulness of his statements or of knowing if they correspond to real life. If he decides to incorporate his statements into the system of science, it is because he has a subjective feeling that they are correct. He believes that these statements are not incorrect, he is convinced that to some extent they satisfy our curiosity and broaden our understanding of reality, that they lead to valuable practical consequences, that they broaden the sphere of human action, or at least that they facilitate action. This pragmatic point of view finds application not only in the empirical sciences but also in the formal ones, e.g. in mathematics, as Poincaré has shown with respect to Euclidean geometry. The question whether Euclidean geometry is real has no significance according to Poincaré:

> One geometry can not be more true than another; it can only be *more convenient*. Now, Euclidean geometry is, and will remain, the most convenient:
> 1) because it is the simplest...
> 2) because it accords sufficiently well with the properties of natural solids...[11]

So we see how the criterion of convenience, a pragmatic criterion, can be of use even in the formal sciences, especially when the criterion of truthfulness cannot be applied. The pragmatistic tendencies of contemporary science are expressed, though in a somewhat exaggerated manner, by Müller-Freienfels. While for Łukasiewicz, universal human needs are basically intellectual needs, Müller-Freienfels considers the needs of the mind and soul in the broadest sense. He argues that the question of satisfying those needs is not about the compatibility of ideas with the world of objects, but that it is a question of life values, of emotional satisfaction and of practical applicability. Man will always substitute the slogan, "Pereat vita, fiat veritas" by "Pereat veritas, fiat vita"[12] (let truth perish, long live life).

The origin of the creative scientific drive still remains steeped in darkness. Psychoanalysts (of the Freudian school) generally believe that the origin of the researcher's curiosity lies in infantile sexual curiosity. The researcher's curiosity is seen as a transformed, sublimed sexual curiosity. M. Chadwick writes:

> Sexual curiosity [...] can therefore be seen as the strongest libidinal source for a later drive to knowledge – it only depends on whether the original instinctual drive is retained, or repressed and substituted by neurotic Symptoms, or successfully sublimated.][13]

---

[11] Henri Poincaré: *Nauka i Hypoteza*, transl. by Maksymilian Henryk Horwitz, Warszawa 1908, 47. ||| This translation follows *The Foundations of Science*, transl. by George Bruce Halsted, New York and Garrison, NY 1913, 65.

[12] Richard Müller-Freienfels: *Das Denken und die Phantasie*, Leipzig 1916, 306. ||| The Polish original gives a translation of the Latin phrase.

[13] Mary Chadwick: "Über die Wurzel der Wissbegierde," *Internationale Zeitschrift für Psychoanalyse*, vol. 11/1 (1925), 54–68: 58. ||| Translation by T.B. The Polish original quotes in German: "Die Sexualneugierde [...] dürfen wir demnach als die stärkste Triebquelle für den späteren Wissenstrieb ansehen – es hängt nur davon ab, ob die ursprüngliche Triebanlage beibehalten oder verdrängt und durch neurotische Symptome ersetzt oder erfolgreich sublimiert wird."

Science would therefore be rooted in the sex drive, whose roots reach back into the child's sexual curiosity. In our opinion the problem with this approach lies in the observable fact that in the child's earliest years, even before the onset of sexual curiosity, the child's curiosity is turned towards a great variety of objects that have no sexual colouring.

However, we must of course admit that scientific creativity and the entire field of scientific interest often exists in a very close relationship with sexual tendencies. This can be seen in the examples cited by Jones, where unconscious sexual complexes have influenced the choice of research topics.[14] Of course it is always possible to construct a relationship to infantile sexuality in such cases. As for other cases, the issue at least remains open to question.

The new opinions that enrich the scientific system, be they certain or uncertain, are of different value to science. Some studies move in tracks made by previous researchers – ruts so deep that it would be difficult to get off track and stray. From time to time, the researcher who moves in those tracks will snatch some unknown little flower along the way, one that his predecessors overlooked. How different is this researcher from one who strides across the huge expanse of science! Such a researcher maps new directions for research, provides the means and methods for their implementation, and outlines the farthest conceptual perspectives, leaving traces on which innumerable less creative successors will tread. This is because there are innumerable degrees of creativity that (at least theoretically) can be arranged in a sequence from the lowest, where we can barely talk about scientific creativity, to the highest, which we call ingenious creativity.

Let us examine this more closely.

II. Degrees of Scientific Creativity

There is no express need to defend the notion of different degrees of scientific creativity. Arranging them in some hierarchical order, of course, presents considerable challenges. It is especially difficult to draw a dividing line determining where we can begin to speak of scientific creativity. Another circumstance that further confounds this hierarchy is the fact that many products, whether they were painstakingly put together or written with great ease, do not flow from real scientific creativity, even though their authors noisily advertise them as scientific works.

A scholar's work is scientific only when it contains an element of creativity. The term "scientific" ought to be reserved for such articles, discourses and products that are the results of creative work. Although this is frequently done, we should not describe as scientific an ordinary article from some book, or a seminar paper that com-

---

[14] Ernest Jones: "Unbewusste Wahl wissenschaftlicher Untersuchungen," *Zentralblatt für Psychoanalyse*, vol. 1/4 (1911), 166–167: 166.

pares the results of a few scientific treatises, or a journalistic feuilleton that discusses scientific issues that are still a mystery to the general public although they are well known in scientific circles. Even a handbook is not scientific unless it displays special creative traits in its presentation of the state of a given field, or in its synthetic account of the material. Merely to accumulate encyclopaedic material and to present it according to an accepted traditional form does not amount to scientific work. Unfortunately we often come across projects that are mere compilations, mosaics composed of tiny stones collected from a greater or smaller number of scientific treatises and works, where the language blurs the origin of borrowed statements, where, therefore, it isn't statements that have been copied, but ideas. To copy statements is to expose oneself to the accusation of plagiarism and to run afoul of copyright law, whereas copying ideas usually goes unpunished. Works of this kind should carry the title of "arrangements" (as in music). But the compiler erases his traces, like a careful Indian, and sometimes even acquires the reputation of an erudite. Another characteristic is common among such arrangers: the tendency to dazzle the reader. One form that reflects this tendency is when the arranger cites the opinions of various researchers and then presents (or even opposes) his "own" opinion, though in reality that opinion is so hackneyed that it has become the common knowledge in scientific circles, a nameless truism that no one would lay claim to anymore. To these arrangers who drape themselves in the togas of scientificity and originality we can apply Nietzsche's sentence: "When grasped they puff out clouds of dust like sacks of flour, involuntarily; but who would guess that their dust comes from grain, and from the yellow bliss of summer fields?"[15]

Graphomaniacs are intellectually related to those "arrangers". The ease with which they write and the greater freedom of their imagination possibly represent the only difference between them and the "arrangers" who work by the sweat of their brow. Mentally and ethically, these graphomaniacs approach the so-called inventors who continually besiege patent offices with their worthless ideas; as Rubczyński remarks, graphomaniacs are pathologically convinced that "their written works, even short-lived articles, contribute something new and useful to scientific or imaginative literature."[16]

Of course the painstaking, dogged work of the "arranger" and the easy, manic work of the graphomaniac, both lie entirely outside the confines of scientific creativity.

But it is easier to tell what does *not* belong to the field of scientific creativity than to determine where the first, even the weakest, glimmers of that creativity occur. We all know projects (e.g. doctoral theses) that contain valuable ideas and propositions – ideas that the professor came up with and the student realized. The project is then

---

[15] Friedrich Nietzsche: *Tako rzecze Zaratustra*, transl. by Wacław Berent, Warszawa 1905, 151. ||| This translation follows *Thus Spoke Zarathustra*, ed. by Adrian Del Caro and Robert Pippin, transl. by Adrian Del Caro, Cambridge 2006, 89. [T.B.]

[16] Witold Rubczyński: *Filozofja życia duchowego* [The Philosophy of Mental Life], Poznań 1925, 445.

published under the student's name, and its results will lead many readers to judge the creative power of the person whose name figures on the title page. If every project included a "biography of creative thought," as A.B. Dobrowolski proposed in the first volume of Nauka Polska[17] then we would know the exact relationship between the work and the author, instead of using the work's results to draw conclusions, often wrong ones, about the author's creative mentality. But Dobrowolski's project, no doubt significant for the psychology of scientific creativity, does not stand a chance of becoming a matter of course among broader circles of creative scientists. This is for the simple reason that creative people have such a strong inner orientation towards the objects they study that the idea of dividing their mental energies between the studied objects and the task of controlling the creative process one would strike them as plainly undesirable.

We are not going to further concern ourselves with these intimate accounts of the origin of numerous, usually lesser, works. Smelling out the irregularities of a work's origin, a Holmesian chase after its spiritual father, is neither pleasant nor does it benefit science. To adorn oneself in borrowed plumes is doubtless both immoral and demoralizing, even if it is widespread among scientists. But we are not concerned with that. Let us assume honesty and examine individual scientific products for the degree of scientific creativity manifest in them.

Rubczyński lists several assessment criteria for the degree of scientific creativity.[18] Every scientific project – we use the term "scientific" in the sense specified above – contributes something new. But that novelty is not the same across all projects. First, let us consider only the novelty of the results, not the novelty of the methods that lead to those results. It is apparent that some research results, though new, have nothing unexpected about them, but are marked by a strong resemblance to results already known. Such projects do not open up new horizons, nor do they suddenly shed bright light onto a previously obscure field of knowledge. They have their value as modest scientific contributions, but *per se* they do not constitute evidence of some greater creative power. The situation is different when an author submits an entirely new result or uncovers facts that had been quite unknown before. The novelty of those facts inspires many other researchers to explore further, which is how they cause a considerable expansion of the scope of our knowledge. Like a comet, a truly new scientific result is often followed by a long tail of ever-fainter ones with an ever-smaller degree of novelty, until it finally loses its power of attraction and ceases to inspire further independent projects. When Boll discovered visual purple in rod photoreceptor cells of the retina of frogs and rabbits in 1876, his result was significant and highly novel within research on the physiology of the eye. Along with Boll's theory that visual purple decays when exposed to light but regenerates in the dark, this result produced strong and long-lasting impulses for further research. But the degree of novelty of

---

[17] Cf. ▸ Antoni B. Dobrowolski: "The Urgent Need for Mental Education in Poland" 1918, 298.
[18] Rubczyński: *Filozofja życia duchowego* (footnote 16), 443.

those later studies continually decreased. Kühne remarked that visual purple is dissolved in bile acids and bile salts, thus revealing some chemical reactions of that pigment. Others simply tried to ascertain whether visual purple is also found in other animals (not only frogs and rabbits); eventually visual purple was discovered in all animals that have rod photoreceptors in their retinas, except pigeons and chickens. These studies have a decreasing degree of innovation compared to Boll's discovery. Based on these studies, visual purple came to be understood as a fundamental component in the process of seeing, until further studies eventually toppled it off its pedestal – this is the usual course of scientific hypotheses and theories.

What I said about the novelty of results applies even more to the novelty of method. The discovery of any fact, such as visual purple in the example analysed above, can promote many further discoveries, which would be less remarkable in terms of their novelty. But for science the discovery of a new method is usually more effective, as it entails the discovery of many facts in all the fields of science to which that method is applied. With a new method in hand, the researcher can arrive at numerous results of different degrees of novelty in all fields of science to which that method is applicable. We feel that method is the vehicle that takes us to new results and we understand the tendency, common among scientists, to consider creative effort that leads to the invention of a new method to be greater than the creative effort of attaining a new result by applying an established method.

Perhaps Fraunhofer's discovery of the so-called Fraunhofer lines in the solar spectrum was in itself no more significant for science than Boll's discovery of visual purple. It is true that Fraunhofer observed a fact that was fascinating and puzzling within the scientific framework of his time, that researchers took a lively interest in it, and that it became a strong incentive to study and explain it. But it is only thanks to Kirchhoff and Bunsen's method of spectrum analysis that physics, and with it many other sciences, gained new fields of study, where research continues until today even though legions of scholars have been exploring them for decades. It is not up to us to track the effectiveness of this simple and ingenious method. Suffice it to say that with its help the inventors themselves discovered two new elements, namely Caesium and Rubidium, while Crookes discovered Thallium, and other researchers identified Indium, Gallium, etc.; even the existence of Helium was established in the Sun fifteen years before it was discovered on Earth, and soon this element came to play the key role in the theory of radioactive bodies. We also know what significance spectrum analysis attained in astronomy, in determining the constitution of the Sun and even of the most distant fixed stars. As I write these words on a day of solar eclipse (29 June 1927), one idea springs to mind: it is the method of spectrum analysis that has made it possible to reveal the bright lines in the spectrum of the Sun's protuberances, and to verify the hypothesis of the Sun as a burning white hot mass.

The discovery of a fact, even a most weighty one, is something finite, something closed within itself. A new method carries thousands of embryos, innumerable possibilities that are realized gradually, leading to ever-new scientific results. Every new method – as long as it is legitimized by its effectiveness in some scientific discipline – is alive, is unbounded in its sphere of influence, and we cannot foresee its end. If its place is taken by other, newer methods, this is not because all its possibilities have been exhausted, but because other methods turned out to be more effective or advantageous in terms of the economy and technicality of scientific work. There is nothing more important in science's struggle for the conquest of reality than finding new, ever more effective methods. They are the fuel that sustains the flame of scientific creativity.

Similar to the above-mentioned hierarchy of results, which depends on how important these results are for science, i.e. what they suggest about the soaring of creativity, drawing on the criterion of effectiveness we can also observe a hierarchy of method. That criterion will never be more than relative, since every method potentially contains countless possibilities that might be realized in the future. We can assume that a greater degree of creativity is involved in developing a method that shows significant effectiveness in scientific practice and opens up vast horizons further along than there is in developing a method that is less effective and does not reach as far into the future.

A fact, once discovered, is a thing of the past, while method lives on in the present and constantly grows into the future.

The discovery of new facts and the development of methods that lead to such discoveries of new facts represent only one aspect of creative scientific efforts. Scientific creativity reaches towards other goals, too, namely towards formulating hypotheses and scientific theories. Here the scientist's creative imagination has a lot of room for manoeuvre, which is why varying degrees of creativity must make themselves felt very strongly. To issue a hypothesis is to construct a cause for the effects observed, but without succumbing to immediate observation. We see a greater deposit of intellectual creativity in a hypothesis that connects a great number of manifold facts, especially if those facts had appeared to be distant from one another until the hypothesis connected them by the thread of similar causes, and also if the hypothesis agrees with the entirety of our knowledge. If someone returned today to the ancient hypothesis of vision, according to which each object emits tiny images of that object that enter the eye, he would be ridiculed for uttering statements so incompatible with the entirety of our knowledge. The hypothesis of *horror vacui* appears worthless to us, and yet it might have endured to this day if vacuum had not been found anywhere in the world.[19] The hypothesis of air pressure that rightly took its place immediately created a clear and harmonious connection with facts that were well known in physics, and it supplanted the bold hypothesis of fear of empty space. Not all aspects of Huygens's wave theory were compatible with the state of knowledge at the time, and its verification showed major gaps. And yet it became a point of departure and a point of reference for Young's and Fresnel's research.

---

[19] Mach: *Erkenntnis und Irrtum* (footnote 7), 246.

This suggests at the same time how difficult it is to judge the value of a hypothesis, since even such an important criterion as compatibility with the current state of knowledge (as e. g. in the case of Huygens's hypothesis) can fail. This is why the assessment of the degree of creativity is both difficult and uncertain, and must lead to different outcomes depending on what criterion is given emphasis (is it the simplicity and understandability of the hypothesis, is it its unusualness, or its compatibility with the current state of knowledge, or what range of phenomena it explains, etc.). What is more, there exist hypotheses that are generally considered to be the efflux of enormous creative power, even though in some respects it simply boggles our carefully fashioned and strictly reasoned understanding of the phenomena and processes of nature. This is the case for instance with the force of gravity, which in an instant acts across enormous astronomical distances, or with ether, to which are ascribed qualities that are not encountered together in any known body. And yet, in those cases the hypothesis is a bold idea that expresses an unusual creative power.

Similar considerations could be made about scientific creativity, a topic that is manifest in scientific theories. Viewing theory as a system of duly justified hypotheses and laws, we could easily demonstrate that the most creative theories are those which are simplest in their entire construction and which – despite the simplicity of their construction – span the greatest range of multifarious facts and dependencies among facts, that show the greatest novelty in relation to previous theories, that are most compatible with other well founded scientific statements, etc. A classic example of such a theory is Newton's theory of gravitation.

All the aspects we just considered separately tend to be structurally combined in the work of the great exponents of science. Great creators are known for their ability to cut new paths, that is to say to create new methods which then allow us to uncover series of new concrete facts and relationships between facts; they are able to express multiform reality in ever more general concepts. They do this by identifying common characteristics in objects where their commonality had been overlooked; by observing previously undetected similarities and differences, by formulating laws that apply to facts and their interrelations, by creating hypotheses and by joining all this together in scientific theories and systems.

We have now arrived at the highest degree of scientific creativity – ingenious creativity. Many psychological studies have addressed the abilities of geniuses, but so far they have not succeeded in identifying the essential mental springs that raise the pressure under which works of genius are produced. Francis Galton made considerable contributions to the study of individuals of the highest ability.[20] He observed that schooling is of secondary importance to them. They learn in every fleeting situation in a manner that is completely incomprehensible to others. We could leave them al-

---

[20] Francis Galton: *Hereditary Genius: An Inquiry into Its Laws and Consequences*, London 1869. ||| The original text gives the English title but cites the German edition *Genie und Vererbung*, transl. by Otto Neurath and Anna Schapire-Neurath, Leipzig 1910.

most entirely to their own devices, giving them very little unnoticed direction. They proceed along the line that nature dictates to them, and often show a simply unbelievable energy in overcoming both internal and external difficulties. An excellent example, according to Galton, is Jean-Baptiste le Rond d'Alembert, who is not only one of the greatest mathematicians but also distinguished himself as an outstanding philosopher. An illegitimate child and foundling taken in by the wife of some glazier who made light of his desire to study, mocked at school, d'Alembert, despite these life obstacles, did not break down intellectually, not even when it turned out that his first scientific discoveries had previously been realized by others. None of these bitter disappointments managed to turn him away from science. On the contrary, he overcame outrageous fortune and at the age of 24 was elected into the Académie.

The fact that an uninterrupted and regular course of study is relatively unimportant for people of the highest abilities is also apparent in the fact that a great number of ingenious individuals can be listed who did not study in their youth, and yet came to achieve exceptional results in later years. Julius Caesar Scaliger, an outstanding physician and naturalist of the sixteenth century, earned his living as a soldier until the age of 29, then lived a transient life for many years, continually changing his profession and subjects of interest. It was not until he was 47 years old that his first publication appeared, and from that moment on his scholarly career evolved very smoothly. A trait that is characteristic of ingenious individuals is expressed here: the plasticity of their mental organization. Claparède wrote of the ordinary, average person: "Adult age is crystallisation, petrification; the aim of infancy is to defer as long as possible that moment when 'being,' losing its aptitude for 'becoming,' congeals, takes its definite form, like a piece of iron which the blacksmith has allowed to grow cold."[21] In a normal person the years between the age of 15 and 22 are above all a time of moving forward, where their impressionable nature absorbs information and is sensitive to the cultural environment's faintest stimuli. But in supernormal individuals, all years of their lives resemble most people's seven youthful years. Recently the Marburg-based psychologist E.R. Jaensch emphasized the similarity between the personality structure of the youth and that of the creative person.

Jaensch argues[22] that even in the most exact sciences, creative thinking is closer to the mentality of the artist and the youth than one might assume based on a logician's perspective on scientific creativity. In scientific creativity we can observe the same identification with the subject of study, the same merging of object and subject, as among youths and artists. The process of creative thinking takes place quite different-

---

[21] Édouard Claparède: *Psychologia dziecka i pedagogika eksperymentalna*, transl. by Franciszka Baumgarten, ed. by Florian Znaniecki, Warszawa 1918, 102. ||| This translation refers to: *Experimental pedagogy and the psychology of the child*, transl. by Mary Louch and Henry Holman, 2nd ed. London 1913, 147.

[22] Erich Rudolf Jaensch: "Die Eidetik und die typologische Forschungsmethode," *Zeitschrift für pädagogische Psychologie, experimentelle Pädagogik und jugendkundliche Forschung*, vol. 26 (1925), 37–55, 202–219, 236–257: 215. ||| For a contemporary English translation see Erich Rudolf Jaensch: *Eidetic Imagery and Typological Methods on Investigation*, transl. by Oscar Oeser, London 1930.

ly from how it is described in textbooks of logic, a discipline that portrays the outcomes of thinking as ordered and resulting from one another. H. Poincaré's personal account, where he analyses his own creativity, corroborates this, as does D. Hilbert, who outlines the allure of mathematical research.

But Galton's views have met with opposition from a great number of psychologists.[23] It has been observed that the environment in which a highly gifted individual is raised often determines the appearance and development of his ability. It has also been stated that the examples on which Galton's argument is based are not always sound. The American researcher L.F. Ward has demonstrated that Galton occasionally drew on dubious biographical material. For instance it turned out that d'Alembert had not been abandoned by his family, that he received a rather good education and that his father left him an annuity of 1200 livres, which was no meager sum in the eighteenth century.

Another American critic of Galton's, Cattell, asserted that a child born in the states of Massachusetts or Connecticut is fifty times more likely to become a scholar than a child born in the south-eastern shores of Georgia or Louisiana. Cattell recognizes of course the factor of individuality, but he also puts a very strong emphasis on the influence of a child's upbringing.

Very valuable for our purposes are the studies performed by Odin, which conclude that 90 % of illustrious writers were born in cities and in their early youth came to great intellectual hubs; 90 % came from rich or wealthy families; 98 % have enjoyed a good education.[24]

In his article "The Conservation of Talent," the American psychologist Lewis M. Terman also suggests that Galton's view is rather too optimistic.[25] Terman argues that geniuses have often found themselves in the profession that allowed them to apply their ingenious abilities only by chance, which is why much genius must have gone wasted.

In his *Anthropologie in pragmatischer Hinsicht abgefasst*, Immanuel Kant devotes a few remarks to the abilities of geniuses. Those remarks have lost none of their value. He claims frankly: "Now the talent for inventing is known as *genius*."[26] But Kant makes a distinction between the invention and the discovery. Discovering a thing we

---

[23] Solid criticism of Galton's views can be found in Georges Poyer: *Les problèmes généraux de l'hérédité psychologique*, Paris 1921, 196–205.

[24] Alfred Odin: *Genèse des grands hommes, gens de lettres modernes* [Genesis of Great Men, Men of Modern Letters], Paris 1895.

[25] Lewis M. Terman: "The Conservation of Talent," *School and Society*, vol. 19 (1924), 359–364 (cited in Otto Lipmann: "Experimentelle und statistische Untersuchungen zur genetischen pädagogischen Berufs- und Arbeitspsychologie" [Experimental and Statistical Research on Genetic Pedagogical Professional and Industrial Psychology], *Zeitschrift für angewandte Psychologie*, vol. 24 (1924), 270–289: 278–280).

[26] Immanuel Kant: *Anthropologie in pragmatischer Hinsicht abgefasst*, Königsberg 1798, 159–162. ||| This translation follows *Anthropology from a Pragmatic Point of View*, transl. and ed. by Robert B. Louden, Cambridge 2006, 119–120. The direct quotation is on p. 120 in the translation, and on p. 160 in the original: "Nun heißt das Talent zum Erfinden das Genie."

assume that it existed before, even if it was unknown, just as Columbus discovered America for example. But to invent a thing is to create something that did not exist previous to the creator's design. To be worthy of being called ingenious, this creation must be original, which is to say it must be neither a simple construction nor an imitation of something that already exists; it must exhibit new and previously unknown values; it must also be exemplary, which is to say it must be a model that fully deserves to be emulated. "So a human being's genius is 'the exemplary originality of his talent.'"[a]

What is creative in people of genius is above all their imagination. Works of the imagination can be original thanks to the fact that the imagination is freer than other mental faculties. But even the imagination must, according to Kant, be subject to certain rules; notably it must ensure that the product matches the idea from which it originated. A product stripped of all rules, one whose presentation, according to Kant, contains no "truth," can be original, different from previously known works, full of subjective innovation, but it can never be a work of genius.

III. Categories of Scientific Creativity and Types of Scientific Creators

We have discussed the degrees of scientific creativity and analysed the products of the creative mind. All scientific products, however, come into being thanks to the mental factors that produced them. This is why we can draw some inferences from the creator's product to reach conclusions both about the activities behind its development, and about the abilities that must have existed in order for those creative activities to come to pass and for new scientific products to come into being. But to infer from the product something about the psychological substructure based on which that product was realized is only possible to some extent and requires extreme caution. A product of scientific creativity, fixed once and for all after the finishing touches have been put to the manuscript, is a result of the creator's entire personality. It comes into being through a collaboration of all his mental "faculties," based on countless mental acts whose effects no analysis of the product will ever be able to work out, not even as a rough outline. Only a "biography of creative thought" could be of some help. The product does not reveal along what paths the creator's mind proceeded from the moment when the problem appeared as a foggy outline until the moment it was solved. Whatever stages the creator's thought traversed on its way toward the final goal, its traces are usually obscured, and the failures and difficulties of the creative effort are passed over in silence. We know that some projects that impress us with their clear and lucid structure, with the lightness and deftness of their pres-

---

[a] This is Robert B. Louden's English translation (footnote 26), 120. B. provides his Polish translation of the passage and then quotes it in the original German. "Also ist das Genie eines Menschen 'die musterhafte Originalität seines Talents'." [T.B.]

entation, are in reality the fruit of the author's painstaking and prolonged labour. A typical attribute of the product is that as it approaches completion, it begins to tear itself away from the living substructure of the creative personality. Once completed, it severs all ties with that personality and becomes objectivized – a work from which the creator has eliminated as many moments of subjective difficulty as possible, viewing the creative struggle as inessential to the scientific work itself.[27]

And yet, even if the creator views his work most objectively and takes care not to let his work reflect the wave-like, often even broken line of his creative effort, the work's appearance will never obscure all ties that bind him to it. The work will always cast a light, though usually a very faint one, onto the degree of his creative effort as well as onto his creative practice and the type of his creativity.

A researcher's scientific talent manifests itself above all in his achievements. That said, Spranger has aptly shown that we cannot treat talent in quite the same way as results, since a given talent does not always lead to corresponding (proportional) results and, *vice versa*, a good result need not always be seen as the outcome of a corresponding talent.[28] There are factors that lower the performance of talent. The work consequently falls short of the natural talent. Sometimes this is due to internal factors such as laziness, an inability to concentrate, a lack of confidence in one's own faculties, and sometimes the cause lies in external factors such as illness, poverty, malnourishment, gaps in education or other factors that do not stem from the organization of the mind. Spranger points out that some factors can also cause the work (*Leistung*) to outgrow natural talent. These factors, too, can be either internal or external. The former include e.g. assiduity or ambition, the latter wealth, background, education, good results. Thus scientific talent always presents itself in combination with a great variety of factors that form a unique structural whole in each individual. This is why a scientific work, being an extremely complex whole, is always the product of an entire creative personality. It is obvious that from one creator to another, different factors take centre stage within this structural whole. For one individual, the dominant factor is a capacity for precise observation, resulting in characteristically accurate and confident descriptions; for another what comes to the fore is an ability to grasp similarities and differences (there are two types: one more focused on similarities, the other on differences); a third again has an unusual ability to generalize, and so on. Spranger is right to argue that the one essential function that permeates all processes of scientific work and unites all types of this work is *thinking*. To be sure,

---

[27] Hermann von Helmholtz for instance writes in this manner: "In meinen Abhandlungen habe ich natürlich den Leser dann nicht von meinen Irrfahrten unterhalten, sondern ihm nur den gebahnten Weg beschrieben, auf den er jetzt ohne Mühe die Höhe erreichen mag." ("Erinnerungen: Tischrede gehalten bei der Feier des 70. Geburtstages, Berlin 1891," in: id. *Vorträge und Reden* (vol. 1), Braunschweig 1903, 1–22: 14. ||| "In my works I naturally said nothing about my mistakes to the reader, but only described the made track by which he may now reach the same heights without difficulty." Leo Koenigsberger: *Hermann von Helmholtz*, transl. by Frances A. Welby, Oxford 1906, 181.

[28] Eduard Spranger: *Begabung und Studium*, Leipzig 1917, 15–27.

thinking is not the exclusive domain of science. But scientific thinking has a unique colouring, since it concerns a conceptual grasp of reality for purely cognitive purposes. As we saw in the introduction, not all new opinions enter into the scientific system, but only those that represent real progress for our conceptual grasp of reality as compared to the scientific system already in existence.

Having isolated thinking – or, to put it more precisely, *reasoning* – as the fundamental function in any creative scientific work, we can begin to try and identify categories of creativity and creative types. It has long been known that thinking takes a more synthetic and combining form in some individuals, and a more analytical and critical form in others. Here we make a clear distinction between two types that probably never occur in extreme forms. On the contrary, each of these types always contains some smaller or greater element of the other. Especially when it comes to the synthetic type, we must assume that his structure necessarily contains an analytic component. Meumann pointed out that the ability to create new and original combinations requires a breaking down of acquired knowledge and commonly accepted into their constituent elements, which will then be combined into new syntheses.[29] But even if every creator of the synthetic type exhibits analytic qualities that are structurally part of the synthetic type, while similarly the analytic creator is unlikely ever to be lacking some traits of synthetic thinking, it is still synthetic mental activity in the former and analytical mental activity in the latter that give a specific colouring to their entire creative output. Only in creators of the highest order do we observe a perfect harmony between the synthetic and the analytic way of thinking.

[...]

IV. The imagination, memory, erudition, the subconscious and their roles in scientific creativity

I. *The Imagination.* – It is beyond doubt that the imagination plays a huge role in scientific creativity. Only people who make no distinction between the concepts of imagination and fantasticality sometimes question this fact. Products of the scientific imagination, even if they are teeming with inspiration, cannot be fantastic. Fantastic ideas characterize pseudosciences such as occultism, alchemy or astrology, out of which, as the history of ideas has shown, true sciences often developed. Even in the most coherent and rationalized sciences, such as mathematics and logic, the imagination provides the ideas that reason then examines and integrates into the system of science. "In a conversation concerning the place of the imagination in scientific work," says Liebig, "a great French mathematician expressed the opinion to me that the greater part of mathematical truth is acquired not through deduction, but

---

[29] Ernst Meumann: *Intelligenz und Wille*, Leipzig 1908, 159.

through the imagination. He might have said "all the mathematical truths," without being wrong.[30]

In *Nauka Polska* Z.L. Zaleski presented a thorough and beautiful study of the imagination in scientific creativity. It seems appropriate therefore to limit our observations to a few key issues, referring the reader to Zaleski's study for further details and the relevant literature.[31]

Wundt, Colozza, Visconti, Ribot, Dugas, Second and other scholars have emphasized the role of the imagination for scientific creativity. But the most important studies, according to Zaleski[32] are found in the works of H. Piéron[33] and E. Abramowski.[34] What Piéron calls superior memory "is already an active imagination, or at least it is the *mechanism of imagination*, well crafted, cleaned, oiled and ready to be set into motion"[35]. The superior memory is characterized by extremely lively memories and a great number of associative connections between those memories, whereby those connections themselves should be as faint as possible. Ordinary memory forces a person to imitate, to repeat; superior memory, meanwhile, multiplies and simultaneously weakens associative connections, thus liberating the mind from the domination of ordinary memory.

But we cannot understand the workings of memory without taking into account the emotional factor, "which sets into motion the apparatus of memory and gives it the impulse – the direction towards new crystallizations"[36]. This is where Abramowski's theory of memory allows us better to understand what significance emotional factors have for the processes of both memory and the imagination. As our mental representations disappear from consciousness, making space for other mental representations introduced by the stream of consciousness, they leave behind an "emotional equivalent," a "nameless feeling" free of all imaginational elements but still perceptible through an act of consciousness. Everything that used to be in the conscious mind slips away and becomes unconscious, while everything that was imaginational loses that quality and is reduced to a nameless feeling. These emotional equivalents of the imagination are stored in the conscious mind; through them we can recall into consciousness those mental representations which we experienced in the past. Thus Abramowski's theory of memory inserts the world of feelings between the domain of images and the domain of the subconscious.[37]

---

[30] Ernest Naville: *Logique de l'hypothèse*, Paris 1895, 27, cited in Théodule-Armand Ribot: *Essai sur l'imagination créatrice*, Paris 1921, 205. ||| This translation follows Ribot's *Essay on the Creative Imagination*, transl. by Albert H. N. Baron, London 1906, 245. [T.B.]
[31] Lubicz-Zaleski: "Rola wyobraźni w twórczości naukowej" (footnote 1).
[32] Ibid., 7.
[33] Henri Piéron: *Evolution de la Mémoire*, Paris 1922.
[34] Edward Abramowski: *Le subconscient normal*, Paris 1914.
[35] Zaleski: "Rola wyobraźni w twórczości naukowej" (footnote 1), 7.
[36] Ibid., 10.
[37] Stefan Błachowski: "Pamięć a świadomość" [Memory and Consciousness], *Przegląd Filozo-*

All that has traversed the conscious mind thus continues to live in the conscious mind in the form of an emotional equivalent. It becomes absorbed into the cenesthetic self as a base of one's general sense of existing. "In this cenesthetic self," Abramowski writes, "everything accumulates that the individual has experienced, but it accumulates in a namelessly emotional form; this is the fundamental non-conceptual content of our self, the deep biological foundation of our character and temperament, in the sphere of thought as well as in the sphere of feelings and actions."[38] Thus Abramowski linked all intellectual processes with the entire biological individuality of a person by the thread of emotionality. Thereby he lifted a hem of the mysterious process of how mental representations and thoughts re-emerge from the subconscious (conceived dynamically and creatively and not as a reservoir of immutable mental representations) through emotional equivalents accumulated in the cenesthetic self back into ego-awareness.

Abramowski was one of the first to highlight the structural unity of memory (and with it, the imagination) and mental life as a whole. After all, the imagination is not, as earlier psychologists believed, an isolated mental faculty that atomizes mental life, but an integral part of a whole that is composed of all mental functions and dispositions. When it comes to the creative imagination in science, it would particularly be wrong to separate that imagination from thinking. The creative imagination, as Müller-Freienfels has correctly observed,[39] has a teleological character, due to which it appears as a form of productive thinking. The creative imagination must be distinguished from the playful imagination (*spielerische Phantasie*), which is active in dreams, feverish delirium, in states of intoxication, in maniacally racing thoughts, etc. It is true that the border between these two kinds of imagination is fluid. How often does the researcher's imagination lead him astray by suggesting ideas whose falsity will be revealed by later testing! There are also scholars of the romantic type, especially in the younger, still rather underdeveloped sciences, who get carried away by their imagination and the deceptive mirages it presents. Other researchers with a more critical and rigorous attitude towards new ideas will expose their work as daydreams, as poetic thought with no connection to reality. That said, the position that official science occupies with regard to new ideas often undergoes radical shifts. The cosmological ideas of German philosophers of the idealist era were taken very seriously for some time, until the flourishing of natural science in the mid-nineteenth century rightfully consigned them to oblivion.

Some of the ideas suggested by the creative imagination are later proven wrong, often following painstaking and prolonged labour. In such cases the imagination has led the researcher astray, and woe betide those who know not how to shake off such

---

ficzny, vol. 16/4 (1913), 484–494: 489. In this text I discuss some of the problems with Abramowski's emotion-based theory of memory.

[38] Edward Abramowski: "Źródła podświadomości i jej przejawy" [The Origins of the Subconscious and Its Manifestations], Warszawa 1914, 146.

[39] Müller-Freienfels: *Das Denken und die Phantasie* (footnote 12), 224.

ideas in good time. That the imagination should lead us off course surprises no one, so much this idea corresponds to our notion of the imagination. Astonishment and wonder are only expressed about those unexpected ideas that later, sometimes after many years of work, turn out to be perfect solutions to given problems. To explain such events we must view them as outcomes of a harmoniously functioning mental organism in which the imagination collaborates closely with a critical approach. This critical approach, moreover, must be engaged at the same time as the imagination, not just afterwards, and it must be creative criticism, rather than a criticism that simply rejects and negates.

We often hear creators say that creative work requires luck, that some ideas are fortunate while others are pertinent and astute but not fortunate.

Subjectively we experience this as if the creative imagination were capricious to give a creator mediocre ideas that might obsess him for many years, even his whole life. I found an interesting observation on this topic in E. Weber's letter to Fechner, dated 12 December 1850. Fechner had sent Weber his work on psychic measurement along with a letter in which he referred to his "fortunate idea" (*glückliche Idee*). Weber replies that he would hesitate to speak of a fortunate idea in this case, since a fortunate idea is one that

> coincides with the discovery of new facts which lend themselves to precise apprehension and which support that idea in a special way. I would call the idea of the wave theory of light as presented by Euler astute and correct, but not fortunate; the same idea, reproduced by *Fresnel* and coinciding with the phenomenon of inference, I would call fortunate. [...] Only through the facts that support them do those ideas really gain a foothold in science.[40]

In general there exist huge differences in researchers' subjective attitudes towards ideas suggested by the imagination. Some easily trust their imagination, not suspecting at all that it might lead them off course, while others are careful. Some immediately have a sense for the scientific value of their ideas, distinguishing more and less important ones, while others do not have such a feeling for the importance of their ideas. Considering that researchers (at least those of a certain type) are extremely tenacious and can remain nailed to their ideas for a long time, it becomes obvious that an erroneous assessment of the value of a product of the imagination can have fatal effects. Richet writes that many inventors – Pasteur, Claude Bernard, Berthelot, Marey – held on to their ideas with admirable tenacity.[41] But tenaciousness must not be blind, since there exist ideas from which we should free ourselves as quickly as possible, with or without regrets about the time we lost on them.

Many have asked themselves if there exists a specifically scientific imagination. For Zaleski[42], the answer to this question must be phrased in very general terms, even if

---

[40] Gustav Theodor Fechner: *Elemente der Psychophysik* (vol. 2), Leipzig 1889, 557 ||| Quotation translated from German by T.B.
[41] Charles Richet: *Le savant*, Paris 1923, 39. ||| This translation follows *The Natural History of a Savant*, transl. by Oliver Lodge, London, Toronto 1927, 40.
[42] Zaleski: "Rola wyobraźni w twórczości naukowej" (footnote 1), 14.

some qualities may represent a special – though not exclusive – predisposition for scientific thinking. Referring to the above-mentioned characterization of memory of the higher order (according to Piéron), Zaleski argues that

> the researcher-creator's memory ought to possess above all an exceedingly sensitive dissociative apparatus [...] In a sense we could also say that if for the artist that essential means of self-expression is to stylize reality according to his own inspiration and emotions, for the scholar such basic activities are the dissociation and simultaneous schematization of the experiment's results. Operating with this moveable material of memory that is split into abstractions and symbols, the scholar constructs, by way of analogy, both theories and hypotheses.[43]

Assuming, as Ribot also does, the existence of a separate scientific imagination (*imagination scientifique*), we can further ask ourselves if individual sciences require specific kinds of imagination. Ribot is certain about that.[44] In his opinion, the scientific imagination falls into a number of types, even genres, depending on the nature of a given science. The mathematician-analyst who operates with the most abstract symbols needs a different imagination than the geometrician. Monge, the father of descriptive geometry, probably had a different imagination than the mathematician who for his whole life dealt with the theory of numbers. The imagination of physicists and chemists must be concrete, since they must constantly use mental representations of the visual, tactile, motor, aural, and thermal types, amongst others. It is not much different in the case of geologists, botanists, zoologists, etc.

II. *The mnemonic (imaginational) type* uses imaginational material in his activities. It has been known for a long time that objects, as we think about them, become present in our consciousness either through visual (or rather acoustic, motor, thermal, etc.) mental representations of those objects, or through words that mostly appear in the visual or acoustic-motor form. Ever since Galton and Charcot, psychologists distinguish various types of imaginational (mnemonic) types, the most important ones being the visualist, audist, and motor types. Psychological studies have shown, however, that so-called pure imaginational types are rare; these rare cases include people who constantly use either exclusively visual or exclusively acoustic mental representations. Usually, one and the same individual will use more visual or acoustic or motor representations, depending on the situation, the object that he thinks about, and his habits. We are mostly interested in the influence that the object that we think about has on the way in which it becomes present in our consciousness. We know that the same person who while thinking about concrete objects uses visual representations, can resort to acoustic representations while thinking about abstract objects, where a verbal approach plays a crucial role. The type of object therefore exerts a strong influence on the manner of mentally representing them. Concrete objects, which we perceive most accurately through the sense of vision, can also be most accurately and most easily recalled into consciousness with the help of re-creative

---

[43] Ibid., 17.
[44] Ribot: *Essay on the Creative Imagination* (footnote 30), 237–238.

visual representations, while abstract objects are best captured in words, elements of speech, which is in the life of every person above all heard and spoken speech, based on acoustic and motor elements, so that we think about abstract things mostly through acoustic-motor mental representations.[45]

The fact that abstract objects are best captured in words was already perfectly clear to Saint-Paul, who aptly observed that the mathematician, thinking $ax^2 + bx + c = 0$ sees, pronounces or hears in his imagination only signs, letters or numbers, and at the same time he thinks through verbal representations.[46] Saint-Paul further argues that it is only in exceptional cases that a mathematician uses no verbal representations in algebra.

Even if the type of imagination that we use in our thinking depends on the objects of our thought, it is beyond doubt that some have a greater tendency to think by using visual representations, other by using acoustic representations; both groups base their thinking on the appropriate visual or acoustic imagination, as using the wrong mental representations would make effective work difficult, sometimes even impossible. In creative scientific work, where the mental organism must function under extreme tension and in full harmony, to force thought to conform to mental representations that are not suited to one's strongest dispositions is to risk slowing down the creative momentum, which might lead to setbacks.

This understanding of visualist and audist types makes it clear that some scholars tend to use acoustic representations of words in their creative work, while others tend towards visual representations of the objects of their thought. Among the first group, the audists, there are those for whom acoustic memory is so important that the idea of thinking without the help of speech strikes them as an impossibility. Max Müller expressed this by saying, "No thought without language", though of course his maxim, which concerns audists alone, is rather too narrow. Among the second group, i.e. the visualists, some are convinced that their entire thinking takes place without the help of words. Francis Galton was one of them.[47]

In the visualist category psychologists distinguish two subtypes, namely one that uses visual representations of objects (this is the one discussed above) and the typographical subtype, certainly less common, where the imagination suggests visual representations of written or printed words and other (e.g. mathematical) symbols. Typographical visualists, as Rignano argues[48], have no need to imagine reality, and think through visually given symbols that represent reality. This type occurs quite markedly among mathematicians. It includes individuals who are fond of symbolic algebra but remain averse to mental representations in geometry. Among analytics,

---

[45] Stefan Błachowski: *Struktura typów wyobrażeniowych* [The Structure of Imaginational Types], Poznań 1924, 50.
[46] Georges Saint-Paul: *Le langage intérieur et les paraphasies (La fonction endophasique)*, Paris 1904, 76.
[47] Eugenio Rignano: *Psychologie du raisonnement*, Paris: 1920, 385–390.
[48] Ibid., 390.

that is to say logicians (according to Poincaré), and especially among formalists (according to Klein)[b] we often experience this phenomenon of avoiding geometry and fondness of algebraic symbols and mathematical logic. G. Cantor, for instance, claimed in Fehr's survey: "I lack the special imagination for geometry"[c], while Hermite said of himself: "I cannot tell you how much effort I must put in to understand anything of the drawings of descriptive geometry, which I detest."[d] Here we should mention the loathing many mathematicians feel towards "the geometrization of the mathematical question". The invention and use of symbols clearly give rise to great progress in science as long as those symbols are applied to concepts and symbolic objects with utmost accuracy, and as long as we remember that symbols symbolize something and that we must go from the signs to that which they mean and signify. Otherwise, as Twardowski argues, we fall into symbolomania and pragmatophobia, which is, presumably, mostly practiced by typographical minds. According to Twardowski, symbolomaniacs

> maintain an unwavering belief in the infallibility of the symbolism they employ. They hold that should the results attained by applying the symbolism diverge from the convictions that are held independently of the symbolism, the latter must yield to the former. Indeed they seldom come to grips with that conflict, for they forget that the symbols symbolize something – occasionally, they even deny it outright. They therefore make no attempt at all to interpret the results shrouded in symbolic garb. The symbols and the operations performed on them, originally the means to an end, become for them an end in itself, an object of ardent affection and the source of great intellectual delight.[49]

---

[b] At this point, B. refers back to a passage from the part of the text that was not included in this edition. In this passage, B. presents Henri Poincaré's description of *logicians* and *intuitionalists* in mathematics (Henri Poincaré: The Value of Science, transl. by George Bruce Halsted, New York 1907, 15–25. He refers to the French *La valeur de la science*, Paris 1911, 11–16) to proceed with Felix Klein, who added the *formalist* to this typology: Felix Klein: "Clebsch," in: id. *The Evanston Colloquium: Lectures on mathematics, delivered from Aug. 28 to Sept. 9, 1893, before members of the Congress of Mathematics held in connections with the World's Fair in Chicago, at Northwestern University, Evanston, Ill.* (reported by Alexander Ziwet), New York 1894, 1–7: 1–2 (B. quotes the Polish *Odczyty o matematyce, miane w Evanston*, transl. by Samuel Dickstein, Warszawa 1899. He then presents Ostwald's distinction of the classic and romantic type (Wilhelm Ostwald: *Große Männer*, Leipzig 1909, 364) and hints at two Polish works on the field: Józef Nusbaum-Hilarowicz: *Uczeni i uczniowie* [Scientists and Students], Lwów 1910, and Jan Tur: *Nauka i uczony* [Science and the Scholar], Warszawa 1917.

[c] *Enquête de "L'Enseignement mathématique" sur la méthode de travail des mathématiciens*, ed. by Henri Fehr in collaboration with Théodore Flournoy and Éduard Claparède, Paris, Genève 1908, 19. Henri Fehr's survey was published in consecutive volumes of the international review journal *L'Enseignement mathématique* between 1905 and 1908. The 11 parts of the survey were finally joined in the single volume quoted here. Although the survey was published in French, Błachowski gave Georg Cantor's quote in German: "Für die Geometrie fehlt mir eine gewisse Vorstellung des Raumes."

[d] B. gives no reference. See, however, *Correspondance d'Hermite et de Stieltjes* (vol. 2), ed. by Benjamin Baillaud and Henry Bourget, Paris, 1905, 41.

[49] Kazimierz Twardowski: "Symbolomania i pragmatofobia" [Symbolomania and Pragmatophobia], *Ruch Filozoficzny*, vol. 6/1–2 (1921), 1–10: 1. ||| This translation follows "Symbolomania and

The symbolomaniac who shuns things themselves, preferring not to think or hear about them, says (and here Twardowski quotes a sentence from Poincaré's *Science et méthode*[50]): "We not only know nothing concerning what these *things* are, but we should not even expend any effort to know anything about them. We have no need for this…"[51]

Symbolomania and pragmatophobia most often occur in mathematics, although they are also observed in other sciences, e.g. the natural sciences, when they use mathematics and its symbolism. Bouasse describes the ideal of the symbolomaniac and the pragmatophobe. Here is Twardowski's translation:

> The mathematician is an apparatus for performing deductions, characterized by an almost complete absence of thought. He cranks the handle of the algorithmic machine. He incessantly derives from this machine truths – or even stupidities – of every imaginable form that he had previously inserted into it. He has highly refined tools at his disposal for carrying out this trivial work. The perfection of these tools liberates him from thinking about what he is doing, and even from thinking at all. He cranks the handle, just as the squirrel does in its cage, the rotisserie's dog in its treadmill, the criminal condemned to *hard labour* in his prison.
>
> Since he lacks intellectual finesse, that is, the most basic common sense, he never examines what he puts into his machine. The results of the chain of reasoning, *always most perfectly consistent*, sometimes manifest a patent absurdity, which grows in direct proportion to the number of terms, so that despite all his non-intelligence it may happen (everything is possible!) that the mathematician suffers pangs of conscience; that, however, is an inordinately rare occurrence. Ordinarily, he does not quit the game until the lamp runs out of oil, both literally and figuratively.[52]

Although Twardowski considers this characterization to be exaggerated, he believes that if the blunt epithets and comparisons were toned down, we would find examples of this type in the real world.[53]

The pathological equivalent of symbolomania and pragmatophobia in the sciences is arythmomania in its many forms, which are completely unproductive as a rule, as for instance the constant and completely aimless permutation of symbols.

But even if the typographic mental representation of symbols apparently plays a major role in these symbolomanic and pragmatophobic tendencies, we must remember that we have no accurate studies yet on the relationship between a researcher's

---

Pragmatophobia," in: id. *On Actions, Products and Other Topics in Philosophy*, transl. by Arthur Szylewicz, ed. by Johannes Brandl and Jan Woleński, Amsterdam, Atlanta, GA, 261–270: 262. [T.B.]

[50] Henri Poincaré: *Science et méthode*, Paris 1908, 156. ||| This translation follows Twardowski: "Symbolomania and Pragmatophobia" (footnote 49), 263. Arthur Szylewicz acknowledges Sarah Cordova for translations from the French. [T.B.]

[51] Twardowski: "Symbolomania and Pragmatofobia" (footnote 49), 263. Italics in the original.

[52] Henri Bouasse: "De la formation des theories et de leur transformation pragmatique," *Scientia*, vol. 14 (1920), 253–268: 264–265. ||| While B. quotes Twardowski's own translation from the French (Cf. Twardowski: "Symbolomania i pragmatofobia" (footnote 49), 4), this edition follows Twardowski: "Symbolomania and Pragmatophobia" (footnote 49), 263–264.

[53] Twardowski: "Symbolomania i pragmatofobia" (footnote 49), 5 ||| In the English version see Twardowski: "Symbolomania and Pragmatophobia" (footnote 49), 266.

imaginational type and his type of work. All we can say about that relationship is merely conjecture based on a number of cases; these are no thoroughly justified claims. Not even the visualism of geometricians has been established precisely, so we must refrain from making categorical statements when we express opinions in this field. Katz even supposes that the elementary types of imagination that psychologists have studied so far have no immediate significance at all for mathematical types and work methods.[54] Until impeccably performed experiments prove that geometricians have an outstanding visual imagination, Katz argues, the question how a mathematician's imaginational type relates to his work method will have to remain unanswered.

Nonetheless, there exist many individual facts that shed some light onto the role of the visual and acoustic imagination in scientific work. Based on them we can – and we should – form an opinion about these issues, even if that opinion will be temporary… just as almost everything in science.

III. *Erudition*. – We often hear professionals say that erudition hampers the imagination, and, by extension, scientific creativity. It appears that they are right, and that we cannot counter with a few examples of polyhisters who were at the same time great creators (e.g. Leibniz), since these rather seem to be exceptions that prove the rule. Of course, reading is indispensable for any scientific work, since science develops organically and grows out of the current system into the future. So it is not surprising that many scholars, as we have seen above, arrive at new ideas thanks to their reading.

What makes the erudite stand out among scholars is an inner need to learn about everything that science has created. He longs to become familiar with the terrain that has already been discovered and prefers to travel in the land of thought with a guidebook in his hand, which must be meticulous and can be boring, like the Baedeker; he is not particularly keen on setting out into the unknown. If despite this he works in science, if a spark of creativity is still alive in him, he will try to read and take into account everything that had been said on the given topic. "This fullness of learning rather paralyzes his initiative," as Richet mildly claims; "for in the immense and confused storehouse of scientific documents, there is scarcely a subject that has not been already broached."[55]

William Ramsay maintained that too much expert knowledge can stand in the way of discovery, while the sailor Cook did not want to take along geographical maps in order to make discoveries on his own.[56] Storing in one's memory an excessive amount of diverse facts impedes their conceptual organization as well as the discovery of unknown relationships between them, while incessantly resurfacing traditional truths impede the search for one's own paths.

---

[54] David Katz: *Psychologie und mathematischer Unterricht*, Leipzig 1913, 66.
[55] Richet: *Le savant* (footnote 41), 42. ||| For this translation see *The Natural History of a Savant* (footnote 41), 43.
[56] Ostwald: *Große Männer* (footnote 48), 364.

It has often been noted that a somewhat limited memory makes it easier to generalize and to create new concepts, and that it forces the mind to apply general concepts, laws, patterns, etc. (see e.g. Helmholtz, Maudsley, Galton, Rignano). Forgetting details, especially less important ones, paves the way for the mind to create concepts that span a great number of objects. Aware that we cannot trust our memory to master too many individual cases, we tend to incorporate them into more general relationships, and in this way a mental type develops that discusses an individual object by becoming conscious of its general pattern, a pattern that includes that given object.

A relatively weak capacity to remember things that have no connection to one another is no rarity among creators. Helmholtz for instance confesses that his weak memory as to unrelated things had made itself felt very early on. Studying languages at school he had difficulties memorizing vocabulary and grammatical exceptions; the same applied to historical facts and passages of prose. Learning them by heart was an ordeal. "This shortcoming, of course, only got worse, and has become the plague of my old age."[57] These words may sound odd coming from an ingenious researcher whose oeuvre spans vast areas of science. But from the psychological point of view we can understand today that his inborn mnemonic disposition destined the author of the *Treatise on Physiological Optics* to become a great researcher. Besides, Helmholtz himself expressed the essential point, claiming that even small mnemonic devices helped him learn and remember. He adds words or great weight: "The most perfect mnemotechnic aid that exists is the knowledge of the law behind the phenomenona."[58]

With the erudite it is different. Victimized by his memory and hampered by that in his creative initiative, when he finally sets out to work, he begins by leafing through the entire bibliography on the subject. This painstaking and usually unproductive work keeps him, however, from the unpleasantness of making discoveries that others had made before – a risk to which true creative minds often expose themselves.

IV. *The Subconscious*. – The subconscious factor in creativity will be discussed in the next chapter, where we will analyse the origin of ideas. Let us therefore content ourselves with a few general remarks. The sudden appearance of scientific ideas after a period of fruitless labour, after we have ceased to be consciously concerned, their appearance at a completely unexpected moment, while we are performing some indifferent task, or early in the morning after we wake up, sometimes even in dreams – all this suggests that a subconscious factor is at play. The creator's mood at such a moment, when he feels the flow of new and unusual ideas, that mood, emotionally charged to the highest degree, is often described as inspiration. According to Ribot, inspiration has the negative quality that it does not depend directly on our will, but

---

[57] Helmholtz: "Erinnerungen" (footnote 27), 6. ||| While B. quotes this passage in his own Polish version, this translation is based on the German source text cited by the author. [T.B.]

[58] Helmholtz: "Erinnerungen" (footnote 27), 7. ||| B. quotes this passage in German. [T.B.]

that we must wait for its appearance; it also has the positive qualities of suddenness and impersonality, as if the thoughts were coming from someone else, as if they were messages from a higher being.[59] Phenomena of inspiration are more often observed in the field of artistic rather than scientific creativity, the latter being dominated by conscious work of the intellect. Most remarkable are those creations that originated during sleep and were written down after awakening. Here are a few examples provided by Bernheim.[60]

Coleridge woke up and felt that he had composed several hundred verses that surged into his consciousness, ready and with no sense of effort. Thus a fragment of fifty-four verses was created, jotted down as fast as the pen allowed. A break in this effort, caused by an hour-long visit, erased the following part. The memories had lost their clarity.

Tartini, having fallen asleep after a vain attempt to finish a musical composition, suddenly saw in his dream the devil offering to finish the piece in exchange for Tartini's soul. Accepting the condition, Tartini hears with absolute clarity how the devil performs the end of the composition on the violin. He wakes up and in joyous exultation leaps out of bed and writes down the entire passage from memory.

But we know that once they are examined under the magnifying glass of conscious reason, the ideas that appear germane and original in dreams oftentimes lose all their brilliance.

For Ribot, inspiration is the result of the subconscious work that occurs in all people, though in some it occurs to an especially high degree. Since that work takes place outside consciousness, we cannot say anything about the essential nature of inspiration. What Ribot is certain about is that inspiration represents either the end of unconscious work – short or long – or the beginning of conscious work.

In *La Logique des sentiments*, Ribot returns to the question of unconscious reasoning, maintaining his earlier statement about the existence of two categories of the unconscious: 1) "the *static* unconscious, which comprises habits, memory and generally all that is organized knowledge" and 2) "the *dynamic* unconscious, which is a latent state of activity, of incubation, of elaboration".[61] Unconscious reasoning, if it exists at all, naturally belongs to the second category. According to Ribot, hypotheses about the essence of unconscious states are liable to serious criticism and do not in fact explain anything. Ribot lists two major hypotheses about the unconscious. The first assumes that unconscious activity is purely cerebral, that is to say it takes place without the psychic factor; the second hypothesis assumes that in the same person there exist "multiple currants of consciousness, only one of which is currently known;

---

[59] Ribot: *Essay on the Creative Imagination* (footnote 30), 51.
[60] Hippolyte Bernheim: *Automatyzm i sugestja*, transl. by Aleksander Arct and Stefan Błachowski, Poznań 1924, 32–35. ||| For the French original see *Automatisme et suggestion*, Paris 1917, 35–38.
[61] Théodule-Armand Ribot: *Logika uczuć*, transl. by Kazimierz Błeszyński, Kraków 1921, 260. ||| For the French original see *La Logique des sentiments*, Paris 1903, 79.

the others, even though they take place in obscurity, do not for that reason change their nature; they remain, essentially, psychic"[62].

However the question of a theoretical approach to the unconscious presents itself, the fact remains that a huge part of creative scientific work takes place outside our consciousness. The conscious life of the moment constantly emerges from the unconscious and we can even agree with Bernheim's saying that all ideas flow from the unconscious. "To think with the intention of creating a piece of work is to concentrate one's attention and to wait for the ideas to emerge"[63].

V. Problems and ideas

The way in which progress with respect to the current state of research takes place is that problems arise in the mind of a scholar and then push towards their solution. Naturally, there are numerous paths towards those solutions. The problem is the question, the idea is the answer. From a psychological point of view, we should be interested in the entire complex process of how problems are formed; we should also be interested in how a researcher attempts to work them out and finally to solve them with the help of an idea that in his view is at that moment the best answer to the question implied by the problem.

Problems ensure that science does not ossify. Phases of stagnation within a given science are always marked by an absence of problems, at least of more consequential ones. Thanks to these problems, the current state of research increases into the future; they represent a bridge between the statements valid at a given moment and whatever the further development of science might bring. They are truly "the arc of the covenant between the old age and the new"[e]. The scholar's creative mind must always embrace a lively consciousness of problems that are consequential for science, a consciousness from which opinions arise that modify the current state of research. All scientific production, even the preliminary function of collecting raw material, is based on scientists' awareness of problems and ability to pose questions. The problem, being an embryo of countless possibilities, might even be more significant to science than the concrete answer rooted in a specific scientific constellation.

All true researchers are students of Socrates, i.e. they proceed, in whichever special area it might be, along the line that leads from ignorance through the desire to know to finding; but even in their awareness of ignorance there lies a secret art of seeing.[64]

---

[62] Ibid., 80.
[63] Bernheim: *Automatyzm i sugestja* (footnote 60), 40. ||| This translation follows the French original. [T.B.]
[e] Błachowski's reference to "the arc of the covenant between the old age and the new" represents a bastardized version of two verses from Adam Mickiewicz's narrative poem *Konrad Wallenrod* (1828). [T.B.]
[64] Spranger: *Begabung und Studium* (footnote 28), 18.

The psychology of formulating, working out and solving problems is still music of the future today. What we are able to provide are but *disiecta membra* of some organism that still awaits its constructor. Within the scarce literature on the subject, Müller-Freienfels's discussions of the problem stand out.[65] He bases his views on those of Richard Avenarius[66], who claims that every mental process is a "vital series" that falls into three parts: the beginning, the middle, and the end.

The beginning of such a vital series is always represented by some content that troubles the mind. That content is something "different," "contradictory," "doubtful," "unexpected," "unusual," "new," "mysterious," "vague," "unreal," "impossible," etc., or at any rate something that arouses anxiety or that threatens the individual's accepted or habitual views and prompts him to react.

The mind, unsettled from its balance, tries to eliminate the anxiety by reacting in such a way as to understand that difference, contradiction, novelty, etc. The middle part of the vital series is the search for such a solution that would eliminate the cause behind the anxiety and tension.

The last part, finally, consists of finding a solution; what occurs here is the reaction that ultimately eliminates the anxiety and tension. The researcher finds what he sought; the strangeness vanishes, doubts and contradictions are solved, what prevails is a feeling of relief, of contentment, of excitement, of freedom.

This general pattern can help enhance our understanding of the psychology of formulating, working out and solving problems – it only needs to be filled in with content taken from analyses of "vital series" of scientific creativity.

So let us begin by having a closer look at the part that is "the beginning," the birth of a problem.

Our ordinary attitude towards the objects we perceive or the words we hear or read is usually such that we just become aware of their existence or simply "acknowledge" them. Normally the things we see and hear do not surprise us or worry us by clashing with what we have become used to knowing. In this state of affairs nothing inclines us to ask ourselves any questions, or to ask them of others, and if we ever do, then those questions are of a conventional nature, superficial and banal, and they do not engage our emotions. Only when something strikes us by being new, when we notice differences and similarities that we had previously overlooked, when something dazzles us with it unusualness or by being a glaring contradiction, when something awakens doubts or wonder, then we begin to be troubled by feelings and by nagging questions that demand to be answered. Sometimes again someone else throws a question our way, and we, once we have understood it, take it on as our own and also try to find an answer.

This is the case in everyday life as it is in the sciences, the only difference being that a true scientific problem must have the potential to lead to an answer that represents genuine progress within the current state of research.

---

[65] Müller-Freienfels: *Das Denken und die Phantasie* (footnote 12), 243–306.
[66] Richard Avenarius: *Kritik der reinen Erfahrung* (vol. 2), Leipzig 1890.

Examining how research problems are formed and what incentives underlie their formation, we can distinguish – as we do in our day-to-day lives – problems that the researcher formulates altogether independently from problems that he adopts from others and subsequently refines more or less independently.

A. Let us take a closer look at the first kind of problem. Even here there exists marked variation. Sometimes a researcher is struck by some detail whose existence he had not suspected in the material of the subjects and phenomena under investigation, or his understanding of that detail is at odds with his current observation. To detect a disparity between reality and the image that one previously had of that reality becomes an incentive, strongly coloured by emotion, to eliminate that disparity, to alter one's understanding in such a way that it would agree with reality, thus eradicating from consciousness the unpleasant sensation of a contradiction.

In other cases the problem has a more abstract beginning. For instance, we know, from previous experience or from the literature, that some factor X impacts a phenomenon Y_1. The question arises if X would not also impact phenomena Y_2, Y_3..., which are similar to Y_1, and next there could be the question of the impact of factor X on the phenomena Z_1, Z_2 ..., which are not similar to phenomenon Y. It is precisely this way of formulating problems and making discoveries in physics that [Czesław Białobrzeski][f] discusses in his "Autobiographical Sketch":

Some researchers, especially early in their careers, tend to make discoveries by testing whether any physical factor might influence a given phenomenon, e. g. if a magnetic field affects fluorescence, or if X-rays affect the spectrum of live steam.

It happens sometimes that experiments of this kind lead to new observations. As a rule, however, one should beware of such simple ideas, unless they are based on well-considered theoretical foundations that augur a positive result.[67]

Oftentimes problems are born not of some incompatibility or contradiction in some unexpected place, but on the contrary, they arise out of a similarity (one that affects perhaps just one characteristic) between things that had previously been considered dissimilar. Such a sudden realization of similarity causes astonishment. It forces the mind to solve the riddle of that similarity between things that had until then been seen as different; it also inclines the mind to search for the reason or cause of that similarity.

A good example of a problem coming together like that – this time from the humanities – can be found in the biographical outline titled "The Trajectory of My Intellectual Development" which [Franciszek Bujak] wrote for *Nauka Polska*:

Preparing a lecture about sources in the history of agriculture I had to familiarize myself with the wonderful edition of the famous *Polyptychum abbatis Irminionis* [the Polyptych of the Abbot Irminon], which contains a very detailed inventory of the possessions of the monastery

---

[f] Complying to the original bibliographical information, Błachowski only used the initials "C.B." for Czesław Białobrzeski and "F.B." for Franciszek Bujak in the following passage.

[67] ▶ Białobrzeski 1927, 309.

of Saint-Germain-des-Prés near Paris from 819 AD, i. e. from the time of Louis the Pious. This inventory lists the names of the monastery's tenants and other inhabitants on its grounds. The editor compiled those names in alphabetical order. Browsing this catalogue of names I was struck by the similarity of some of those names with Lithuanian names I was familiar with, above all Germund and Witold.[68]

Later, looking in the *Słowinik Geograficzny* [Geographical Dictionary], [Franciszek Bujak] found such a large number of Lithuanian and Germanic names that were identical or similar that he was "quite shocked".[69]

The emotional shock of suddenly discovering a similarity between things (names) that had previously not been considered to be similar is clear from the words above, and this is probably why the author kept returning to this problem. But here we already enter the phase of working out the problem, which will be discussed later.

According to [Czesław Białobrzeski]'s remarks from his "Autobiographical Sketch," most researchers find ideas that lend themselves to scientific elaboration by browsing the latest literature in search of topics that their authors had not considered. [Czesław Białobrzeski] does not provide any more details on this effort to find new topics through reading. It probably amounts to such moments as we described above, or similar ones. But the way in which he describes his personal method of identifying problems (topics) is so interesting that we cannot but quote some of his sentences:

> Reading scientific papers, usually only minor ideas would come to me. If they are good researchers, authors present well-considered issues that have been investigated as comprehensively as possible, so that following their train of thought we will find it difficult to find a way out of it. My search for problems ready to be solved happened in a different way. Considering a broader set of phenomena I tried to find those that had not yet been explained. Of course, this method requires a good command of the subject, better even than when we identify research problems by examining the literature. Furthermore, the horizon of the unknown broadens and deepens in the mind's eye as the researcher gains a better understanding of the field. The hardest part is to select from the realm of the unknown those problems that allow for connections with what has already been studied, and that lend themselves to existing methods of experimental and mathematical research.[70]

As we can see, the author recommends direct thinking about some broader field of phenomena and searching for unexplained phenomena in order to find explanations for them, rather than studying specific works from the most recent scientific literature. Such thinking – or perhaps such reasoning, to be more precise – is what Rignano would call a series of subsequent operations or experiences that have only been thought, not carried out materially, because we already know, thanks to previous experiments that had been carried out materially, the result of each individual experiment. This is in fact the thought experiment that Ernest Mach discusses in his *Erk-*

---

[68] Franciszek Bujak ["F.B."]: "Drogi mojego rozwoju umysłowego" [The Trajectory of My Intellectual Development], *Nauka Polska*, vol. 6 (1927), 77–136: 132.
[69] Ibid., 133.
[70] ▸ Białobrzeski 1927, 311.

*enntnis und Irrtum*⁷¹. Every thought experiment is about imagining objects in given conditions, also merely imagined, and then to consider how those imagined objects will behave in those conditions. A thought experiment is therefore nothing else than a series of experiments performed in the mind.

Expecting little from perusing specialist scientific treatises in the search for new problems (topics) – doubtless a strictly individual tendency – [Czesław Białobrzeski] usually turned, as we read in [...] his "Autobiographical Sketch," to good synthetic works in some broader discipline, trying to identify questions that had not been sufficiently explained as well as new problems. Thus, e. g. while reading Henri Poincaré's *Leçons sur les hypothèses cosmogoniques*⁷², the author of the "Autobiographical Sketch" became aware of the role that the pressure of radiation plays in the equilibrium of the sun and stars. This is how he came to write the text "Sur l'équilibre thermodynamique d'une sphère gazeuse libre," which was published in the international Bulletin of the Cracow Academy of Learning (May 1913) and which showed (and demonstrated mathematically) "that the pressure of radiation equal to the pressure of matter must determine the equilibrium of the sun and stars understood as huge gaseous spheres."ᵍ

In fact, reading good synthetic works serves only to introduce the scholar to some broader field of phenomena in which he then performs various purely imaginary experiments. As he experiments in his thoughts, problems are born whenever he notices in a thought experiment some previously overlooked similarity or difference, some ambiguity or contradiction in judgment, etc.

B. We made a distinction between problems that the researcher identifies independently and problems that he takes on from others in order to then work them out more or less independently. In those cases, in fact, creative abilities only manifest themselves in the phase of working out the problem.

One factor that determines whether another person's problem will become the focus of one's own research is the intellectual resonance, and above all perhaps the emotional resonance, that the problem evokes in the researcher. We will pass over the numerous cases where a professor distributes research topics (e. g. for doctoral dissertations), which the students then work out in a manner that is more craft than crea-

---

⁷¹ Mach: *Erkenntnis und Irrtum* (footnote 7), 183.

⁷² Henri Poincaré: *Leçons sur les hypothèses cosmogoniques*, Paris 1911.

ᵍ In this paragraph Błachowski refers to a section of Białobrzeski's "Autobiographical Sketch" (▶ Białobrzeski 1927), specifically to pages 65–66 of the original Polish version, which are not included in this edition (Ibid., cf. the editorial introduction). Białobrzeski refers to his paper "Sur l'équilibre thermodynamique d'une sphère gazeuse libre," *Bulletin International de l'Academie des Sciences de Cracovie (Série A: Sciences Mathématiques)*, vol. 5/1913, 264–290. *As common in the Bulletin, the titles of the articles which were usually printed in French and German were also given in Polish, in this case:* "O równowadze termodynamicznej kuli gazowej swobodnej". *Białobrzeski published the paper as Tcheslas Bialobjeski, a phonetic transcription of his name for international readers which he employed at various occasions (Cf. Andrzej Kajetan Wróblewski:* "Polish Physicists and the Progress in Physics (1870-1920)," *Czasopismo Techniczne: Nauki Podstawowe*, vol. 1 (2014), 255–273: 269, 273).

tivity, not worrying about it too much and working only with their examination in mind. What concerns us are only those works where the topic, suggested from outside, encounters an accumulation of highly responsive mental representations, a material that is ready to form new systems; this topic will galvanize the researcher's emotionality and become the emotional centre of a longer period of activity.

Problems that enter one's mental structure from outside have the same capacity as independently formulated problems to hold the mind captive for years, even for one's whole life. In this respect there is no difference between Robert Mayer, who spent his whole life working on the independently formulated problem of energy conservation, and J.-J. Rousseau, who chanced on a newspaper announcement of an essay competition; the proposed topic stirred him so deeply that he stayed with it for years.

It would be difficult to find in literature a more accurate and beautiful representation of the entire mental process that takes place when some topic that suggests itself from outside enters the unusual mental organism of a great creator, moving him deeply. In 1749, during the hot summer months, Rousseau would often visit Diderot, who was then imprisoned in Vincennes. The journey on foot was arduous for the delicate Rousseau, and so, exhausted with the heat and effort, he frequently lay down on the ground to recover enough strength to keep walking. To force himself to slow down his pace he would take a book along. This is what he writes in his *Confessions*:

> One day I took the *Mercury of France* and while walking and glancing over it I fell upon this question proposed by the Academy of Dijon for the prize for the following year: *Has the progress of sciences and arts tended to corrupt or purify morals?*
> At the moment of that reading I saw another universe and I became another man. [...]
> What I do recall very distinctly on this occasion is that, when I arrived at Vincennes, I was an agitation that bordered on delirium. Diderot noticed it; I told him its cause, and I read him the prosopopeia of Fabricius written in pencil under an Oak. He exhorted me to give vent to my ideas and to compete for the prize. I did so, and from that instant I was lost. All the rest of my life and misfortunes was the inevitable effect of that instant of aberration.
> With the most inconceivable rapidity my feelings raised themselves to the tone of my ideas. All my little passions were stifled by an enthusiasm for truth, for freedom, for virtue, and what is most surprising is that this effervescence maintained itself in my heart during more than four or five years to as high a degree perhaps as it has ever been in the heart of any man.[73]

Of course the emotions that occur when a suitable problem is found are rarely so exuberant; their vehemence rarely entails some kind of revelation and sudden transformation in one's worldview and experience of one's personality ("I became another man"). It is rarely going to be an "agitation bordering on delirium" or a "fermentation" in the heart that is maintained "for more than four or five years as intensely perhaps as it has ever worked in the heart of any man on earth". Nonetheless, wheth-

---

[73] Jan Jakób Rousseau: *Wyznania* (vol. 2), transl. by Tadeusz Boy-Żeleński, Kraków 1918, 93. ||| This translation follows Jean-Jacques Rousseau: *The Collected Writings of Rousseau* (vol. 5: The Confessions and Correspondence, incl. the Letters to Malesherbes), ed. by Roger D. Masters and Christopher Kelly, transl. by Christopher Kelly; Hanover 1990, 294–295.

er a problem arises within the researcher's mind or is adopted from someone else, its birth and further elaboration are always marked by an emotional experience. As an essential element, this emotionality is proportional to the temperament of the researcher, who might be more or less susceptible to such excitements. The notion of a type of scientific researcher whose mind creates and solves problems in an entirely unemotional manner seems unlikely, and we ought not to waste too many words on the ideas as researchers rising above all passion, the sharp and cold scalpel of reason, or the threat that emotional stirrings pose to scientific work – the fact is that analyses of creative thought processes always confirm that feelings inspire and maintain thought. The role of feelings is certainly different in art than in science, since the latter attempts (in the final phase of a project's completion) to eliminate from the system of statements everything that does not belong to logical deduction. And yet, in the germination of both scientific and artistic projects, the influence of feelings is analogous. It is also quite certain that Tadeusz Grabowski's thesis, published recently in his *Wstęp do nauki literatury*, applies to the phase of a scientific project's conception: "The germ of the work, born of an ecstasy of feeling, generates a multiplicity of surrounding images and assimilates whatever seems fit. This is what brings about the unity of the work [...]."[74]

Rousseau's example showed how a problem taken on from outside can become the germ of an excellent work, as long as it falls on ground that has a vivid resonance, intellectual and emotional – that of an exceptional personality. Besides such fortuitous formulations (e.g. topics for competitions), some problems are important or even fashionable at a given time; they can interest or trouble whole legions of researchers for years, decades or even centuries. Müller-Freienfels correctly observed that sometimes this happens because those problems lead to the next stage that is attainable departing from the present state of research[75]. Sometimes again these are problems that arouse a general interest in a given epoch, so that every thinker falls into their thrall. The problem of universals and particulars, which was debated throughout the Middle Ages; the problem of substance and the relationship between the body and the soul in the seventeenth century, the problem of knowledge in English philosophy – these are perfect examples for how entire epochs of human thought can be marked by a given problem. This phenomenon occurs particularly in the history of philosophy. Take any handbook from that discipline which organizes the material into groups of questions, e.g. Janet and Séailles's *Histoire de la philosophie*[76], or Windelband's *Lehrbuch der Geschichte der Philosophie*[77], and you will be convinced that certain questions excite many generations of philosophers, perhaps precisely be-

---

[74] Tadeusz Grabowski: *Wstęp do nauki literatury* [Introduction to the Science of Literature], Lwów 1927, 62.

[75] Müller-Freienfels: *Das Denken und die Phantasie* (footnote 12), 262.

[76] Paul Janet and Gabriel Séailles: *Histoire de la philosophie. Les problèmes et les écoles*, Paris 1887 (Cf. *A History of the Problems of Philosophy* (2 vols.), transl. by Ada Monahan, London 1902).

[77] Wilhelm Windelband: *Lehrbuch der Geschichte der Philosophie*, Tübingen 1903.

cause solutions to their problems do not satisfy the curiosity of the questioning human mind. But just as interest in certain questions wanes over time in individuals, so it does also across the history of sciences, especially in the history of philosophy, where this phenomenon is generally more pronounced than anywhere else. As a consequence, there is a return to old, worn-out problems – something that is again especially noticeable in philosophy, though it also happens in other sciences. At any rate we should take note that there exist problems that have long been solved and have ceased to be problems, such as the relationship between the hypotenuse and legs of a right triangle, as well as problems that were considered solved but might suddenly reappear as problems. In general we must be very careful if we want to announce that some problem has ceased to be a problem (by virtue of having been solved), as we can see in the problem of parallel lines and their behaviour at infinity.

The return to older problems, to a seemingly worn-out past, should be of much interest to biographers of creative thought. This is the category of problems that are taken on from others in order to be elaborated independently. But these are problems that are rooted in older, sometimes very distant generations of problems, so that to approach them, to adopt them intellectually, must demand a different intellectual attitude than adopting current problems that are *en vogue* in the sciences. E.R. Jaensch's book Über *die Wahrnehmung des Raumes*[78] presents interesting remarks on this topic. In his research, which is of outstanding quality, Jaensch deliberately calls for a return to fundamental experiments (*Die Forderung der wiederholten Rückkehr zu den Fundamentalversuchen* [A Call for a Repeated Return to Fundamental Experiments][h]), which implies a return to fundamental problems. His call cannot be met in equal measure in all the sciences. In physics, for instance, it would be entirely impossible in Jaensch's opinion, while in psychology it could lead to exceedingly interesting results. Jaensch asserts that to repeat the experiments described in Gilbert's book *De magnete*, published in 1600[79], would not be likely to benefit today's studies in electricity. In psychology, however, a straightforward progression from simple issues to more complex ones risks leading to a state where to continue in the same direction will fail significantly to promote knowledge in the field. In that case it is a sudden return back to the issues that appear most simple, back to fundamental experiments, that can lead to the desired results.

Finally, to conclude this part about the inception of problems, let us address the well-known situation, which occurs just the same in the arts, where certain problems whose nature makes them exceedingly significant to the further development of science, are met with incomprehension or indifference at the moment in which they are presented. This is either due to the fact that they actually are well ahead of the current state of research – hence the incomprehension – or it is due to the fact that

---

[78] Erich Rudolf Jaensch: *Über die Wahrnehmung des Raumes*, Leipzig 1911, 2.

[h] B. quotes the title of the introduction to Jaensch's above-mentioned publication. [T.B.]

[79] William Gilbert: *De Magnete, Magneticisque Corporibus, Et De Magno magnete tellure. Physiologia noua, plurimis et argumentis, et experimentis demonstrata*. London 1600.

science is too preoccupied with something else – hence the indifference. An example of such a problem is the problem of heredity as formulated by Mendel in 1865[80]. At that time Mendel's work did not create any effect. Only when analogous facts were discovered in 1900 by H. de Vries in Amsterdam[81], by C. Correns in Tübingen[82], and by E. Tschermak in Vienna[83], the problem of Mendelian heredity fell onto fertile ground.

Moving on to the "middle part" of working out and solving problems, we must mention again that methods and means that researchers use more or less consciously in this phase are so numerous and variegated that it would be difficult to frame them in one category. The researcher's originality manifests itself not only in the processes of posing problems and of obtaining results, but also in the very effort to solve those problems. Here the question of method, which we already discussed in previous chapters, moves to the fore again. To find a suitable method is to get onto a beaten track; at any rate it does much to facilitate the process of solving research problems. But the pursuit, adaptation or indeed the development of an appropriate method is usually an extremely complex activity whose psychic structure is complicated and difficult to access by way of analysis. What is more, we must remember that the phase of solving a problem is never limited to simply applying a method. It also includes collecting material and anticipating results that are provisionally accepted and that one then attempts to test or justify.

The phase of working out and solving a problem – at least where the subjective, psychological aspect is concerned – aims ultimately to eliminate the inner anxiety and emotional tension and to satisfy the curiosity that were all aroused by one's awareness of the problem. This resolution occurs through a reaction which manifests one's tendency to express oneself through statements that represent progress in relation to the state of research in a given field.

It would be hard to imagine any sort of scientific work that is done indifferently. On the contrary, we should assume that this work must be accompanied by a constant and vivid interest, often intensified to the highest degree. There is no doubt that the effect of factor on scientific work is of utmost significance, and studying it would certainly enhance our understanding of the psychology of scientific creativity. But this interest has not yet been elaborated in any detail, and knowing so little about its very nature, we cannot directly apply data from general psychology in the field of the psychology of scientific creativity. According to G.E. Müller, the factor of interest works in three directions: first, negatively, by limiting interest in other fields to a smaller or greater degree; second, it intensifies the individual's attention onto the

---

[80] Gregor Mendel: "Versuche über Pflanzen-Hybriden," *Verhandlungen des Naturforschenden Vereines in Brünn*, vol. 4 (1866), 3–47.

[81] Hugo de Vries: "Sur la loi de disjonction des Hybrides," *Comptes Rendus de l'Académie des Sciences*, vol. 130 (1900), 845–847.

[82] Carl Correns: "Mendels Regel über das Verhalten der Nachkommenschaft der Rassenbastarde," *Berichte der Deutschen Botanischen Gesellschaft*, vol. 18/1 (1900), 156–168.

[83] Erich Tschermak: "Ueber künstliche Kreuzung bei Pisum sativum," *Berichte der Deutschen Botanischen Gesellschaft*, vol. 18/6 (1900), 232–239.

given material; third, it makes him exercise constantly, as he is obsessed with perfecting himself.[84] This notion of interest aptly emphasizes some aspects, but is unsatisfactory above all because it fails to underline the affective-volitional factors that mark such interest. F. Malsch's definition therefore seems more pertinent: "Interest is the will to possess a thing, broadly conceived, in connection with an awareness of the thing's value and a feeling of pleasure."[85] We know what a negative influence fading interest can have on the course of one's work. Examples are easily found – every research worker knows some. I will cite only one, relating to the above quotation from "The Trajectory of My Intellectual Development"[86]. Having discovered the similarity of Lithuanian and Germanic names, [Bujak's] interest endured for quite a while; admittedly not in a continual manner, but surfacing into his consciousness like a wavy line. But then, even though accumulated a considerable amount of promising material, his interest faded and finally fell to such a low level that he was unable to overcome the obstacles in his way.

In other cases, however, the researcher's interest in some problem can be so intense that it comes to dominate his whole being and to limit his interest in other fields; all other tasks and goals vanish from his intellectual horizon; he values his work so much, finds it so exciting, and is so determined to find a solution, that his feelings about the problem become all but ecstatic. There is no better description of that enthusiasm than Rousseau's above-quoted passage from his *Confessions*. There is of course significant individual variation from one creator and another, but Rousseau is in any case an extreme example of a passionate creative sensibility.

The work that goes into solving a problem also comes in very different forms. Sometimes the idea that answers the question implied by the problem suggests itself immediately and effortlessly, in which case the researcher's further work is about testing or justifying that idea. Sometimes again one must toil away for quite some time, collect material, perform experiments and thought experiments, somehow to construct the idea that will eventually conclude one's work on the problem; in that case no further testing or justifying is necessary.

Experienced researchers, having already completed a certain number of projects, work out a technique and economy of creative work, or at least they know how to create internal and external conditions that favour the "spontaneous" appearance of ideas. Sometimes such work can be a terrible struggle. Again Rousseau can serve as an example when in his *Confessions* he recounts how he went about his essay for the competition:

I dedicated the insomnias of my nights to it. I meditated in my bed with my eyes closed, and I shaped and reshaped my passages in my head with unbelievable difficulty; then when I had

---

[84] Georg E. Müller: *Zur Analyse der Gedächtnistätigkeit und des Vorstellungsverlaufes* (vol. 1), Leipzig 1911, 245.

[85] Fritz Malsch: "Das Interesse für die Unterrichtsfächer an höheren Knabenschulen," *Zeitschrift für angewandte Psychologie*, vol. 22 (1923), 393–441: 395.

[86] Bujak: "Drogi mojego rozwoju umysłowego" (footnote 68), 69.

succeeded in being satisfied with them, I deposited them in my memory I could put them on paper: but the time it took to get up and get dressed made me lose everything, and when I had applied myself to my paper, almost nothing of what I had composed came to me anymore.[87]

Later Rousseau got the idea to dictate the results of his night-time meditations from bed – a work method he stayed with for a long time.

Extreme enthusiasm "that bordered on delirium" and the torment of creating have come together to form an inexorable whole.

In his "Autobiographical Sketch"[88], [Czesław Białobrzeski] explains that his work method resembles a mosaic. Having arrived at a leading idea that will become the axis of the entire project, he jots down everything that enters his head and seems valuable in his meditations on the subject. Finally, based on these loose notes, there comes into being a whole, organized in his mind, and a more or less detailed plan emerges, according to which [Czesław Białobrzeski] usually writes the work directly in its final form.

The most important element in this whole phase of elaborating a problem is the emergence of a concept, a leading idea. We have mentioned already that that idea can appear in a number of ways. Some arrive at it after a long time of focused effort, after painstaking consideration of the problem, after having searched for analogous problems to study how they had been solved; to other creators, meanwhile, ideas occur so suddenly that the solution almost comes as a shock. Besides these two extremes, there are of course all kinds of other possibilities.

How the unfolding of ideas happens for ingenious creators can be learned from their personal reflections. On his seventieth birthday Helmholtz gave the following account of his creative scientific work:

Yet as I have often been in the predicament of having to wait on inspiration, I have had some few experiences as to when or how it came to me, which may perhaps be of use to others. Often enough it steals quietly into one's thoughts and at first one does not appreciate its significance; it is only sometimes that another fortuitous circumstance helps one to recognize when, and under what conditions, it occurred to one; otherwise it is there, one knows not whence. In other cases it comes quite suddenly, without effort, like a flash of thought. So far as my experience goes it never comes to a wearied brain, or at the writing-table. I must first have turned my problem over and over in all directions, till I can see its twists and windings in my mind's eye, and run through it freely, without writing it down; and it is never possible to get to this point without a long period of preliminary work. And then, when the consequent fatigue has been recovered from, there must be an hour of perfect bodily recuperation and peaceful comfort, before the kindly inspiration rewards one. Often it comes in the morning on waking up [...] as Gauss also noticed [...]. It came most readily, as I experienced at Heidelberg, when I went out to climb the wooded hills in sunny weather. The least trace of alcohol, however, sufficed to banish it. Such moments of fertile thought were truly gratifying, but the obverse was less pleas-

---

[87] Rousseau: *Wyznania* (footnote 73), 94. ||| This translation follows Rousseau: *The Collected Writings* (vol. 5) (footnote 73), 295.

[88] ▶ Białobrzeski 1927, 313.

ant when the inspiration would not come. Then I might worry at my problem for weeks and months, till I felt like

> [...] a brute upon a barren heath, with tether bound
> By some wicked spirit, and in a circle driven,
> While pastures green and beautiful are lying all around.

Sometimes nothing but a severe attack of headache could release me from my spell, and set me free again for other interests.[89]

This is how Gauss describes the sudden appearance of ideas, also preceded by a period of intensive but fruitless searching, in his letter to the astronomer Olbers:

> This shortcoming spoiled everything else that I found; and hardly a week passed during the last four years where I have not made this or that vain attempt to untie that knot – especially vigorously during recent times. But all this brooding and searching was in vain, sadly I had to put the pen down again. Finally, a few days ago, it has been achieved – but not by my cumbersome search, rather through God's good grace, I am tempted to say. As the lightning strikes the riddle was solved; I myself would be unable to point to a guiding thread between what I knew before, what I had used in my last attempts, and what made it work.[90]

Finally we must recall the exceptionally subtle observations made by Henri Poincaré, whose three great publications in the theory of science (*La science et l'hypothèse, La valeur de la science, Science et méthode*[91]) demonstrate both his interest in and his deep understanding of the problems of scientific creativity. The ingenious mathematician describes how on an excursion, just as he was getting on an omnibus, putting his foot on the step, it occurred to him that the transformations that he used to define Fuchsian functions were identical with the transformations of non-Euclidean geometry:

> I did not verify the idea; I should not have had the time, as, upon taking my seat in the omnibus, I went on with the conversation already commenced, but I felt a perfect certainty. On my return to Caen, for conscience's sake, I verified the results at my leisure.[92]

Another time, after fruitless efforts to solve some arithmetic problem, Poincaré went to the seaside and occupied his mind with other thoughts. And then one day, as he was walking along the cliffs, an idea come to him:

---

[89] Helmholtz: "Erinnerungen" (footnote 27), 15–16. ||| This translation follows Koenigsberger: *Hermann von Helmholtz* (footnote 27), 208–209. The translation of the verse from Goethe's *Faust* follows Johann Wolfgang von Goethe: *Faust. A Dramatic Poem*, transl. by John Wynniatt Grant, London 1867, 58. [T.B.]

[90] *Wilhelm Olbers. Sein Leben und seine Werke* (vol. 2: Briefwechsel zwischen Olbers und Gauss, Erste Abteilung), ed. by C. Schilling, Berlin 1900, 268–269. ||| This translation follows Samuel J. Patterson: "Gauss's Sums," in: *The Shaping of Arithmetic after C.F. Gauss's 'Disquisitiones Arithmeticae,'* ed. by Catherine Goldstein, Norbert Schappacher and Joachim Schwermer, Berlin et al. 2007, 505–528: 507.

[91] Henri Poincaré: *La science et l'hypothèse*, Paris 1902, id. *La valeur de la science* (footnote 48), id. *Science et méthode* (footnote 50).

[92] Id.: *The Foundations of Science* (footnote 11), 388.

The idea came to me, with just the same characteristics of brevity, suddenness and immediate certainty, that the arithmetic transformations of indeterminate ternary quadratic forms were identical with those of non-Euclidean geometry.[93]

Continuing to think about this result and trying to draw all possible consequences from it, Poincaré again encountered significant difficulties. At first all his efforts only led him to a better understanding of the difficulties he would have to overcome. This entire labour took place very consciously.

Thereupon I left for Mont-Valérien, where I was to go through my military service; so I was very differently occupied. One day, going along the street, the solution of the difficulty which had stopped me suddenly appeared to me. I did not try to go deep into it immediately, and only after my service did I again take up the question. I had all the elements and had only to arrange them and put them together. So I wrote out my final memoir at a single stroke and without difficulty.[94]

For Poincaré, this sudden appearance of a good idea is typical for his work. Other mathematicians, as suggested by the survey in the monthly *Enseignement Mathématique*, have similar experiences. The key moments in the above-cited personal reflections by Helmholtz and Gauss overlap with Poincaré's exceptionally interesting account of the course of creativity. This probably is one of the major ways in which valuable scientific ideas emerge. But we will only be able to create a typology of the appearance of scientific ideas some time in the future, once we will have collected a greater number of examples.

It is not easy to explain this sudden appearance of good ideas, which puts creators in mind of something like inspiration, in psychological terms. Conscious work and conscious effort do not always lead to good results. It seems beyond doubt that in the subconscious layers of the soul some processes take place that betray their existence at an appropriate moment, e. g. when the mind is not preoccupied by anything else or when it is well rested, by suddenly appearing to conscious thought, understandable to us only as the result of a longer chain of reasoning.

Sometimes again the researcher deludes himself that the whole idea grew out of the depths of the subconscious, because he never noticed or forgot how often his thoughts had delved into the problem. Research on conscious but unnoticed experiences has many followers among psychologists today, even though it faces many difficulties and ambiguities. If we suddenly catch ourselves thinking about something, we can explain that phenomenon by assuming that conscious but unnoticed thoughts were suddenly noticed; nothing forces us to believe that those thoughts had been subconscious, suddenly became conscious and only then were noticed.

Another frequent phenomenon is the blotting out of intermediate conceptual links. Clearly, the final links in any conceptual chain impress themselves on the conscious mind, which is why they are more easily remembered than intermediate ones.

---

[93] Ibid.
[94] Ibid.

It is rather inconsequential by what succession of thoughts we arrived at a given result, and this is precisely why those fleeting intermediate thoughts generally fail to settle in our memory, and only the more important stages in the course of thought will be accessible to later reproduction. Every scholar has experienced those fruitless attempts to remember the entire stream of thoughts that lead to a result in its original, authentic form. Trying to reproduce from memory the whole interior situation that preceded the moment when the problem was solved, we are too easily led to a logical construction instead of a faithful reproduction of our reasoning. Many of us have felt like Rousseau[95] who, as soon as he sat down to confine them to paper, could not remember them anymore.

It is also possible that sometimes (e.g. when the mind is very fresh) the psychic processes that are active in solving a problem take place so quickly that their effect is that of something unexpected, especially when the mind had previously spent a long time struggling over that problem in vain.

But we must not forget that scientific creativity is always based on a great store of information and experience. A scientific researcher, no matter on what areas he works, and even if those areas are very wide apart, always begins by preparing the ground for his constructions. This involuntary foresight is rooted in the unity of the mental structure that unifies experience in a tightly joined whole. This is why at moments when we become aware of a problem that touches our entire mental structure, our accumulated imaginational material, fragments of various chains of reasoning, our intensified ability to perform mental experiments, and our practiced ability to concentrate all come together to form a reaction that can solve that problem almost like a lightning. Some individuals react so quickly that they seem to come up with the answer as soon as the question is posed, without even the briefest moment of reflection. Others (such as the ingenious physicist Pierre Curie[96]) are characterized by exceptionally slow reaction. Some are very well aware that that quickness to react results (considerably at least) from previous preparatory work, which sometimes of many years, but which never goes to waste, as the mind instantly puts it to use at a suitable opportunity. Thus for instance Napoleon – though admittedly he was no genius of science – had an exceptional ability to react instantly to the most complex questions. He told Roederer:

> I am always at work; [...] I meditate a great deal. If I seem always equal to the occasion, ready to face what comes, it is because I have thought the matter over a long time before undertaking it. I have anticipated whatever might happen. It is no genius which suddenly reveals to me what

---

[95] Rousseau: *Wyznania* (footnote 73), 94. ||| This translation follows Rousseau: *The Collected Writings* (vol. 5) (footnote 73), 295.

[96] Henri Piéron writes: "There is no doubt that our great Curie had difficulties with certain experiments because he was terribly slow, while everywhere we feel some tendency to value quickness… But sometimes depth takes the place of quickness, and originality replaces the capacity for quick assimilation." ("Le rôle de la psychologie dans l'orientation professionnelle," in: *Dix conférences sur l'orientation professionnelle*, ed. by Arthur Fontaine, Paris 1923, 42–75: 65–66. ||| Translation from French by T.B.

I ought to do or say in any unlooked-for circumstance, but my own reflection, my own meditation. ... I work all the time, at dinner, in the theatre. I wake up at night in order to resume my work.[97]

Here we see how the personality of this ingenious leader and organizer, rooted strongly in the real ground of life, categorically rejects all inspiration, all revelation of truth from without, and emphasizes his thorough preparation. H. Fehr's survey of the work of mathematicians also emphasizes the key role of preparatory work, as it highlights "the necessity of study, of reflection, of patience, in short, of work undertaken to prepare or to perfect the gifts of chance or of inspiration."[98] The mathematicians who participated in this survey agree that mathematical discoveries do not happen of their own accord, but require a mind that is well prepared, like fertile soil, by dint of conscious work. We can also refer the testimony of that masterful intuitionist Poincaré, who proposes as a condition for the efficiency of unconscious work nothing less than a period of conscious work before and after:

> These sudden inspirations [...] never happen except after some days of voluntary effort which has appeared absolutely fruitless and whence nothing good seems to have come, where the way taken seems totally astray. These efforts then have not been as sterile as one thinks; they have set agoing the unconscious machine and without them it would not have moved and would have produced nothing.[99]

It will not harm to add that in general, creative scientific work proceeds on long and windy paths; we often lose the right way and must return to the roads that are certain, to well known points of orientation. Sometimes we walk past the goal, unknowingly, off by one step. It would be hard to find a more suggestive image than the one pained by Gustav Theodor Fechner in his account of his search for the psychic measurement:

> imagine someone standing on the periphery of a circle; he is looking for some object, which lies a step away from him, but he is standing with his back towards it and so he must travel along the entire circle until, having overcome many difficulties, he finally reaches the object he had sought and realizes with astonishment that he would only have had to turn around in order to have it at once, and of course to have in it not exactly what he had expected. This is what happened to me in my search for the psychic measurement. But I must not regret the distance I travelled, since that journey made me aware of the very scope of the principle of the measure, which the shorter path, from Weber's law and Euler's formula to the general principle of the psychic measurement, would not have been able to do. As far as I had to travel backward for it, as far forward it took me.[100]

---

[97] Pierre-Louis Roederer: *Histoire contemporaine* (= Œuvres complètes, vol. III), Paris 1854, 544–545. ||| This translation follows Hippolyte Taine: *The Origins of Contemporary France* (vol. 1: The Modern Régime, pt. 1), transl. by John Durand, New York, NY 1890, 19.

[98] *Enquête de "L'Enseignement mathématique"* (footnote 48), 39. ||| B. does not refer to the whole survey, but only to vol. 8 (1906) of *L'enseignement mathématique* without indicating a page number. As the reference is vague, B.'s Polish translation is difficult to trace in the French original, which, however, contains a similar passage by George Fontené.

[99] Poincaré: *The Foundations of Science* (footnote 11), 389.

[100] Fechner: *Elemente der Psychophysik* (footnote 40), 553 ||| Quotation translated from German [T.B.].

Fechner's comparison does not apply only to the concrete case he describes.[101] On the contrary, this is the typical picture of the spirit's progress along the unknown paths of science.

The moment an idea springs to mind and strikes us as suitable to solve a problem, or when we choose one among many different ideas and reject others, the creative scientific process has been completed in principle. Our belief in the idea's felicity may be based on the feeling that we have committed no errors in our reasoning, but even more perhaps on a purely emotional element, on a sense of relief, of contentment, exaltation, freedom. Some researchers decide on their solutions with incredible self-assurance, as if they were guided by some instinct. Others see different possibilities that seem to them equally good and struggle to find an inner partiality for either of them. These are often people who are described in German as Projektenmacher; they can create some ferment and inspire others to do research. In life they would have no success; despite their ingenuity their indecisiveness would make them into sluggards. In science we cannot negate their significance, though we must add that what characterizes great researchers is an almost infallible surety when it comes to feeling that a certain solution is right, and also when it comes to choosing ideas.

We will pursue the history of the scientific work no further, passing over efforts to bolster the theoretical resilience of research endeavours (testing them and bringing them into harmony with other known statements) and to find the most suitable way of presenting them. This would go beyond the framework of our problem of scientific creativity. We will only note briefly that there is a whole spectrum of researchers between the two extreme of the one who writes down the outcome of his thoughts in one sitting, and the one who constantly improves and edits his manuscript, who will even pull the sheets out from the press, becoming the bugbear of the editor and printer. In his preface to the second edition of *Die Lebensformen*, Spranger mentions: "With a festive occasion in mind, the first version of this text was designed and completed in fourteen days of fortuitous inspiration."[102]

[Franciszek Bujak] confesses:

Editing is burdensome drudgery to me, I moan and swear that I will never again undertake this kind of work. [...] I usually leave the left half of the page blank for additions and inserts. The manuscript usually has so many things crossed out that it is almost unreadable to anyone but me, so I am forced to copy it out myself, or, as I do now, to dictate it to a typist. A rather

---

[101] A similarly beautiful account is provided by Helmholtz: "Erinnerungen" (footnote 27), 14: "I am fain to compare myself with a wanderer on the mountains, who, not knowing the path, climbs slowly and painfully upwards, and often has to retrace his steps because he can go no farther then, whether by taking thought or from luck, discovers a new track that leads him on a little, till at length when he reaches the summit he finds to his shame that there is a royal way, by which he might have ascended, had he only had the wits to find the right approach to it." ||| This translation follows Koenigsberger: *Hermann von Helmholtz* (footnote 27), 180–181. [T.B.]

[102] Eduard Spranger: *Lebensformen. Geisteswissenschaftliche Psychologie und Ethik der Persönlichkeit*, Halle 1922, VII. In contrast, it took Spranger seven years to work on the second edition of this work.

small portion of my manuscripts to date was ready to be sent to the typesetter in its first version.[103]

Incidentally, the agony of elaborating and finalizing a work is familiar even to the greatest creators (in particular those of the classic type), who with hard and relentless work drape their ingenious ideas into the robes of words that make up their oeuvre. A good example is Helmholtz, who confesses in his anniversary speech that he finds the task of expressing his scientific research in writing wearisome to the utmost degree.[104] Many parts of his writings went through four to six versions; he kept changing the structure and did not cease to work until he had succeeded in formulating his research with the greatest precision and with no gaps in the logic.

---

[103] Bujak: "Drogi mojego rozwoju umysłowego" (footnote 68), 131–132.
[104] Helmholtz: "Erinnerungen" (footnote 27), 18.

The Man of Action and the Student

Franciszek Bujak

FRANCISZEK BUJAK (16 August 1875, Maszkienice near Brzesko, Austrian Galicia – 21 March 1953, Cracow, Polish People's Republic) economic and social historian. Studied history, historical geography and law at Cracow Univ. (1894–1900). Prof. of history at Univ.s in Cracow (1909–1918), Warsaw (1919–1921), Lviv (1921–1941) and Cracow again (1946–1952). B. was an active organiser of scientific activities, for example as president of the Polish Historical Society (*Polskie Towarzystwo Historyczne*, 1932–1933, 1936–1937) and Lviv Scientific Society (*Towarzystwo Naukowe we Lwowie*, 1933–1937), and also engaged in politics – he was an expert in the Polish delegation at the Paris Peace Conference (1919–1920) and very briefly served as a minister of agriculture (1920).

B.'s text was originally published in vol. 1 of Organon in 1936. It was, in fact, a contemporary translation of a Polish version originally published in vol. 11 of Nauka Polska nine years before. B. had presented a version of this text in the second session of the Science Studies Circle on 10 December 1928 (▶ **IN THIS VOL.** Sprawozdanie 1929, ▶ List of Sessions). This edition follows the 1938 version except for a first footnote to the author's name introducing him as "President of the Sciences and Letters Society in Lwów and Professor of Economic and Social History at the University of Lwów." Linguistic errors have been corrected without indication.

ORIGINAL REFERENCES
Franciszek Bujak: "Działacz i badacz," *Nauka Polska*, vol. 11 (1929), 11–23.
— "The Man of Action and the Student," *Organon*, vol. 1 (1936), 20–32.

FURTHER TEXTS MENTIONED
▶ "Sprawozdanie z działalności Koła Naukoznawczego," *Nauka Polska*, vol. 11 (1929), 353–355.

CONTENTS: Definition. Objects of action. Character. Means and ways of action. Organization of work. Conclusion.

Definition

Human life as well from the social as from the biological point of view consists of an uninterrupted flow of reactions which might be broadly termed «activities». Since consciousness takes part in the reactions of a human being, they might be connected with the notion of studies. A human being perpetually adapts itself and its organs to circumstances and environment, through verifying, foreseeing and, shortly, through studying. Every human being must be, at the same time, in the biological sense a man of action and a student. Similarly, every human being, living even in the smallest human community such as a family or clan, acts and reacts to actions of other human beings, speculates, viz. chooses time and means of action, foresees counteraction of others and prepares for the same.

We are here concerned with social types of men of action and students, otherwise with human beings, in the life of which action or studies dominate and obtrude themselves to us as the chief factors of their characters.

The man of action directs social life, creates it and organises it, more or less independently, to a greater or lesser extent, from small societies and even temporary associations up to great political parties and trade-unions, from a rural community, village school and village church up to a Ministry or even a State; from the management of a small business up to the directorship of a great international trust. This category comprises also as well antisocial activities such as banditry, forgery, as double dealing (espionage). When we mention men of action, we have in view, in the first instance, their typical representatives such as military leaders, statesmen, promoters of industries, creators of schools of thought or of religious movements.

The student aims at the knowledge of the intrinsic meaning of phenomena, at deepening the study of nature and technical sciences, at the knowledge of the mechanism of the human mind and social life psychological processes. He is therefore, in the first instance, a scientist and an inventor.

The type of a man of action is by far more numerous than that of a student, which evolves at a comparatively later stage from the former.

Though the coming into being of these types results from the tendency to differentiation in human society and to division of labour, this division does not decide about the difference between the two types, for every action and especially an action of a higher grade requires a certain speculative activity, and, on the other hand, every investigation is necessarily connected with more or less widespread and intensive activities. The difference between these two social types is therefore based on the quantitative relation of action and speculation respectively. Their mutual relations could be defined figuratively by calling the man of action a practical student and the student a theoretical man of action. The man of action is an earlier social type and therefore the most vital species of the normal, average man. The derivative character of the type of the student is manifested by the fact that very often, not being satisfied with mere theoretical studies, i.e. with the creation of ideas or principles or projects, he passes to action, trying to apply his theories in practice by making propaganda, promoting industries, directing institutions or certain spheres of state life. It happens but seldom, on the other hand, that a man of action should turn to studies. Even the cases of a former student's return from the sphere of action to speculative work are rather infrequent.

Objects of action

The man of action's objects are, firstly, other human beings, secondly, technical appliances. The student deals foremost with nature, human culture and instruments. Human beings are to him but of secondary importance and taken into consideration only when help or collective studies are concerned. The passive universe cannot cooperate with the student but does not actively oppose him. A man of action's case is a different one. In his ease an active resistance of human beings is possible and often essential (war, politics, the police, penal prosecution). The man of action directs all his efforts to overcome this resistance. Human passivity may also become an important obstacle for him (passivity of pupils, for instance).

The man of action must choose his collaborators or treat human beings as objects for his activities. He must therefore prepare them to be easily managed. He must employ his collaborators with great skill and treat them methodically.

The man of action, and the student as well, solve certain problems which must be formulated in a concrete and practical manner in order to be solved. The character of problems varies, however. The man of action aims at immediately influencing human destinies viz. at ordering their lives or labours or inculcating upon their minds certain useful ideas or practices, as well as supplying them with certain advantages or causing them certain damages. The student aims also at influencing human beings, but not directly. He wishes to impart certain knowledge of new facts or principles or of new technical possibilities. He therefore aims at influencing indirectly their conduct or actions.

While the man of action turns to life and orders the same around himself, the student isolates himself and, so to speak, turns his back on life, even if life is the immediate object of his studies.

Theoretically speaking, the objects of interest are limited enough to the man of action, for they are confined to a certain limited number of his contemporaries in a limited space whilst the objects of interest to the student, however limited in a different manner, are usually not so confined as far as the sphere of phenomena and their relation to time and space is concerned.

There are of course many points of contact between both fields of activities and they even overlap. The student might and sometimes must formulate his problems in connection with actual interests or necessities, in trying to discover, for instance, some germs or vaccines, or inventing new asphyxiating gases or means of paralysing their influence. The legislator, who belongs to the group of men of action, when trying to influence directly the destinies of a country, is, as far as the method of his action is concerned, rather a student than a man of action. The efforts of a man, who amasses means for scientific research, are an action, which benefits studies.

Character

The man of action and the student do not differ much in the essential features of their respective characters. Either of the two social types may manifest all four essential temperaments, especially the active and receptive one. Active temperaments dominate of course over the passive (receptive) ones, especially among men of action, but the same applies no doubt to students. On the whole let us say that the more active the character, the better the results and the higher the position in social hierarchy. A passive temperament usually keeps a human being within the confines of an auxiliary part, which consists of carrying out orders. The man of action with a passive temperament usually works on the account of somebody else and remains usually a subordinate even, if he realises his own conceptions. The lack of authority amongst his fellow-workers or lack of self confidence and an aversion to assume responsibility causes such a man of action, even if very gifted, to remain anonymous, or, at least, hidden. The student with a passive temperament usually carries out somebody else's conceptions of minor importance, supplying the shortcomings of the studies, by means of minute diligence.

The active temperament of a man of action manifests itself in daring to assume responsibility and to solve problems without any help. The active temperament of a student makes him aim at mastering the field of his research and assuring for himself independence of judgment and freedom for his labours.

The combative temperament is one of the essential features of a man of action. This is however not a rule, for many of them prefer rather to conciliate people than to conquer them. They prefer mild methods and slow achievement of their aims by devious routes rather than by attack which provokes opposition and vigilance of their adversaries. The student's is an opposite case. Quiet temperament prevails in accordance with the nature of his activities. The fight for truth and progress requires impartiality of judgment, patience and devotion to the possibly highest degree.

Both types must possess an active ambition i.e. a higher aspiration than that of breadwinning for themselves and their family, especially the tendency to distinguish themselves and to outrival their competitors. The ambition of a man of action may grow into greed for power and honours, or into greed for money. The ambition of a student assumes more or less noble forms of aiming at a career and culminates in the greed for recognition and popularity. Students are very often disinterested professional men or women actuated by their propensities and rewarded by self satisfaction. This is very often based on the sense of social responsibilities. Such types are however not lacking also amongst men of action, viz. the type of an idealist, who disinterestedly devotes himself to a profession, but such individuals are less often encountered among men of action.

Willpower, endurance, fidelity to purpose and even stubbornness are the usual and almost indispensable features of the character of a man of action as well as of a student. Both must pay for the great successes achieved with great expenditure of energy, they must therefore possess a great capacity for intensive efforts and work during prolonged periods of time. The greater the efforts, the more inevitable periods of exhaustion and inactivity. These are particularly dangerous to men of action, especially when they coincide with decisive moments in their action, as for example a military or electoral campaign. Willpower extended sometimes to astounding proportions sustains the physical forces of men of action and students for longer periods during their decisive struggles, but causes ruin of those forces and often ends in a catastrophe.

The combative temperament of students finds often its expression in the tendency to controversy. An inclination to propagandism occurs with comparatively few individuals. In this case, there exists an evident aspiration to a man of action's part, an aspiration, which is able to balance or even overweigh, in time, the addiction to study.

The essential psychic basis of both social types is, on one hand, discontent with the existing conditions, resulting in a critical attitude and opposition, and, on the other hand, confidence in one's own brain power and technical abilities. This urges such individuals to search for new forms of life and for new ways, better and more perfect than those prevailing. Alternatively, they try to replace others, less capable and less suitable for a place.

There are great similitudes in the respective mentalities of men of action and students. Both have usually a great susceptibility, which enables them to receive and absorb a great quantity of impressions, i.e. they possess a strong ability for observation and a capacious memory. Very active and plastic imaginative power makes a very valuable contribution to their power of observation, provided that this power is not excessive and does not tend only to create highly aesthetic visions, but has a realistic tendency and coordinates the known parts of reality, filling only with the most possible amount of probability the existing gaps. These mental features are connected with the capacity of quick thinking, i.e. of associating ideas and notions. These processes are often subconscious and, in consequence, only some parts of the same; especially their final stages, reach consciousness as a manifestation of intuition. Thanks to intuition, the brains of such men easily produce new conceptions, which facilitate the solution of capital problems or that of the difficulties which emerge during action and pertain to details.

Plastic imagination is the source of constructive speculation and of the capacity for exactly foreseeing possible circumstances and consequences. Aims assume, then, usually a clear and concrete form. Plans for their realisation are also well constructed and defined. With a great part, perhaps with the majority of men of action, as well as of students, plans of execution swiftly follow conceptions. Those plans are almost ready in every detail or at least ready to such an extent that later additions do not alter them radically, but only supplement them or alter details. The minority develop and shape their aims slowly and gradually, according to the development of the problem, and similarly their plans are long and laborious, and are sometimes formed with the help of an apparatus of detailed studies and assistants.

The lack of a clear aim and a preconceived detailed plan reveals a would-be man of action as a mind, which belongs rather to the type of a student. The vagueness of purpose and especially of proceeding, which may be described as groping in the dark, happens, of course, more often to the student, who deals with things yet unknown, has aims, sometimes situated beyond the explored reality, and must look after means not easily defined beforehand, often even non-existent. A military commander, or a surgeon during an operation are sometimes in the same position, although they are concerned with presence or absence of known objects in certain places and with the choice of the manner of proceeding, fitting only the foreseen circumstances.

The capacity for a quick alteration of plans and their adaptation to changed circumstances is of vital importance to men of action dealing with situations, which require immediate decision, such as a contest with human beings or with elements, viz. war, political action, gambling on the stock exchange, surgical operations. The student does not need this capacity, for the success of his does not depend on instantaneous decisions and he can alter his plans after having thought out the matter quietly and thoroughly. Among both types there are people of one idea (monomaniacs), who are imperceptible to other psychic stimuli and often appear unintelligent for lack of interest in other problems and aspects of life. The leading idea absorbing their mentality becomes, in time, an obsession. The concentration of all mental powers upon one single purpose very often decides the issue. A human being too much absorbed in one single idea becomes mentally dull. This morbid phenomenon usually kills every success.

Means and ways of action

There is a great difference as to ways and means between a man of action and a student. A man of action who deals, in the first instance, with human beings, must be a very good practical psychologist. He must have a knowledge of the human soul generally, as well as of all its varieties and shades. He must know very well the customs of all social strata and ethnical groups. We have of course in view a purely practical knowledge and its application to individuals, which are to be influenced. This knowledge is not to be acquired at a moment's notice; usually a long experience is essential. Practical psychology requires constant studies of human beings and changes in their character. There is neither a pattern nor a formula available; means, which gave good results in the past, might in the course of time prove totally inadequate.

The man of action must not only practice his psychological knowledge in respect of individuals. He must also be acquainted with social psychology, in a practical sense of course, and in particular with mass-psychology; he must know, how to use it. Success is achieved by military leaders, prophets, statesmen and agitators of every kind through skillful influencing of public opinion, especially through the influencing of crowds and the rousing of emotions, which are needed at a certain moment.

The man of action conciliates his supporters, followers and admirers by various means. The most powerful factor, effective especially with societies, which occupy lower stages of development: or belong to the emotional type, i. e. are ruled rather by emotion, than by reason, is personal charm. It evokes feelings of sympathy, amounting to worship, even to what is called blind devotion, which eliminates the advice of reason, moral considerations, legal principles. This devotion is above other social ties. It is difficult to describe this way of securing devotion as the matter has not been adequately investigated. It seems to me that its higher forms or manifestations show a certain similarity with suggestion and hypnosis, for it drives human beings into a state of subconscious psychical subservience amounting to a state of mental servitude.

The popularisation of the man of action's personal charm by his followers and admirers, sometimes, the exaggeration of this charm, is of vital importance, for they preach his glory, the greatness of his actions and his soul. In this process of creating the spiritual authority of a man of action there is, no doubt, a blending of elements of mass-suggestion with mass-interests: moral or material ones.

The man of action exercises his personal charm through his eyes, voice, captivating behaviour in personal relations, ravishing eloquence and characteristic public appearance. Sometimes these means are used consciously, sometimes they are subconscious with the man of action and his followers. There are various other means of influencing human willpower and reason. The force of argument enhanced by eloquence and persuasiveness in written messages is of great importance, particularly at the beginning of a man of action's career.

A very popular and effective means is flattery or praise, doled out in proper doses and at the right moment. They enable the man of action to induce misers to generosity, and people lazy by nature to efficient hard work. The chief and seldom failing trick is the offering or promising of material gains to people, from a rich reward for real and honest services up to bribes for services of mean capacity.

Terror and physical control, applied in various forms and proportions, are also a very effective means of action. Most often they assume the shape of the State's executive power, which, no doubt, is the most powerful form of coercion and fails only in exceptional circumstances. This power extends from pure order through every form of direct propaganda up to the exercise of indirect influence by means of most noble cultural factors such as education, church organization, literature and science. This power can deprive human beings of fortune, freedom and life, but can also lavish on them all kinds of grants and profits, distributed to all citizens, or meant only for the few.

As we said before, human beings are the object of action for the man of action. He selects them and groups them for his own purposes. The more the thoughts and willpower of people who are the object of his action cooperate with his willpower and thoughts, the more they understand his purpose and the more they are personally interested in it, the surer and quicker the results.

Though human beings are the chief object of his action, the technical aspect also plays an important part, such as knowledge of the country, of roads and weapons in war, knowledge of production, credit conditions, markets and possibilities of sale in peace. The wider and more thorough the mastery of the technical aspect of the problem by the man of action, the greater his independence and his personal influence on the results.

The student has very little to do with human beings and he little needs these ways of influencing them. His capacities in this respect are very often nil or very undeveloped. When a student addresses his fellow-creatures in order to make them adhere to his plans or supply the necessary means for their realisation, he does it only incidentally and in a clumsy way. Students, who are exceptionally higher endowed in this respect and score successes of this kind, become, in the course of time, heads of learned societies or institutions for research, and, being busy with the work of organization, become men of action since they have no longer time to spare for studies.

The student devotes all his attention to methods of investigation and technical means for applying them. He looks for new and more perfect methods of investigation or tries to perfect the old ones by coordinating them or applying them in a different manner from before. The criticism of the results of former investigations consists in pointing out mistakes made by predecessors in the appliance of methods or in demonstrating the inefficiency of such methods, or inadequacy of technical means for obtaining more satisfactory results. In order to avoid the mistakes of his predecessors the student aims at a more accurate handling of instruments, at a more strict selection of materials, at a greater precision of observation and a wider scope of the same, at a greater accuracy in drawing conclusions. The conviction as to the inadequacy of methods and technical means causes him to think out new methods of ways of investigation, to construct new instruments, or at least, to improve or perfect them.

While for the man of action quickness and steadfastness in action is of prime importance, the aim of the student is attained through constant and patient preparation and study of thousands of specimens, analyses, measurements and notes. The trying out of various means is sometimes for a student a necessity, usually a profitable one. It is obvious that the energetic and constant following of a once chosen and approved way of proceeding, is always very profitable for the student and in most cases awards him a first place in the scientific race. On the other hand, there are spheres of action, where quickness and determination might be characterised as damaging to results, for in such cases a slower and more patient proceeding and winning of confidence is essential. Otherwise, it is possible that one might alienate sympathies and thus delay or make even impossible the attainment of the purpose in view.

The faith in the importance of one's work and the chosen direction of the same is a vital element of success for the man of action and for the student. No less important factor for both of them is self-criticism. By way of the same they are prevented from over-valuing their own forces and merits, delivered from dangerous illusions and given the opportunity of profiting through their own mistakes and saved from repeating them. This quality is more often observed in scientists than men of action, who being compelled to act with authority lose in the course of time self-criticism and become conceited and too self-confident. This always causes failures. On the other hand, experience acquired prevents the same mistakes. being committed again, facilitates action at later stages and makes subsequent successes more easy.

The profits gained by experience are usually paralyzed by routine. This habit is acquired through applying the same method in a mechanical or automatic way. Routine is the result of the economy of work or tendency to obtain results with a minimum of effort. No wonder that even gifted individuals succumb to this lure. It is very difficult to avoid routine after having pursued the same direction of studies for a considerable length of time, for the coordination of movements and thoughts soon becomes automatic in our nervous system. As it is difficult to move a cart from deep ruts on a muddy road, so it is still more difficult to free one's thoughts from automatic associations. Routine leads men of science towards atrophy and sterility. Their pursuits, when standardised, tend to be superficial, even if the object of the same is changed. They are in such cases outrivalled by their competitors, who attain better results through more detailed and penetrating studies, the method being, on the whole, unchanged, and are higher valued. Routine for men of action is likely to become even more dangerous. The appliance of the same methods leads to disaster as the foe studies these methods and finds means of counteraction. The appliance of the same methods usually betrays criminals and delivers them into the hands of justice to be punished for their last crime and the former ones.

In some forms of action an important part is played by surprise, and turning to one's advantage of the lack of foresight and vigilance on the part of opponents. The achievement of success by surprise causes more hatred and thirst for revenge in the opponents betrayed, than in opponents defeated in a loyal contest. This makes the position of the victor more difficult for the future. Striking craftiness and deceit applied by a man of action make him ill famed as a crafty man and justify similar methods on the part of others. This also has a detrimental effect on his subsequent fortunes.

In the sphere of studies hypothesis is, to some extent, the equivalent of craft. Hypothesis is the adoption of a thesis possible, but not verified. Hypothesis may also bring undesirable consequences; sooner or later mistakes and self-assumed facts are discovered and the hypothesis collapses. It is worth remembering that hypotheses cannot be excluded from the arsenal of weapons in the fight for truth and that great achievements of science usually begin in the form of hypotheses. Similarly deceit cannot be excluded as the means in the struggle for existence, being a kind of corrective which gives to intellectual force some chances in the struggle with the physical force.

Organization of work

The proper organization of work is very important to men of action as well as students. It is indispensable to a man of action, especially when he is dealing with a great number of helpmates and great distances. Without a proper organization of work, viz. without a proper division of actions and their succession, he cannot concentrate his forces and energy in a proper time and place, in order to achieve the desired end. The man of action must be careful about the proper utilisation of his time and its proper exploitation, he must accurately foresee the results of his actions, classify them and, according to such a classification, divide them between himself and his collaborators. The consequence of classification is a distribution of functions amongst collaborators (hierarchy); thus a great and complicated human organization comes into being. The creation of such an organization and its management is a science in itself. The man of action must be born with such a capacity or acquire the same through arduous labour and at the cost of painful experiences or he must find a cold-blooded specialist who is fit for replacing him in the organization of such an efficient apparatus and its running.

Good organization of work is not a condition as essential to a student as to the man of action. Firstly, only a comparatively limited organization of his own work comes into question (the division and proper succession of activities). Secondly, he has to use the work of but a small number of collaborators properly. It is however evident that good organization of work spares the efforts of a student, makes them more efficient and therefore makes him to achieve his results sooner, multiplies their quantity, or enables him to lead a many-sided personal life (i.e. family and social life), while otherwise he would be a prisoner shut up in his study or laboratory.

Conclusion

Men of action and students are born, but not all such men are allowed to use their natural faculties in their respective proper fields of action. Through circumstances or social position many people might be unaware of their talents or hindered to apply them through infirmity. The natural gifts of a human being develop during its lifetime. Education might influence them, especially by accelerating their development. It might, however, easily happen that education is directed by parents or teachers, consciously or unconsciously, towards the suppression of those germs of action or study. On the whole, however, we may assume that great capacities are able to overcome obstacles created by education and circumstances and find expression in deeds.

With these remarks we conclude this short survey of similarities and differences existing between the social types of a man of action and a student. Our remarks aim to a certain extent at provoking a discussion and at attracting attention to this very wide subject which deserves to be treated in a detailed work. I shall consider myself happy, if my remarks will incite anyone to undertake this task.

Science and the Forms of Social Life: Issues at the Intersection of Sociology and Theory of Science

Paweł Rybicki

PAWEŁ RYBICKI (30 September 1902, Janów, Austrian Galicia – 15 July 1988, Cracow, Polish People's Republic), sociologist and historian of social sciences. R. received a doctoral degree from Lviv Univ. in 1926 before continuing sociological studies in Berlin, Cologne and Paris. From 1934–1939 and 1945–1957 he was director of the Silesian Library (*Biblioteka* Śląska, former Parliamentary Library of Silesia), but also was involved sociologically at the Scientific Institute of Silesia (Śląski *Instytut Naukowy*) and at Wrocław and Cracow Univ.s before becoming an assistant (1957) and eventually full prof. at Cracow Univ. (1972).

R.'s text was published in vol. 11 of Nauka Polska after a presentation of the material in the third session of the Science Studies Circle on 7 Feb. 1929 (▶ **IN THIS VOL.** Sprawozdanie 1929, ▶ List of Sessions). The author engages strongly with Florian Znaniecki's earlier outline of a science of knowledge (▶ Znaniecki 1925) and seeks to add an anthropological strain. Starting from Ludwik Krzywicki's work, he includes German sociology (esp. Wissenssoziologie, for ex. Max Scheler, Max Weber, Leopold von Wiese, Paul Honigsheim, Wilhelm Jerusalem) as well as French sociology and anthropology (Émile Durkheim, Lucien Lévy-Bruhl and Marcel Mauss) that he had studied after leaving Lviv. This edition includes the first and fourth section of R.'s original Polish text (24–32, 60–64) dealing with general thoughts on theory and sociology of science and omits the historical-anthropological considerations in sections two and three. Translation by Tul'si (Tuesday) Bhambry.

ORIGINAL REFERENCES
Paweł Rybicki: "Nauka a formy życia społecznego. Kilka zagadnień z pogranicza socjologii i teorii nauki," *Nauka Polska*, vol. 11 (1929), 24–64.

FURTHER TEXTS MENTIONED
▶ "Sprawozdanie z działalności Koła Naukoznawczego," *Nauka Polska*, vol. 11 (1929), 353–355.
▶ Florian Znaniecki: "Przedmiot i zadania nauki o wiedzy," *Nauka Polska*, vol. 5 (1925), 1–78.

CONTENTS: 1. The theory of science and approaches to science as its subject. The sociological current in the theory of science and manifestations of sociologism[a] in

---

[a] "Sociologism" appears in the Oxford English Dictionary as a "frequently depreciative" term denoting "the (excessive) tendency to explain matters in sociological terms, or to ascribe a sociological basis to other disciplines." Rybicki's use of the term *socjologizm*, however, is neutral. [T.B.]

explanations of cognitive phenomena. A general description of the problem. 2. The question of science in primitive groups. The characteristics of social groups in which science manifests itself. The conditions for the existence of science in a social group. 3. Processes of social integration and differentiation and their impact on the character of science. Medieval knowledge and modern empirical science. The social group's attitude towards scientific activity and the transformation of that attitude as a result of processes of social integration and differentiation. 4. A glance at some other problems of the sociology of science. Closing remarks on key issues. The sociology of science and contemporary concepts of scientificity.

1. We hear more and more often nowadays that a new field of theoretical research ought to be established, nay, that it is already beginning to take shape. Its focus is to be the study of all forms of the sciences.[1] The call for such research is rooted in the assumption that science represents a distinct field within our cultural reality, which is why this field is not and cannot be wholly covered by any of the existing sciences. Questions of science are tackled by cognitive theory, logic and methodology. But neither the normative judgements of logic and methodology nor the philosophical approaches to knowledge from within cognitive theory are tantamount to studying scientific activity and its outcomes, and they cannot take the place of such study.[2] What is more, while representatives of various branches of knowledge frequently share their ideas on science, its trends, goals and values, such contributions tend to arise from loose reflections rather than from a conscious position with respect to the subject under investigation. Meanwhile, people's activities in the field of science – like any other kind of human activity – represent a complex of problems that demand empirical research and theoretical approaches purposely designed to solve them.

Many different contributions and attempts suggest that the call for a science of science or a science of knowledge is beginning to be answered. However, the subject of this field of research has not yet been established. We intentionally presented the two terms "science" and "knowledge," since the subject's limits and the concerns of this discipline lie somewhere between their usual meanings. What we call science today no doubt belongs to the subject of the new branch of science. But we must recognize that it is not possible to limit our research to the contemporary definition of science only, since in that case research in the science of sciences would exclude analogous cultural phenomena in other periods and social groups. Our understanding of

---

[1] Cf. ▸ Znaniecki 1925, Kurt Lewin: "Über Idee und Aufgabe der vergleichenden Wissenschaftslehre", *Symposion*, vol. 1/1 (1925), 61–93. ||| Rybicki frequently uses the expression "życie naukowe" – literally "scientific life". I translate this expression as "the sciences" because in English that expression evokes the concept of the scientist's vocation to devote their life to science, i.e. to lead a "scientific life". Rybicki also touches on this question in the second part of his text, so a literal translation of "życie naukowe" might cause confusion. [T.B.]

[2] Znaniecki convincingly argues this point in ▸ Znaniecki 1925, 114–116.

science rests on logical and methodological criteria and rules, but such a normative approach cannot define the subject of empirical science. On the contrary, the science of science must include cognitive operations and their outcomes irrespective of any criteria pertaining to their value. The concept of knowledge, however, appears rather too broad for our purpose if we understand it to comprise all kinds of cognitive operations and outcomes. For instance, we understand knowledge to mean religious cognizance, which is part of the whole of religious life and can be understood only through the study of religion; we also take knowledge to mean a collection of practical information within the domain of practical activity, which is why we need no separate definitions in scientific research. Today we are able to point out these difficulties in determining the subject of the science of science, not to solve them. Only developing research will be able to draw the boundaries of the subject, though it is very likely that peripheral problems pertaining to different scientific fields will always continue to exist. There is nevertheless one criterion that promises to help determine the fields of research, at least provisionally. We see this criterion in the awareness of the distinct nature of cognitive activity above all with respect to religion and the entire sphere of practical life. Where there is no such awareness, where all information, all elements of cognition and worldview are contained either in practical life or in art or religion, there can be no question yet of science as a separate cultural field. Not until certain complexes of cognitive operations and their outcomes break away from other cultural domains, and only when these operations begin to represent a distinct domain according to those who practice them, does science come into being as a separate sphere of cultural reality and the proper subject of research in the science of science becomes visible. Such a concept of science is narrower than the general concept of knowledge, but it still spans a great diversity of cognitive activities and their outcomes. The concept we retain for the present considerations naturally differs from concepts of science and scientificity used today, when certain logical and methodological rules are applied to describe scientific research and its ideal characteristics.

To study science means to study all manifestations of cognitive activity that occur as separate cultural phenomena in the mental life, broadly conceived, of individuals and of society. Such research can be either historical or systematic in nature. The call for historical research on science is realized to a great extent by histories of individual abilities. Besides this type of historical account (which often comes in the form of an introduction and therefore relates to the systematic presentation of a given science), the history of science as a whole is particularly interesting from the point of view of the science of science. Here we stand before a research endeavour that appears to be almost unfeasible, especially today, given the advanced separation and fragmentation of the sciences. But all that eludes the frame of history, which is an individualizing science, becomes a subject for systematic study, a cognitive opportunity. The systematic study of science is referred to as the theory of science. The theory of science, similar to other systematic fields in the humanities, such as economics and sociology, is not concerned with unique phenomena and does not attempt genetically to retrace their course. In systematic research, the researcher's attention is focused on recurring phenomena, typical facts and constant characteristics of subjects; the goal of the research is to investigate the structure of a cultural field, to recognize the conditions determining the occurrence of facts, and to describe permanent relationships between them. The theory of science, therefore, does not target the very core of creativity or go into the individual creative act, but aims to grasp the whole complex of conditions, situations, dependencies, and – if such assumptions can be made – the laws that circumscribe the development of cognitive operations and to which manifestations of creative and original thought are bound.[3]

The systematic study of science is still in its infancy today. Among the efforts in this field, a considerably notable one is the sociological current. Initiated in France with regard to general cognitive operations in Durkheim's school, and recently revived in Germany, especially by the well-known philosopher Max Scheler, this current refers to social facts, above all to forms of collective life in the social group. It is in these facts that it seeks to elucidate cognitive phenomena. Followers of this current emphasize the significance of a sociological interpretation of knowledge. Even if this significance has not yet been wholly determined and must be subject to debate, an interest in this current itself is fully justified. It is warranted by the importance and the interesting nature of the questions that emerge when we study the relations between social life and manifestations of knowledge.

Above all, we must realize that our everyday thinking and framing of phenomena usually overlooks these relations. When it comes to science, it is commonly seen as the result of individual work, rather than as the result of any social effects. It is true that social groups devoted especially to science (universities, academies, etc.) play quite an important role in the sciences, but the general opinion is that those institutions represent auxiliary resources for the cognitive creativity that is in fact carried out by individuals. More important than these opinions, whose partial truthfulness no sociological interpretation will undermine, is the broadly accepted and widespread belief that scholars and their work exist in isolation from current social life. Their relations appear loose here, and society does not appreciate its own influence on the development and character of scientific activity.

If sociological research is not likely to corroborate such an opinion, the opposing viewpoint does not arise directly from approaching issues of cognitive creativity in separation from other fields of cultural life. This viewpoint is based on research on human mentality and the elementary intellectual operations that are manifest in any cultural field. In this context, sociologists have been particularly interested in primitive mentality and the intellectual operations of primitive social groups. Having studied forms of classification among primitive peoples, Durkheim and Mauss realized that those forms correspond to those peoples' social organization, the system of clans in particular. Hence they suggested that primitive forms of classification – clas-

---

[3] Znaniecki gives programmatic instructions in the above-mentioned article (▶ Znaniecki 1925).

sification in general – depend on the structures and character of the social group.⁴ Even if their essay on primitive forms of classification does not directly concern science, it does suggest certain lines of thought and indications. Concluding their sociological reflections on manifestations of intellectual life in primitive groups, Durkheim and Mauss propose that scientific classification is related to changes in the relationship between the social group and the individual and to the expansion of individual freedom. Thereby they consciously raise the profile of sociology in accounting for the logical operations and scientific concepts and posit the sociological perspective as a new and valuable point of view in research on the operations and outcomes of cognition.⁵

Assessing the views proposed by Durkheim and his school in the above-mentioned work as well as in later ones, we must keep in mind the attribute that has come to be accepted as that school's general tendency in explaining cultural phenomena. That trait is sociologism, a current that seeks to determine all kinds of cultural phenomena through social factors. In Durkheim's work, sociologism appears in relation to cognition as much as it does in relation to religion or morality. The outlook proposed by the French sociologist and his school allows for a broader definition of the social fact⁶, but it also facilitates a consistent rootedness in research on primitive groups, where the collective's relationship with the individual is always particularly significant. Concerning studies on the intellectual life of primitive groups, today the most well-known – at least as a topic of debate – is L. Lévy-Bruhl's work on primitive mentality and the primitive psyche.⁷ At present we are not interested in those studies themselves, but only in the implied possibility of a sociological approach towards cognitive activity in all its stages. When it comes to the sociological interpretation of knowledge, the French school has found a special supporter in the German philosopher Jerusalem, whose short treatise "Die soziologische Bedingtheit des Denkens und der Denkformen"⁸ presents an important and expressive outline of such an interpretation. Referring to the work of Durkheim and Lévy-Bruhl, Jerusalem argues that primitive groups have no actual concept of objectivity, only an awareness of things that are important for the group. The individual is only led to grasp objective truth once certain transformations occur in social relations, and above all once the necessity arises to perform specialized work within a more differentiated social order. A

---

⁴ Émile Durkheim and Marcel Mauss: "De quelques formes primitives de classification. Contribution a l'étude des representations collectives," *L'Année Sociologique*, vol. 6 (1901–1902), 1–72.

⁵ Ibid., 72.

⁶ Émile Durkheim defines this in: *Les règles de la méthode sociologique*, Paris 1919 (7ᵗʰ ed.), ch. 1. ||| Cf. *The Rules of Sociological Method*, transl. by W.D. Halls, ed. by Steven Lukes, New York e.a. 1982.

⁷ Lucien Lévy-Bruhl: *Les fonctions mentales dans les sociétés inférieures*, Paris 1910, id.: *La mentalité primitive*, Paris 1922, id.: *L'âme primitive*, Paris 1927. ||| Cf. *How Natives Think*, New York 1925.

⁸ Wilhelm Jerusalem: "Die soziologische Bedingtheit des Denkens und der Denkformen," in: *Versuche zu einer Soziologie des Wissens*, ed. by Max Scheler, München, Leipzig 1924, 187–207. ||| Wilhelm Jerusalem was in fact an Austrian Jewish philosopher and pedagogue.

further stage is reached when a group in which individuals are already aware that certain judgments are objective comes into contact with other social groups, thus coming closer to the notion of humanity. In these conditions the recognition of objective truth leads to universalism and cosmopolitanism, since that which appears objective begins to strike individuals as significant not only within the domain of their own group but also for all groups; in other words, individuals recognize the universally human significance of their objective judgements. Consequently, Jerusalem points out three factors of cognition: the social one (in the form of primitive collective representations and a reciprocal social impact), the individual one, and the universally human one. He considers all three of them to be dependent on social factors.

It is easy to critique these views, especially by pointing out their one-sidedness. But that does not undermine the questions that the sociological current has raised concerning phenomena of knowledge, or the new points of view introduced by that current. An expression of the growing interest in these questions can be found in two edited volumes published by the Research Institute for Social Sciences [Forschungsinstitut für Sozialwissenschaften] in Cologne, the first of which deals with sociological issues relating to a certain type of teaching and the popularization of science,[9] while the second groups together studies that directly concern the sociology of knowledge.[10] Max Scheler, the editor of that second volume, outlines in his *Probleme einer Soziologie des Wissens*[11] the very complex set of issues and proposes a series of bold and broadly conceived hypotheses. Assessing this work we must keep in mind that Scheler was above all a philosopher, and that his sociological arguments are based on general assumptions that belong in fact to the philosophy of culture and present potential material for philosophical discussion in their own right. Besides, Scheler's considerations of the ties connecting social life and manifestations of cognition are based on extremely broad concepts. In his discussions of knowledge and society, knowledge includes all types of cognition e.g. myth, religion, metaphysics, positivist science and technology, while his use of the term society sometimes amalgamates the concept of a social structure and the concept of the culture contained in that structure. Such a broad understanding of these issues opens many possibilities for studying and identifying putative relationships. Scheler's work suggests many lines of thought and indications, which makes it a rich but often one-sided and arbitrary expression of those possibilities. If we continue to investigate the relationships between science and the forms of social life, the methods of this work will not seem a fully adequate base.

---

[9] Leopold von Wiese (ed.): *Soziologie des Volksbildungswesens*, München, Leipzig 1921.

[10] See Max Scheler (ed.): *Versuche zu einer Soziologie des Wissens*, München, Leipzig 1924.

[11] *Probleme einer Soziologie des Wissens* is the first part of Max Scheler: *Die Wissensformen und die Gesellschaft*, Leipzig 1926, 1–229). Both represent expanded versions of the introduction to the edited volume *Versuche zu einer Soziologie des Wissens* (footnote 10). ||| Cf. *Problems of a Sociology of Knowledge*, transl. by Manfred S. Frings, London 2014.

What we need above all is a certain limitation of fundamental concepts. This is why instead of the concept of knowledge, which covers all manifestations of cognition, we focus on the concept of cognitive activity as a distinct type of human operation. Thus we study not all phenomena of intellectual life but only cognition as a special function – in other words science as a particular field within cultural life. This narrowing down of the subject of study allows us to confront a concrete sphere of phenomena, rather than a general process that affects life as a whole and can be felt in almost any cultural operation. However, the concept of society also demands either to be defined more precisely or to be substituted by some other concept. Instead of presenting society as a subject of study, the more recent currents of sociology propose to study social phenomena understood as a distinct sphere of cultural phenomena.[12] Social phenomena include all facts of one individual's relationship with another or with a group, as well as every fact of collective organization. Considering science and the forms of social life we must concern ourselves above all the social group, its structure and the processes of social action occurring between the group as a whole and its members. What relationships exist between the group's structure and those processes on the one hand, and engagement in cognitive activity on the other hand – this is one of the main questions of what we call the sociology of science.

That the relationship between science and social life exists as a problem cannot be denied. However rigorously we separate the sphere of social phenomena, they will always be associated with every form of cultural life – technics and economy, religion, art and science. This is why research on science, just like research in all other humanistic fields, must confront the question what significance social factors have in the creation and development of cognitive activity and its outcomes, and how this activity impacts the relationships between people and the life of social groups.

In this work we hope to undertake the first of these questions: the question how forms of social life influence science. We do not aim to exhaust this issue, but only to propose a few special questions that seem particularly interesting and characteristic with regard to the general problem. The state of research today does not allow us to propose ready and fully justified conclusions in this field. What we have in mind is above all to propose an approach to these issues and to analyse them somewhat more closely. To identify these questions seems to be an important task, since on the one hand we often come across complete ignorance of the role that social factors play in the sciences, while on the other hand currents are emerging that tend to be biased and exaggerate the significance of the sociological outlook when it comes to explaining phenomena of knowledge.

[…]

---

[12] Florian Znaniecki: *Wstęp do socjologii* [Introduction to Sociology], Poznań 1922, 240–254. Noteworthy is also Leopold von Wiese's theory of social relationships: *Allgemeine Soziologie als Lehre von den Beziehungsbedingungen der Menschen* (vol. 1), München, Leipzig 1924.

4. Discussing the sociology of knowledge we only touched on a few questions, though the border area between sociology and the theory of science contains a great many of them. We are merely able to point them out within the limits of this study. For instance, the complex of important and interesting issues arises from the fact that individuals who engage in a scientific activity organize themselves in special groups dedicated to that activity. These groups appear to be rooted in the relationship between the teacher and the student – a particular social relationship that does not exclusively belong to the world of science, but that plays a key role in that world. This relationship and the intellectual collaboration of individuals of equal rank constitute the basis for many scientific groups, which we can identify already in the original and distinctive philosophical schools of Greece, and which include a broad spectrum of types, from the loose intellectual milieu to institutions such as our academies and universities. These groups present a rich source of material for the sociology of science. A very characteristic case here is the scientific group, as this is where the society's influence on science meets with its opposite – science's influence on society. The influence of society on science is expressed in the form of opinions, needs and demands, and above all through various ways of supporting research. Scientific organizations and institutions, being the focal points of society's influence on science, are where society's influence on individual scientific work becomes most intense. However, scientific groups have their own existence, which makes them centres of independence and resistance against the influence and pressure of society; in some cases these centres even have an active impact on social life.

In the course of these considerations we only signal the existence of scientific groups, without undertaking to reflect on this issue. We should only add that studying scientific groups promises a certain completion for the problem of society's fundamental attitude toward science and that attitude's influence on it. It is apparent that social processes of differentiation could not have influenced the development of science as much as they did in the nineteenth century if it had not been for scientific institutions, especially universities, which provided the conditions and a suitable organizational framework for the increasing division of labour and specialization. But here a series of further issues arises that demand to be examined separately. One of them deserves to be highlighted. European universities, especially in the second half of the nineteenth and in the beginning of this century, are not only centres of scientific specialization, but also centres where knowledge is created and augmented. This corresponds to society's view, which acknowledges the importance and independence of specialist scientific goals, while at the same time asking of science that it continually develops and expands. In order to fully appreciate the influence of society, its formations and currents, onto the nature of science in the nineteenth century, we must take into account at least these two aspects of society's outlook and the simultaneous action of those two aspects. Particularly interesting is the question in how far a quantitative expansion of science, sustained by popular demand, impacts the character, even the type, simply, of scientific work.

And so, wherever we turn, there emerges a multitude of issues for the sociology of knowledge, not easy to grasp and all the more difficult to exhaust. In the present work we wished only to present a few basic questions that introduce the problem of society's relationship to science. These questions have been tackled by sociologism, a trend that aims to determine science and all other cultural phenomena in social terms. A careful and critical take on the relationship between the facts of social life and science must result in an unambiguous and discernible delimitation of the social determination of cognitive phenomena, quite at odds with sociologism. That determination is not only partial, but it is also indirect; in other words, social facts are not even in part the direct causes of any scientific manifestation. In as far as they even exist beyond individual creativity, a sphere that eludes causal approaches, such causes seem to exist only in the very facts of cognition. It is social life that represents the conditions for the development of scientific activity. Social life does not create anything itself, but it makes the appearance of certain scientific operations and outcomes possible by facilitating some research trends and hampering others.

Viewing the influence of social facts upon the development of science in this manner we enter a territory whose realness cannot be questioned. The individual devoted to cognitive activity is a member of society[d] and is subject to the processes and tendencies of its social environment. Society is interested in science as a special kind of activity that is performed by some of its members, and it assumes a certain attitude towards it. The group's attitude, opinion and demands are of great significance on account of the many ways in which a group renders an individual dependent on it. These opinions determine society's influence upon the direction of scientific activity.

But the problem of social conditions has a more fundamental significance for science. Does every social system that constitutes a society – or, in the case where there is no system, does a single group, like a horde – contain the necessary conditions for the existence of science? Science represents a special kind of cultural activity, separate from religion, art, or any sphere of practical life. But even though science thus understood need not be the special and exclusive function proper to a few individuals, empirical material does not suggest that scientific activity is possible in societies where, like in the primitive group, all members are engaged in one kind of shared productive work. On the contrary, science is found in social groups that are characterized by differentiation and division of labour, where, consequently, there exists a somewhat greater sphere of individual life. This allows us to posit that the social conditions enabling the existence of science as a distinct cognitive activity are absent in societies whose organization precludes clear differentiation and division of labour, and where there exists only a very narrow sphere of individual life free of a strict relationship to the collective.

---

[d] In this part of the text, Rybicki often uses the expression "grupa społeczna" – the social group. For simplicity's sake I render this as "society". [T.B.]

Those societies whose organization does enable the necessary foundations for the existence of science are able to create conditions varying in degree and in quality. Here we pointed out the two fundamental social processes of social differentiation and integration as two complementary and contradictory factors of social life. The predominance of one or another kind of process gives these groups their specific social character. At the same time this predominance is not inconsequential for society's attitude towards scientific activity; it influences popular opinion on science and shapes society's needs and the demands that it makes of learning. In the present reflections we were only able to touch on the question of what social processes have an indirect influence on science, and what the results of that influence might be. That question no doubt represents one of the fundamental problems of the sociology of science. It is possible to point out at least in general terms how the forms of social life, identified by processes of integration and differentiation, are connected to some research trends and some traits of science.

All these are theoretical issues and conclusions with theoretical implications. But if these conclusions are correct, their meaning will also be practical and orientational for the sciences today. This applies in particular to certain contemporary phenomena related to the so-called crisis of the sciences.[13] These phenomena are addressed particularly in German scholarship, which tackles this particular anxiety and tendency to revise accepted beliefs. Such very characteristic concepts of science as the German economist and sociologist Max Weber proposed in his famous sketch, *Die Wissenschaft als Beruf* do not satisfy those anxieties and tendencies. Weber does not defend scientism, the belief that science can solve all the questions of existence. And yet he rejects the possibility of individuals reconciling a scientific outlook with a complementary spiritual one. By approaching science as a necessarily one-sided outlook on reality, Weber attributes the following characteristics to real scientific activity: it is fragmentary, because fundamental scientific work is the work of the specialist who advances the state of research in a small area of knowledge, and it is impermanent, because every scientific result exists in order to be replaced by newer and better results.

This approach contains what are no doubt the key traits of the dominant understanding of scientificity – traits that doubtless pertain to ideas on the crisis of the sciences. The sociology of science adjusts this approach, at least partially: we have tried to demonstrate that it is partially due to particular social processes within the environment in which science develops that specialization is accepted as the main, ideal and even exclusive type of scientific work. If it is correct, then Max Weber's singular concept must also be tied up with the environment and the epoch, and does

---

[13] Max Weber: "Die Wissenschaft als Beruf," in: id. *Gesammelte Aufsätze zur Wissenschaftslehre*, Tübingen 1922, 524–555, Erich von Kahler: *Der Beruf der Wissenschaft*, Berlin 1920, Theodor Litt: *Erkenntnis und Leben*, Leipzig, Berlin 1923, Alex von Muralt: *Zur gegenwärtigen Krisis der Wissenschaft*, Zürich 1926. ||| Cf. Max Weber: "Science as a Vocation," in: *From Max Weber: Essays in Sociology*, transl., ed. and intr. by H.H. Gerth and C. Wright Mills, New York 1946, 129–156.

not present unconditional criteria of scientificity. Changes in cultural conditions, especially in social conditions, enable certain transformations in views on the science's relationship to culture and in the understanding of the essential traits of scientific activity.

On the Skills of a Researcher

Tadeusz Kotarbiński

TADEUSZ KOTARBIŃSKI (31 March 1886, Warsaw, Congress Poland – 3 October 1981, Warsaw, Polish People's Republic) philosopher, logician and ethicist, central figure of the neopositivist Lviv-Warsaw School. Left aside mathematics, physics and also architecture to finish philosophy and psychology at Lviv Univ. (1912), K. became a lecturer and later assistant and full prof. at Warsaw Univ. (1919, 1929). Actively engaged in political discussion, for example against antisemitism. From 1945 to 1949 rector of the newly established Lodz Univ. before returning to Warsaw. President of the Polish Academy of Sciences (1957–1962), where he established a laboratory (later department) of praxeology.

K. published "On the Skills of a Researcher" in vol. 11 of Nauka Polska after presenting it at the fourth session of the Science Studies Circle on 7 April 1929 (▶ **IN THIS VOL.** Sprawozdanie 1929, ▶ List of Sessions). Translation by Tul'si (Tuesday) Bhambry.

ORIGINAL REFERENCE
Tadeusz Kotarbiński: "O zdolnościach cechujących badacza," *Nauka Polska*, vol. 11 (1929), 1–10.

FURTHER TEXTS MENTIONED
▶ "Sprawozdanie z działalności Koła Naukoznawczego," *Nauka Polska*, vol. 11 (1929), 353–355.

CONTENTS: 1. An attempt at a negative case analysis of researcher material: the star pupil[a], the ineffective, the aboulic. 2. A few observations from life on the characteristics of true researchers. 3. Are there any characteristics unique to researchers? A practical and a psychological understanding of the term "researcher".

---

[a] The Polish word designating the best student in class is *prymus* (from Latin *primus*, the first). I have found no exact one-word equivalent in British English that would not be an anachronism. Depending on the context, *prymus* can mean anything from straight-A pupil to teacher's pet. *Prymus* appears over a dozen times in this section. Rather than settling on any one term in English, I use more or less synonymic expressions to convey different nuances. Kotarbiński also coins the noun *prymusostwo* (with the suffix *-ostwo* denoting a condition or quality), which I translate as "exemplary pupilhood". [T.B.]

1. Ostwald maintains that most great researchers used to be bad pupils, and he attributes this mainly to the verbalism and schematization of ordinary school teaching. This opinion ties in with common views on *star pupils*, namely that they are not at all the best material for outstanding researchers of the future. Here's one of life's thousand paradoxes: we demand of our sons that they do as best they can at school, we praise academic record-breakers, but deep down we're hiding a secret wish: "I'd rather you didn't stand out quite so much, because top students do not grow into leading minds." This lack of trust in straight-A pupils is not entirely unfounded, as this type, usually recruited from among the bright ones, carries the germs of future mediocrity. Why so? We should give this some thought, as the flaws revealed might suggest what virtues are necessary. Seeing what hinders a generally capable person from achieving outstanding results, we will perhaps find out what a competent researcher must aim for. We're interested in the mental characteristics of the individual. Social and organizational conditions will concern us only in as far as they shed light on those mental characteristics.

The high achiever is extremely well adjusted to the school system. He is loyal (at least in his deeds), docile, able to resist his own whims, and, like a liquid, he takes on the shape of the vessel. How should someone like that search for new paths... Especially when those new paths are forged through protest against the old routine. Generally, there are two approaches to establishing a balance between oneself and one's environment: adjusting oneself to the environment, and adjusting the environment to oneself. Average people excel in the former, outstanding people in the latter approach. Star pupils of today's ordinary schools are individuals with an exceptional ability to fashion themselves after the mould of average people. – Thus the head boy is exemplary in his systematic way of doing his homework every day according to the timetable. But there are different ways of being systematic. One comes from without, the other flows from within. The one that is imposed from outside is the regularity of the engineer, who performs his activities according to a plan based on the organization's requirements, a plan prescribed by the management. It is the regularity of a cog in the engine, the regularity of an automaton, set by the will of another. Meanwhile, the regularity that comes from within originates from a feeling for and awareness of the method needed to forge new paths. The pioneer imposes that regularity on himself. It is autonomous and flexible. The element of methodicalness predominates over the element of automatism (of which a certain measure must exist in all systems). Here we understand that methodicalness as the tenacity to abide by methods approved as purposeful. These, if they are good, must be economical. And herein lies the crux of the paradox that those who are perfectly well adjusted to the schedule of the school will have trouble on the road towards their own research goals. They fail to take the shortest route and lose time stopping at places of no consequence. A chess player would say that they lose a tempo. They are lousy managers because, firstly, they waste energy on unnecessary things, and secondly, they do not focus their energy intensely enough to overcome major difficulties.

But all interest is a matter of choice. Someone who has a passion for one thing (and only such a person can become a productive researcher), does not have the same passion for other things. School, meanwhile, demands steady work, and the typical top grade pupil feels comfortable with this. Once we have grown passionate about something, it is difficult to tear ourselves away from the beloved subject. We launch ourselves, like ambitious players, ignoring the strain, taking obstacles that crop up along the way with this trick or that catch. We forget about time, about regular hours for sleeping and eating... The teacher's pet who is comfortable in the school system lives according to the clock. He will eventually hand in his papers at a quarter to the hour and close up shop at a regular time, while the genuinely passionate researcher will hold out until the crack of dawn and maybe then will achieve his "eureka". The swot used to do mathematics on Tuesdays from this time until this time, then pored over his Latin every Friday from this time until this time, and so forth. He studied and learned to leave the campfire before it even was properly ablaze. In such people, who are lined and checked like their notebooks, work kills action. They will come in useful as executors of a plan, as subalterns, for training the masses, to relay information from leading explorers to the general public. They are needed in organizations as an antidote to the sowing of toxic discord. Slaves to minor responsibilities, they have no time to accomplish great designs. They unlearn what is perhaps most important: initiative, and in particular the ability to take on big and ambitious tasks and to do with utmost intensity only what leads to that chosen goal. They work in science, but they do not create it.

The star pupil, well set in his role, develops a taste for the kind of learning that is about accepting the results of other people's research as lectured; he gets used to seeing teaching as a transmission of such information through lectures. If he becomes a professor himself, this will be this teaching method, and this will be the kind of learning he will demand. That will make him useful, to some extent, but he will never become a teacher of the highest class and on the highest level, that is to say one who teaches how to create knowledge, and who does that with the thrilling image of his own example. – What is more, with all his learning, the star student – this is the reward he gets for his hard work and diligence – will remain a dilettante. Even if he specializes later, he will maintain a tendency towards dilettantism... within his own specialization. His dilettantism will not be that of an insufficiently educated amateur, but that of encyclopaedists who have a grasp of their subject, broad horizons, erudition and a wealth of knowledge – and do nothing new in it. – Finally, the model pupil loses the ability to approach tasks boldly. In this arena, he fears naivety. Above all, he wants to know whether the problem that he wants to tackle hasn't been posed yet, and if yes, then what has already been done to solve it. Before making the first independent step, he will try to undertake a broad preparatory study to acquire all available knowledge that others have produced on the subject. This usually ends in hypertrophy of book learning; more courageous people, meanwhile, can be criticized for their inadequate knowledge of the efforts of others and often end up reinventing the wheel, but they meet their problems face to face and at first hand (with a reasonable amount of preparation, of course). Thus, they have an easier road towards the deserved title of pioneer.

The above analysis of exemplary pupilhood derives from reflections on the bitterness of my own experience. Hence the certainty that these are no empty phrases, not even those statements that – sadly – are truisms. And if we rightly recognized what hampers the typical valedictorian's growing up to become an outstanding researcher, then we could try to see what these considerations might teach us. The accomplished researcher must work independently, that is to say he must be able to resist external constraints that impose on him a choice of tasks and a course of action; instead, he must know how to bend his surroundings to his own tasks and work methods. He must be able to focus with complete abandonment. He cannot be interested in all things to the same moderate extent, but must be passionate about one thing, becoming indifferent – at least for a while – about the rest, like an enamoured youth. He must be able to attack problems head-on, without erring on the side of unnecessarily broad subsidiary or preparatory chores.

Let us now turn to another category of person that seems promising for negative case analysis, namely the category of the *ineffectives*. Here, too, we have in mind capable people who are well educated in their field, but who remain unproductive. We have all encountered them in our work environment, though apparently they are especially easy to find among the Slavic peoples. What is it that brings them down to such a state? Not all can blame it on material conditions, though there's no lack of such poor souls, either. Physicists and biologists without labs, geologists and anthropologists without funding for fieldwork… We're not concerned with these, the homeless and unemployed of science, but only with those who have nothing but themselves and their own vices to blame for their unproductivity. – This class consists above all of a considerable body of aboulics of various sorts. Usually they are either lazy or undecided. Among the lazy ones many avoid exertion to guard themselves and their nerves, which are already shattered from excessive work during their youth, from the use of stimulants and drugs taken over extended periods to maintain or rouse energy levels, due to toxins, or as a result of an intemperate lifestyle. Such a person does not necessarily just sit around doing nothing. On the contrary, he is always doing something; he is active, even lively. But he only engages in easy things. He's walking, but on level ground, not uphill. The creation of valuable things, however, usually only begins when drops of sweat appear on your brow and your mental faculties reach a tension that is almost hard to bear. To read a little about this or that, to throw in a detail on a random topic without going deeper into it, to fill long hours by rewriting and repeating things that are already well known, with storytelling, etc. – this is the "work method" of such a lazy person. – The undecided person, on the other hand, is unable to cope with the wealth of possibilities, constantly lets himself be distracted – and doesn't produce anything valuable, either; he gladly changes his subject of interest, begins this and that, doesn't bring it to an end, moves on to something else and again abandons it halfway. Then again, the aboulic cannot decide to publish his results, restrained as he is by an overly scrupulous scientific conscience: ashamed that his lamp is smoking, he hides his light under a bushel. As a result, the entire research apparatus of such a scrupulous person falls into apathy. – But besides aboulics, there are ineffectives of another kind: people of a social bent. They let themselves be elected into boards and directorates, either out of a penchant for leadership and organization, or out of a feeling of responsibility (because someone's got to do it, because you can't shirk from participation, put at risk the survival of the institution, etc.), or for other reasons. And those who once got stuck in the sand of organisational life, they will be pulled in deeper and deeper, until the sand gets to their brains and causes barrenness. The lives of those ineffectives who neglect their research as a result of an overly strong teaching instinct take a similar course. This is contemptuously referred to as pedagogification.

Having considered the faults of the ineffectives, we can register the following desired antitheses: intensive work, vigour, firm decision-making, an ability to stick to the chosen research topic, resistance to temptations and persuasions coming from others who wish to burden their fellow with organizational functions (which appear to be powerful – but are in fact service positions), and finally a certain dose of disregard for that apparently high prestige of organizational work.

2. So far we have explored the characteristics of the competent researcher through the defects that prevent certain people from becoming competent researchers. We will now examine *true researchers*, asking what makes them such, and what mental qualities might be at the root of their success. The author of the present remarks must draw on his observations from the world of his colleagues. Amongst leading logicians, the qualities that come to the fore include: an extremely fervent interest in certain problems; a passion for reasoning; an aversion to inaccuracy, ambiguity, incomprehensibility, incorrectness in justifying and expressing views; an ability and enthusiasm for prolonged and focused immersion in the chosen question, while the heart beats slower for peripheral concerns; independence in the choice of problems and in ideas for their solutions; an ability to impose one's repertoire onto one's research environment and to crystalize their work around that repertoire; a certain, shall we say, intellectual pluck that expresses itself in a spirit of dispute and criticism, not in a readiness to compromise; a belief in the broader scientific significance of one's work and a nourishing joyfulness that comes from the knowledge that one is doing something exceptionally difficult exceptionally well, perhaps not without "feeling victorious," "feeling one is the first" in a given field. – Anyhow, the biographies of outstanding researchers in all sorts of other specializations reveal an entire spectrum of virtues that determine success. Thus, there are masters of patience who are able to record, for months at a time, for many hours a day, the shape and direction of clouds, from which they infer general statements in meteorology. There are fabulously capable lucky devils to whom ideas for experiments come pouring down as if from a horn of plenty. Many of us experience sudden realizations at a moment of

apparent rest: unexpectedly, an improvised solution appears to a question you'd been poring over for a while and on which your mental apparatus had continued to work subconsciously. As usual, inspiration is a matter of solid preparation. Precious bounty comes from such revelations, but these have to be earned with long hours of raw concentration. Some people stand out even in childhood by their universal curiosity, plaguing their surroundings with incessant questions. Others have a gift for unusual associations. Yet others are devoured by a hunger for knowledge that in turn makes them voracious readers. There are fanatics who fiercely believe in their own idea, who attend to it with a reverence appropriate to what is most valuable, and who combat any hint of opposition. There are solitary thinkers who work in the sole company of paper, pencil and a few books. There are also directors of laboratories, where the work is done in teams, feverishly, always with an eye to the latest findings of other equally factory-like centres.

3. The actual difficulty is to identify, among this multitude of abilities, *the virtues proper to any researcher*. Careful reflection should lead to the following conclusion. The attribute to which we are drawing attention – if it is at all something that competent researchers have in common – does not mark them out individually, but characterizes either those people who are exceptional in their work in general, or those who do exceptional work in highly intellectualized disciplines. A passion for complex and abstract reasoning is a great strength in mathematical research. But now an eminent historian testifies to his plain dislike for this. What he does like is to make facts cohere, to think about them one by one and to make short observations on each. Who knows, perhaps a conscious observer must not delve too deep into speculation. Instead, a researcher who is poring over a formula or antinomy might benefit from not gazing at the world too much while hatching up his own idea. Darwin brought back from his voyages oversees a wealth of facts; [Alfred G.] Mayer, struck by two facts, was in no great hurry at all to get to the mysterious mainland of Java, preferring to sit on deck and think up a formula for the mechanical equivalent of heat. That state of inspiration! Perhaps not all serious researchers experience it. But those who do, it doesn't make them special as researchers, since creative inspiration is also known to artists and designers. Patience helps with every kind of tedious work, not only the work of the researcher. An intense interest in the subject, an ability to become absorbed in it… oh, not only the researcher can become absorbed like that! You can get similarly carried away by a fight, a game, sports, musical works, rhetoric, and who knows what else. – The same can be said of the properties of the researcher, which we have tried to bring out by considering negative examples. Independence, confidence in choosing focus points, intensive work, etc. – these are by no means distinct trademarks of the researchers' estate. They are virtues from the catechism of all able people. – In one word, it seems likely that there are no important and yet distinct abilities proper to the researcher per se. Please note that we are not questioning the general abilities that every researcher needs. We could recite an entire litany of those: perceptiveness, attentiveness, skillful handling at least of simple forms of reasoning, etc.,

etc. What we are calling into question is the specificity, the distinctiveness of abilities that are both important and necessary for researchers.

We assert (or rather, we assume) this with reference – naturally – to a certain *definition of the term "researcher"*, a *practical* definition that we've had in mind throughout this text, and that can be summarized as follows: "A researcher is one who successfully strives to discover and account for new truths in the domain of theory". The class of researchers is remarkable by what they do; what they do, meanwhile, is remarkable by the content of the task to be solved (the task is theoretical, not about active engagement, not utilitarian, not from the realm of considerations about what is to be done); finally it needs to be taken into account whether what they do is done well, if their work is working out. [Leon] Petrażycki cautions against misguided attempts at constructing a psychological theory about an arbitrarily chosen class of individuals, that is to say a theory that would adequately and validly attribute some mental characteristic to all these individuals and to them only. It appears that the class of "researchers" as defined above is a good example of a class where it is not easy to detect specific abilities that would at least be important to its members. Let's mark out a tract of land contained for instance between the 20th and the 30th meridian, and then ask geographers to name the unique climatic characteristics of all the countries in this area, from the North Pole to the South Pole. They will surely reply that we've asked the wrong question, since the portion of the Earth thus demarcated covers a huge variety of climate zones, and the countries in it share a climate with countries in areas that do not belong to our section. Something similar is happening in our case. Whether someone is toiling over solutions to theoretical or practical problems is insignificant when it comes to the mental forces needed for efficient work. After all, in theoretical as in practical fields there are simple and straightforward tasks as well as complicated and mentally challenging ones. In advanced engineering, construction problems come so close to questions treated in pure mathematical research! The latter are solved by "researchers," the former by "technicians," whom some would see in sharp contrast to researchers. At the same time, in practice as well as in theory some problems are solved through pure reflection, some are solved through observation, some are solved through experimentation. Perhaps it is these methods that require specific corresponding abilities. But how to characterize researchers generally in a way that would include those who do easy qualitative observations in their field (as this profession makes it possible to work in such a way) just as much as those who as a matter of course tackle problems by way of exceptionally difficult, intricate, abstract mathematical reasoning…

However, to negate the assumption that researchers should stand out by some important abilities specific to them is by no means to negate the crucial role of special skills indispensable to the researcher who undertakes a given task. We agree in fact that this or that task requires not just anyone but a specially qualified person, and that those necessary qualifications include special aptitudes. But these aptitudes must be different in different cases, depending on the nature of the research. To excel at working out the axioms of set theory you need a different kind of ability than to explain the main causes of the French revolution. I would be inclined to believe that those mental dispositions even stand in each other's way.

We might have to modify our answer to the question posed if the concept of the researcher was defined differently. – For instance, the term researcher could designate a person – and such a person only – who enjoys all permanent occupations less than he enjoys posing and solving problems. This *definition of the "researcher"* could be called *psychological*. Can we detect a distinct psychological feature for this class, and if yes then what would it be? And above all, are there people who have a passion for posing and solving problems in general? a passion for problems as such?... some let's call them problemophiles or even problemomaniacs? Let psychologists answer that question. Real-life observation suggests rather that problem lovers have more particular tastes. One has a passion for chess endings, another is terribly curious how people live in faraway countries, yet another is dying to know how evil could have come into a world created by a good being, if the world has always existed and if it is reigned by spirit or chance, etc. Naturally, a certain intellectualism is common to them all – they are all burning to solve some problem. But what problems? And what motivates them? And what characterizes them in their attempts to come to grips with those problems, what mental powers come into play specifically? It seems that there are more differences than commonalities... So we ask whether with researchers it is e. g. like with men in general, who, as such, are attracted to individuals of the opposite sex, which is exceedingly important for their masculinity[b], though different men are particularly drawn to different types of individuals of the opposite sex; or is it as with travellers, each of whom is on the road for a different reason, with different motives and with a different disposition, – but since each must for his own purpose get from one place to another, somehow they all end up on the road, all are travellers. So is there a deep specific psychological alikeness among researchers, psychologically speaking, as there is among men, or is it a superficial alikeness as among travellers? It is hard to even project an answer to this question... My private speculation tends to tip the scale towards the side of the travellers.

---

[b] Kotarbiński literally describes men's sexual attraction to individuals of the opposite sex as "exceedingly important for their characterization as men precisely". While his outright eclipse of same-sex desire among men is most likely due to a certain naivety, I replaced the literal translation of "their characterization as men precisely" with the term "masculinity," which, familiar from studies of gender performance, should bolster the possibility of an ironic reading of this potentially offensive passage. [T.B.]

Investigation and Teaching

Bogdan Suchodolski

BOGDAN SUCHODOLSKI (27 December 1903, Sosnowiec, Congress Poland – 2 October 1992, Konstancin-Jeziorna, Republic of Poland), historian, pedagogical theorist, philosopher. Studied pedagogy in Cracow and Warsaw before graduating in 1925, and later in Berlin and Paris. Prof. of pedagogy at Univ.s in Lviv (1938–1939) and Warsaw (1946–1970). S. later held high representative posts in national and international societies and academies and was involved in founding UNESCO. He became chairman of the National Council of Culture (*Narodowa Rada Kultury*) in 1982 and was a member of parliament from 1985 to 1989.

The text "Investigation and Teaching" was published in vol. 21 of Nauka Polska. S. had presented the paper at the twentyfifth session of the Science Studies Circle on 20 November 1935 (Sprawozdanie dziewiąte 1936; ▶ **IN THIS VOL.** List of Sessions). This edition follows the English translation from vol. 2 of Organon (1938) except for a footnote to the author's name introducing him as "Ph. D., Docent at the Piłsudski University of Warsaw." Linguistic errors have been corrected without indication.

ORIGINAL REFERENCES
Bogdan Suchodolski: "Badanie a Nauczanie," *Nauka Polska*, vol. 21 (1936), 1–44.
— "Investigation and Teaching," *Organon*, vol. 2 (1938), 43–78.

FURTHER TEXTS MENTIONED
"Sprawozdanie dziewiąte z działalności Koła Naukoznawczego," *Nauka Polska*, vol. 21 (1936), 300–324.

CONTENTS: Extent of the Problem. Programme of Analysis. Science and the Scholar: Objective Conception. Education and the Teacher: Objective Conception. Science and Education: Objective Conception. The Conflict Between Science and Education. Results of the Conflict between Science and Education. The Humanistic Point of View. The Process of Learning. The Essence of Investigation. Investigation, Learning, Teaching. Intellectual Processes and the Life Situation. Science and Education as Factors in Forming of Intellectual Culture. Social and Cultural Significance of Both Theories. Psychological Significance of Both Theories. Evaluation of Both Theories.

Extent of the problem

The problem of the mutual relationships which appear in investigation and in teaching is a difficult and many-sided one. Its more exact analysis would require the differentiation and analysis of several larger groups of problems. First of all, that psychical process which appears in investigation and again in teaching may interest us. The discovery of qualities common to and differing in both of these processes would permit us to decide, whether there exist the definitely separate attitudes: that of the investigator and that of the teacher, and also what is their connection with other characteristics of man, with his character, his style of living, his plans for the future, his relation to reality and to other men. In this way we should approach the solution of interesting typological questions, namely, whether and in what sense we may speak of the investigator and of the teacher as a stable type of personality.

Another group of problems appear when, less interested in psychological analysis, we attempt a phenomenological definition of investigation and teaching, when, independently of concrete subjects, we examine the essence and structure of both these functions. From this point of view their mutual relation is presented as the relation of science[1] to education.

A third great group of problems appear when we take the social and cultural viewpoint. Then we are interested in such questions as sources of the investigating impulse and the teaching impulse, the bases for the social recruiting of investigators and teachers, the importance which different epochs have attached to science and education. Then we see the rich heterogeneity of the conception «investigation» and the equally complicated heterogeneity of «teaching». The changing social and cultural functions of these processes, their evolution from practice to theory, from the concreteness of life to the abstraction of books, would deserve here an especially detailed analysis. In connection with this there appears the problem of the intelligentsia as a social class, which makes the development of investigation and teaching its life mission. A revision of this opinion, a comparison of it with the values of other classes would throw considerable light on the social and cultural sources of science and education.

The fourth and last group of problems are concerned with methods and means of purposeful influence on the intellectual culture of people. To this group belong questions of the organization of education on different levels: the problem of separating or combining, within the bounds of one institution, tasks of investigation and teaching, in other words, the problem of institutes for investigation and colleges, questions of the methods of teaching in schools, and methods of extra-curricular educational work. What teaching really is, under what conditions it can be effective, how to make

---

[1] The terms «science» and «scientist» are used here in their broadest sense, «science» including both natural sciences and humanities, and so practically covering the whole field of organized knowledge.

it permeate the personality, how it should be organized, these are the essential questions concerning education from the point of view of organization.

Programme of analysis

These specified groups of problems should be treated in a certain order. It would scarcely be advisable to begin with matters of organization and methods – in fact such a beginning explains the failure of many practical discussions concerning the mutual relation of science and education, of investigation and teaching. The evasion of basic analyses on the pretext that life demands quick and concrete decisions, and that theoretical speculation is long and tiresome, is not justified, since every organization and planning, every concrete regulation imply opinions on the essence of science and education, as well as a certain idea of the process of investigation and teaching. These opinions and ideas are usually very dangerous, unless they are sufficiently clearly thought out and definitely grasped as basic propositions.

Therefore, it is necessary to begin with those considerations which concern the essence of investigation and teaching, science and education. When, however, we try to discover the best founded judgments from the large literature on this subject and to construct a distinct and consistent system of opinions, we are soon convinced that such an attempt is impossible of realization. The ideas are too heterogeneous and conflicting. But it does not seem hopeless to try to find some order in this heterogeneity. We can find it when we realize that the phenomena of investigation and teaching may be looked at from the objective point of view, as actions characterized by objective production, as also from the humanistic point of view, as actions which form man. The first of these points of view prevailed in the epoch of positivism, the second is being formulated only to-day. And on these historical bases, considering the opinions of the epochs of yesterday and to-day, it is possible to analyze these two standpoints more exactly. Of course, in each of these two epochs, as in each previous one, both points of view have appeared and do appear, often intermingled and undifferentiated. The aim of our analysis therefore will be, above all, an exact differentiation of both standpoints and a characterization of each in such a purity of idea as they never attain in life. For, only in this way may we attain a clearer orientation in the complicated and compound processes of reality. By defining in isolation each of these two standpoints, we shall find the direction of the solutions to which these standpoints lead and the kind of one-sidedness which is apprehended.

Science and the scholar: objective conception

From the point of view of the objective conception of science – and this is one readily taken not only by laymen but by scholars as well – science is acquisition of knowledge which is the truth about existence. Knowledge is a great and faithful mirror of reality and, thanks to this, extends beyond the private convictions of individuals and groups, breaking the bounds of time and space. For, truth is eternal and universal. We are constantly seeing more and more of this truth. Although the history of science reminds us that often what through centuries belonged to knowledge has fallen to the level of superstition, and what was looked upon as the visions of madmen has entered the domain of knowledge, yet – and the belief in this is strong – the treasure house of knowledge is constantly being enriched.

This objective, universal, durable knowledge is the ideal goal of investigations, their ambitious aim. The right of entrance to the domain of knowledge is thus acquired by not all the results of investigations but only by those which possess, according to scientific opinion, qualities of certainty and permanence. Investigations should therefore be so conducted that they should permit the attainment of such results. Upon this depends the whole value of investigations. Their function is the enrichment of knowledge, the increasing of its acquisitions.

According to such a conception of knowledge do we define the ideal of the investigator. His chief care should be the acquisition of results which might occupy a lasting place in the domain of knowledge. In order to attain this, he should break with the world, and even in a certain sense with his own self. He should put aside everything which would hinder him from a direct and objective view of reality. He should rise above his country and his epoch, forget himself as an individual, make himself free of everything that is concrete and incidental and become the ideal «objective man». Only then may he face reality, listen to her voice, and faithfully register it. Essentially the investigator is only a tool, whose value is judged according to the results of his activity. J. W. N. Sullivan, characterizing the type of the scientist, refers to the above expression of Nietzsche: an «objective man», and argues that it perfectly suits the description of a scholar, who is and ought to be exclusively an «instrument», exclusively a «mirror». Sullivan illustrates this point of view with the examples of Cavendish and Huxley. Huxley was deeply interested in political problems, religion, social life, he took an active part in quarrels and polemics. Cavendish kept apart from these matters. Huxley was a good husband, a solicitous father, a faithful friend. Cavendish had no wife, he had no friends, he had no enemies. Cavendish did not care for the victory of his views in public opinion, Huxley fought passionately with theories which he considered false, and which had won popularity. Huxley passed successively through states of hope and doubt, Cavendish always kept the same tranquillity. For each of the above reasons Sullivan concludes that Cavendish realized the ideal of the scientist more completely[2].

The ideal so defined stresses the values of the objective function of the enrichment of knowledge, but is not concerned with the subjective side of this work. The influ-

---

[2] John William Navin Sullivan: "The Scientific Mind," in: *Science and the Scientific Mind*, ed. by Leo E. Saidla and Warren E. Gibbs, Great Necks, NY 1930, 113–127: 125.

ence which scientific occupations have on the scientist, his personal relation to them, the creation within himself of a certain attitude towards science and truth – all these are secondary. A certain degeneration of the ideal is easy in these conditions. Training oneself to be an «objective man», making oneself exclusively an instrument of investigation may mean not only freedom from one-sidedness and blindness, but may also become a resignation from the deepening of one's personal relation to scientific work.

Education and the teacher: objective conception

To the treatment of science from the point of view of its objective results corresponds to education considered as propagation of knowledge among people. The test of the value of pedagogical work is the permanence and exactness of the picture of knowledge which arises in the minds of the pupils. Their effort on the other hand is composed of the mastery and committing to memory of that knowledge which is offered to them. Of course it is impossible to give anyone the whole of existing knowledge, since it greatly exceeds the comprehension of an individual mind. But the selection which must be made can be only a quantitative selection. Depending on the age and the profession one has chosen, each receives a certain amount of knowledge. Just as there are large and small encyclopaedias, each having the same structure and the same material, so, although the different amounts of knowledge vary among themselves quantitatively, nevertheless each of them, even the most modest, is a diminutive copy of the same, single, universal, great knowledge. Teaching and learning are understood in an objective manner, as a handing down and perpetuation of knowledge, just as in an objective way investigation is defined as an enrichment of knowledge.

The ideal of the teacher and that of the pupil have qualities similar to the ideal of the scientist. Each of them, if he wishes to occupy himself with knowledge and education, which is the reflection of knowledge, should become an «objective man». The teacher has to be faithful to the truths which scientists have passed on to him. The joys and sorrows of discovering new truths are closed to him. The dramatic battle for their acquisition he knows only from descriptions and studies, – he does not take part in it. He has to use his whole energy for making knowledge accessible to other people. He must translate the difficult language of the scientist into simple everyday speech for the laity. He must make the abstractions of difficult reasoning accessible by putting them into the form of easy examples. He may even substitute simpler statements for those too complicated for the average mind, in spite of the fact that scientific precision might suffer as a result. He should lay the greatest stress on that which in science is most thoroughly investigated, on that which is sure and basic, and he is not permitted, as a scientist is, to be interested in doubtful problems. As a good intermediary and skilful dealer in knowledge he should forget his own self, his own convictions and persuasions, and devote himself, completely and faithfully, to the propagation of that which has been assigned to him. As the scientist reflects reality, so the teacher reflects knowledge. He is the tool of its popularisation. From this objective point of view, it is of no consequence what his relation to knowledge is in his personal life. He ceases to draw from knowledge for his own use, to use it for his own inner improvement.

The pupil is defined in a like manner. Knowledge acquired by scientists, reflected by teachers, is to be perpetuated in pupils. Learning is not to be a process of personal growth, but a method of acquiring and perpetuating information. From this standpoint the individual heterogeneity of interests does not possess a great deal of meaning. The store and structure of information which is to be appropriated are designated in advance: it is the reflection of universal; objective, unchangeable knowledge. The best pupil is he who most quickly and most firmly grasps the store of knowledge presented to him, who is capable of being its most objective receiver.

Science and education: objective conception

In this way there comes into being a close connection between the conceptions of investigation, teaching and learning from the objective, matter-of-fact point of view. Each of these processes is understood independently of its subjective, personal meaning. The acquisition, handing down, and perpetuation of the results of knowledge mean corresponding functions for the scientists, teachers, pupils. The more objective they become, and the more exactly they make just tools of themselves, the better they perform their tasks. The relation of science to education is to be understood, as it were, according to the principles of mirror perspective. In knowledge the scientist seeks to reflect reality, the teacher to reflect its ready and lasting acquisitions, and the pupil to perpetuate them in his own mind. According to the measure of abilities and education, different people, located in various points of this mirror perspective, see a larger and more distinct or a smaller and less distinct picture of reality. The relation of education to science is, therefore, one of subordination. Thanks to science we obtain a faithful picture of reality, while education, on the other hand, is only a picture of that picture. Therefore, every step forward in the field of science is progress in the sphere of education.

The conception of education as a function of the propagation of knowledge is connected with the conviction that education should entirely be determined by the content of the acquired knowledge. All other factors: religious, political, social, which seek to influence the shaping of education, are, from this standpoint, usurpers. They should be fought in the name of the neutrality of education. This neutrality is an expression of the objectivism of knowledge. Just as knowledge forms a field of its own, free from passions of the flesh, independent of the chimeras of individuals and groups, so education, being the reflection of knowledge, should maintain its universal, objective value and stand above the present disputes of various world-outlook camps. The essence of educational work is popularization.

There arises in this way a peculiar dictatorship of science and scientists. For they are to have the exclusive right of dictating to teachers and pupils that which is to be learned. The scientist, the teacher and the pupil form three links in the process of creating, propagating and perpetuating knowledge, just as in economic life there exist producers, middle-men, and consumers. More important still, the producer-scientist has almost the supreme voice. The border between science and education is not to be crossed, scientific investigations, no matter how modest, are an enriching of knowledge, and differ essentially from educational work, which is only intermediary. Therefore, scientific activity is, on principle, valued more highly than any educational work. Therefore, greater respect should be paid to the scientist than to the educationalist: the first must have outstanding creative abilities, the second is only a recreator. Therefore, in scientific circles the popularizing tendencies of scholars are not well received: they are considered an unnecessary and dangerous debasement, a descent to another, worse class of occupation.

This objective world of information, gathered by science, popularized by education, becomes in this conception an independent reality, built, as it were, over the concrete world, related only with «objective people». It is only with that «objective» part of his personality that man can take part in it. Other factors of his personality must either seek satisfaction elsewhere or be exterminated. Science and education are essentially above the individual and the society. The former are not to serve the latter but to be served by them[3].

The conflict between science and education

An analysis of the contents and methods of work in school and extra-curricular work showed more and more clearly that between the needs of education and the structure of knowledge there occurs a disharmony which – if we do not give up the conviction that the aim of pedagogical activity is only to give a diminutive picture of the state of knowledge – will constantly increase.

This disharmony is based, above all, on the fact that the pedagogical need is a need of understanding reality in those concrete complexes of phenomena which we observe, while, on the other hand, science provides us with information and explanations concerning complexes which are singled out and built artificially. This difference was noticed first in elementary teaching. Children are interested in the world in its concrete form and cannot understand why various branches of sciences should

---

[3] In its extreme formulation this standpoint treats education as a means for recruiting scientists and scholars as priests of knowledge. Along with the entire society they have to serve knowledge and make the greatest sacrifices for its benefit.

speak of such a living entity as what is for them just a forest, for instance. The division of studies into objects definitely separated from each other is, as the child feels, something dead and artificial. Therefore, various types of combined studies seek to overcome these dangers of schematism and to organize teaching on foundations entirely different from any scientific classification of knowledge. The school programme should not in its construction correspond to the various fields of knowledge, but it should supply information grouped on the basis of the interests and experience of the children. The teacher, taking as a point of departure concrete events or concrete entities, discusses them from different points of view, teaching at the same time arithmetic and spelling, nature study and drawing, geography and gymnastics. The teaching of separate subjects in the higher grades takes the place of this combined teaching.

But psychological and pedagogical observations made on children have a more universal sense. For not only children, but also the youth and grown-ups like to understand things as a whole. Only specialists are accustomed to directing their interests along the beaten track of separate branches of science, while on the other hand, the laity seeking education is interested, just as children are, in the concrete complexes of reality, without consideration as to how they are divided and classified from the standpoint of the various branches of knowledge. If the problem of genius interests one, then one wishes to see it from all sides and tries to get information from the various branches of science, that is, from psychology, psychopathology, sociology, history, etc. If one is interested in the problem of unemployment, or the problem of nationalism, then likewise one will not find sufficient answer in the works of one branch of knowledge. For the very reason that pedagogical interests are usually concrete and integral, while knowledge is composed of various branches investigating certain separated and purposely grouped kinds of elements, there often arise disappointments expressed by the feeling that science does not answer the questions which are asked of it.

The second misunderstanding between science and pedagogical interests concern the basic heterogeneity of the points of view. Every independent and developed branch of science creates a system of conceptions and problems completely alien and without any interest to the laity. To this belong, above all, questions of methods of investigation and that of tracing the border lines between particular sciences. For instance, the layman interested in social reality and taking up a textbook on sociology, will feel almost completely indifferent about a large portion of it devoted to considerations on the essence and methods of this science. Likewise for one who is interested in literary works but not in the science of literature, it is not important how specialists define the tasks and difficulties of the investigator in this field. Things which are important and interesting for the specialist usually seem to the laity to be not understandable and strange, while those which are important for the laity are in the eyes of the scientist uninteresting or futile. This happens because in scientific activity the criterion of importance of a given subject is based on its value for investigation. Hence the lively interest of scientists in uninvestigated questions, even though they concern matters apparently secondary. Hence the lack of interest with which specialists treat questions apparently important but which are uninteresting from the point of view of investigation. The layman has a different outlook. He is interested in the thing itself, and not in its value as a material for investigation. Therefore, he demands of scholars the explanation of things important and great, he desires to see reality in bold lines of a synthesis, and he is bored with reading analytical items and trifling polemics. In science synthesis is an unattainable ideal, in education an ever-present need.

Further differences between the world of science and educational interests and needs are concerned with their relationship to truth. Science – from the objective point of view – is the acquisition of knowledge, reflecting reality. Everyone wishes to become acquainted with this knowledge and therefore confidently turns to the scientist. But the development of scientific investigations strengthens more and more the conviction that the relation of science to truth is absolutely different from the attitude towards truth of the average man. This attitude is characterized by impatience and the refusal to compromise: people wish to know how it «really» looks and that without any reservations. The scientist also seeks the truth but, in fact, he is half sceptic, realizing that he will not find it neither quickly nor once for all. This disharmony between the needs of education and science has led to the breaking up of the old hierarchy of one-sided dependence of pupils on teachers, and of teachers on scientists. The realization of the self- dependence and validity of educational interests became the expression and reinforcement of the conviction that the general public eager for education may not be treated as a market for the supply of mechanically engrafted scientific information, but its interests must be satisfied. This was – to use the previous comparison from the field of economics – the revolt of the consumers against the dictatorship of the producers. People began to demand such education as they really needed and looked for such education that would satisfy their interests and give them the answers to their questions. They became indifferent towards an education which was only the reflection of knowledge and did not penetrate into their personalities.

On this basis a new conception of education began to develop. Education began to be looked upon not as a function of the propagation of knowledge, but as a function of forming the intellectual culture of man. In connection with this, the tasks of the teacher also began to change. More and more distinctly the teacher was passing from the function of handing down knowledge to the function of enlightening people. It became necessary to choose between these tasks which formerly seemed identical. The types of teachers began to be differentiated: some were more interested in knowledge, others in pupils; the first stood on the side of science in the growing conflict of pedagogical needs and science, the second supported education. But as the realization that education is not only a reflection of knowledge took firmer root, there followed the realization that a teacher cannot be a mere salesman of the goods of knowledge, but he should have a certain independence in relation to the scientists.

All this bore witness to the fact that the paths of knowledge and the paths of education are not parallel, and the attempt at their permanent harmonization on the basis of the conviction that science creates knowledge and education propagates its acquaintance among people – is a delusion.

Results of the conflict between science and education

The fruits of misunderstanding between science and education are bitter for both parties. For, when educational needs cease to be well satisfied by knowledge, and scientific work ceases to find its echo in education, there is no field for the development of science and the growth of education.

Scientific work cannot be organized at will, nor is it the result of material facilities. It arises and develops from personal needs of those who believe that its hardships are worthwhile. This faith needs a friendly atmosphere. It must draw its sap from the earth. The indifference of the masses towards these efforts destroys science sooner or later. Not supported and not understood, in lonely heroism, the race of scholars perishes. Sometimes geniuses are able to create even in these conditions, but how many people of genius have been broken by them, this shall we never know. It is so in all fields of culture, but this social recognition is especially important in science, since science is not born in inspiration and improvisation, but requires continued and systematic efforts. In a society which does not appreciate science, scientists live – to say in the language of modern psychology – with a feeling of inferiority. For, they are occupied with something which is only tolerated. They find themselves beyond the limits of true life. Their queerness is forgivable, although it deserves no praise. In response, they keep aloof, absorbed in their own affairs, ceasing to care for society. By making use of a language and terminology not understandable to the laity, avoiding all popularization, they begin to feel superior and thus find a compensation for the inferiority complex, which social indifference has created within them. If, in spite of this indifference, they keep pursuing their work, they do it often as a kind of protest against social inferiority. However, a deeper and valuable creative science cannot develop in these conditions.

The social appreciation which science requires may have various forms. The most superficial and dangerous one is the appreciation of science out of gratitude for the technical facilities which it bestows upon us. In such an atmosphere science falls to the level of utilitarian speculation. But when science satisfies educational needs it is valued for what is essential in it – the search for truth. When it is able to give answers to questions, when it helps in the organization of our intellectual life – then it is cherished in a manner valuable and stimulating to the scientist.

But the disharmony arising between knowledge and education is depraving not only for science and scientists but likewise for education and teachers. Education which does not satisfy the needs and interests of the individual, becomes for him something artificial and compulsory. He learns to treat it exclusively from an utilitarian point of view, and loses his ability to understand and seek education for its own sake. The danger is very marked in schools with a too heavy syllabus designed for the purpose of handing down knowledge, and not for the purpose of educating the pupil's personality. A man growing up in these conditions retains, during his whole life, a dislike for the broadening and deepening of his intellectual culture.

It is no better when education, desiring to keep pace with life, breaks its connection with science. For, then it easily becomes a means of propaganda. A disdainful treatment of science leads to superficiality. For, no matter what we think about the objective value of science, the fact must remain indisputable that it is a most zealous seeking after responsible and justified opinions about reality. A disdainful attitude towards science turns into a general disregard for human endeavour, a conceited and absolutely unjustified desire to begin everything over again. Science teaches the continuity of effort, and lays stress on the value of method. Education which renounces obedience to science, easily falls prey to the danger of a too speedy formulation and becomes merely a training. Science teaches care and precision in the defining of laws and facts, and does not recognize dogmatism and haste. Therefore, the spirit of science guards education against being mechanized.

The humanistic point of view

There must be found some way out of this conflict between science and education which is highly harmful for both of them. Both scholars and pedagogues are searching for it, attempting to define the requirements which science and education must meet, if their paths are to draw close to each other. In this search, conducted by various people representing various points of view and various fields, there appears, however, a certain uniformity. It is the result of a new, maturing attitude towards the whole of culture. This attitude is often defined by the old name of humanism, and, although there exist very great differences in the exact definition of these humanistic points of view, the basic standpoint is nevertheless distinct and common to them.

The present epoch – in the opinion of the humanists – is suffering from an overgrowth of the objective world. Humanism is a protest against the stifling of man with objective culture, against treating him exclusively as a tool in its development. Both in the sphere of economics, and in the spiritual and political spheres, it values that which shapes man and aims to transform the conditions of his life in such a way that he might best and most fully develop his personality. Its care is first of all for the man. The humanistic point of view – according to this conception – is obviously not a negation of the standpoint of natural sciences. It is an attempt at organizing the whole of reality from the point of view of its importance in the education of man.

In various fields a criticism of the existing state of affairs is being conducted, today, from this standpoint. So, for instance, in the sphere of economics, humanism is opposed to treating people as a labour-producing force and as a market for consumption. Professional work should not be organized exclusively from the point of view of the good of production, but from the point of view of the working man. Similarly, culture with its traditional and present acquisitions should be evaluated according to its reaction to people. Therefore, e.g., the museum and the library should not only preserve past treasures, but should serve actively in the process of shaping man.

From this standpoint, described in outline, the conflict between science and education shows itself in a new light. It is due, to a considerable degree, to the conception of science as an acquisition of objective knowledge and of education as the popularization of it.

When we have freed ourselves from this conception, when we make science and education something really important and valuable for *man* – things which serve him in his inner development, their deeper union will be possible. Now a question arises which we should consider carefully: how can we overcome this objective point of view in the conception of science and education, how should we, adopting the point of view of humanism, define them as functions in the shaping of man? What is necessary for this? The answer to these questions demands first of all an analysis of the problem of learning, investigation and teaching.

The process of learning

We use the expression «I am learning» in order to signify different processes. Thus we say that someone is learning to write on the typewriter, that someone else is learning a foreign language, we say that youth learns in school, and students are learning for their examinations; we say finally that a man learns much from life and the experiences he has undergone. There are certain characteristics common to these processes, which permit us to give them a common designation, but at the same time there are different ones as well. Learning may mean the acquisition of a certain efficiency, considered by us as useful but not very important. If we learn to swim but have no ambitions for distinction in sports, or if we learn a foreign language with the single aim of being able to make ourselves understood abroad, then such learning does not absorb us entirely, does not contribute to forming our personality. The process of such learning is usually burdensome, boring, or, at best, indifferent. We are, above all, interested in the result which we may make use of. A person who has learned to use a typewriter, to ski, to drive a car, to swim, etc. possesses more or less permanently a certain efficiency in those fields which may give certain advantages or pleasure.

Somewhat different, but of a similar character is such learning which supplies us with certain information which we need either for public opinion, or for schools and examinations. The result of this learning is usually not some permanent acquisition, which we may use, as we use an acquired efficiency. This learning usually assures us certain definite advantage, thanks to examination certificates and diplomas given on the basis of evidence that the candidate has mastered a certain group of information. From this standpoint it is by no means important to discover what role this knowledge played in the candidate's life, how it shaped it, what it is for him to-day, what it may become in his further development. Thus, this process of learning also does not go deeper into our personality. The store of information acquired is something external in relation to ourselves. We may make use of it at times, but its ability to shape personality is very small.

But there is still another process of learning. We say that we learn from our life's experiences, our relations with people, our professional occupations, our mistakes, successes and misfortunes. Such learning has the character of inner development, it improves our personality. It is not a burdensome effort to attain intended and expected results. It leads to a development of our character, mind and heart. Its results are not acquirements which might possess an external and additional character in relation to ourselves. They are in us and they compose our history.

In this way we may learn not only from our life's experience but also from science, if our relation to it is a living and personal one and if we treat it as an element in our inner development.

Thus there exist three forms of learning: learning as a process of acquiring efficiency, learning as a collecting of information and committing it to memory, and learning as a process of personal development. The subject of learning may belong to various fields: to sport, technics, ethics, social life, politics, religion, science. Each of these fields may awaken the process of learning in any of its three forms, although certain of them have the ability of awakening one form of this process of learning and others another. Our relation to knowledge was usually expressed in the first and second forms of learning. Learning was based either on the acquisition of efficiency (e. g. in speaking, writing, counting, and the like) or on the memorizing of information.

Modern didactics, which demands a reform in teaching, wants to make learning a learning in its third and deeper form. It is of the opinion that the weakness and faults of former didactics were due to learning being treated exclusively on the level of a superficial memory mastery of the subject. Therefore, learning was conceived first of all as memorizing. When in the year 1909 F. M. McMurry questioned students and teachers as to what is the proper meaning of the word «study», he received the answer in 90 % of the cases, that it was based on the process of memorizing. But the conception of learning as a process which exclusively concerns the intellectual sphere, in isolation from the other spheres of our personality, and is based on the passive mastery of information, is a harmful and obsolete one. For, the process of learning proves to be lasting and yielding results only when it has an active character and develops the personality.

Therefore, in modern didactics we find an entirely different definition of the process of learning than in the old didactics. A.R. Palmer[4] defines learning as a process

---

[4] Anthony Ray Palmer: *Progressive Practices in Directing Learning*, New York 1930.

of development, whose result is a new manner of behaviour of the individual. Its newness may be based either on ability to act, or on knowing how to understand phenomena, or on the personal experience of the value of certain things. The fruits of learning are permanent and valuable changes in the man. There are no other fruits of learning, and, above all, it is a mistake to look for such fruits in the collecting and memorizing of information. Learning engages the whole personality. One must not imagine – writes Walter S. Monroe[5] – that thanks to the process of learning, knowledge passes from the textbook to the mind of the pupil.

Learning is a process of development which requires a certain activity, and not the mastering of certain information. The mastery of knowledge means that knowledge has become an element of our personality, it has ceased to be an external massing together of facts and opinions. Learning – says Noble Lee Garrison[6] – is one's own reaction to situations. It demands the awakening of interests, their satisfaction in independent work, out of which arise new interests. In this way learning ceases to be a short-lived means used for the attainment of desired aims, and becomes a miraculous perpetuum mobile of the spiritual development of man.

On the basis of such a conception of the process of learning, modern didactics is organizing, in an entirely new manner, the work of acquiring knowledge. The point of departure must be some definite interest which impels the search of a new and one's own answer. Situations in which we react in old and mechanized ways do not possess the power to impel us to learning. Only when old habits do not suffice do we begin our independent search. Learning therefore should be a solving of the problems which arise from our interests. These problems may be practical, for the course of actions or the planning of future activity always give rise to very interesting questions whose solutions require intellectual work. Learning stimulated by our interests and plans becomes the expression of the needs and aims of our personality. We learn most permanently and at the same time most quickly that which is necessary and important to us from this personal point of view.

The process of learning so understood is an independent search. It has its defeats and its victories. It is characterized by distrust towards all which has not been proved by our own observations and analyses, and a confidence in the strength of our own mind. This kind of learning gives us as result the intellectual satisfaction which is always present when we arrive to put in order our opinions and to formulate views which formerly we were unable to formulate. But this state of equilibrium does not suffice for long and there soon begins a new search.

In understanding learning as a process of personal development, we stress its active and individual character and, by the same, limit the importance of teaching. Now, teaching serves learning by creating such situations in which the process of learning might most intensively develop. This has an important influence on the re-

---

[5] Walter S. Monroe: *Directing Learning in the High Schools*, Garden City, NY, 1927.
[6] Noble Lee Garrison: *The Technique and Administration of Teaching*, New York, 1933.

lation of teaching and learning to investigation. In liberating learning from a passive imitation of the process of teaching, in making it an important and active intellectual process, we approach the investigative process. Investigation becomes the model of learning. The more the pupil tries in the course of his work to follow the path taken by the investigator, the better his results will be.

The essence of investigation

Scientific work, like learning, may be conducted in various ways. There are investigators who are interested in knowledge but do not give themselves up to their work. They are objective men· and perform their professional functions. They often perform them very scrupulously but without deeper interest. They are interested, before all, in objectively important results, sometimes because these results win them fame, sometimes because they are «feats», or «acquisitions». This conception of the investigator is, as we have seen, characteristic of the objective theory of science.

But there is another type of an investigator. He belongs to the restless and never-satisfied type of person. He is steeped in a passion for the acquisition and broadening of intellectual order within himself. Results obtained in this way are not the basic incentive in his work. Investigative work is his life, the only valuable and desirable way of using the days and hours of his existence. The results he obtains are like ripe fruit falling from a tree which lives and grows constantly, unconscious of the harvest. Only his own development, the overcoming of difficulties he encounters, his struggle with the unyielding reality and unthought-out thoughts are important for him. In this devotion to self-education through the effort of investigative thought, he differs from the investigator who is thinking exclusively of the enrichment of knowledge, of the attainment of results, of the enlarging of the list of scientific publications. All these attainments seem to him rather minor and secondary. It is not for him to appreciate whether he has enriched the store of objective knowledge, but he has a distinct feeling that his own life, devoted to investigation, he has lived well. And that is why the scepticism, common among great scholars, as to the possibilities of attaining the truth does not keep them from investigative efforts and sacrifice. From this point of view we may not consider as a tragic misfortune the characteristic confession of Lord Kelvin:

One word characterizes the most strenuous of the efforts for the advancement of science that I have made perseveringly during fifty-five years; that word is failure. I know no more of electric and magnetic force, or of the relation between ether, electricity, and ponderable matter, or of chemical affinity, than I knew and tried to teach to my students of natural philosophy fifty years ago in my first session as Professor.[7]

His life, however, either for himself or for his pupils was not wasted.

---

[7] From Kelvin's address at the time of his jubilee, cf. "Celebration of Jubilee of Lord Kelvin, 15th, 16th, and 17th June, 1896," in: Lord Kelvin, Professor of Natural Philosophy in the University of Glasgow 1846–1899, ed. by George F. Fitzgerald, Glasgow 1899, 31–73: 70–71.

An investigative work of this kind, engaging the whole of our personality, also engages its moral side. That is why the requirements which we demand of investigators from an intellectual point of view, are at the same time moral requirements. We demand of them conscientiousness and perseverance in work, impartiality and freedom from prejudice and bias, criticism and sacrifice, courage to reveal the truth. These requirements can be met and fulfilled only by a man with a strong character and high intelligence.

Therefore the conviction that science arises exclusively from the intellect, the scientist being only its instrument, is a great mistake. From the new point of view, the greatness of the scholar is measured not so much by his objective attainments as by the degree of his personal union with science. The more himself is the scholar in his investigations, the more they contribute to his own development – the greater is his value as a scientist and a man. In the light of this dynamic conception of the process of investigation, the ideal of the «objective investigator» proves no more satisfactory. Renunciation of one's self in order to become a «mirror» or «instrument» of knowledge, dictated by the desire to get rid of prejudices and individual limitations, becomes harmful rather than useful when it involves the separation of scientific occupations from the inmost part of man. For, only the deep union of self with investigative work, the impregnation of it with one's personal dreams and aspirations – only such a dynamic personal conception of investigation gives value to it. Therefore, one should not be afraid of this connection with one's personality. As an artist, even though he speaks about himself, is able to gain a hearing because he shows in his work the common fate of man, so the scholar who engages his whole self in his investigations is not by any means farther, but perhaps closer to the ideal of «man the investigator».

On the other hand, the ideal of the objective man limits the possibilities of investigation and shuts them up within an isolated sphere of intellectualism. This is avenged not only on the investigators but on their investigations, which become a superficial collection of facts. The separation of intellectual activities from the deeper forms of personal activity makes the scholar of this type function well enough within the field of investigations, but as a man, live apart from them. Therefore, in the critical moments of his own or the national life he is suddenly and unexpectedly showing a complete misunderstanding as to the essence of science and easily betrays the great idea of serving truth. That which appears clearly in such critical situations must, of course, exist also in times of quiet. In fact, a more exact analysis of scholars of this type always shows their weaknesses and faults, their inability to a deeper and more responsible search of the truth. It also uncovers their hidden motives in choosing the scientific profession for the sake of a career and the satisfaction of ambition, and not for the satisfaction of intellectual needs.

Investigation, learning, teaching

Taking the process of investigation and learning from the dynamic point of view, we discover their deep relationship or even perhaps identity. If we wish to learn something well and deeply, we must investigate into it, otherwise the knowledge acquired will be only a mechanical one. We must listen to the inner echo which it awakens within us. If investigation is treated as an acquisition of objective results and learning as a memorizing of results achieved by others, the difference between them becomes immense. The more we are inclined to define these processes not as functions of the enrichment and fixing of objective knowledge, but as functions of the broadening of personal intellectual culture, the closer these two processes become.

The deeper form of learning and the deeper form of investigation may be characterized exactly in the same manner. Born from unrest, from personal needs and interests, they are alike, in their search and trials, in the impermanency of their external attainments and the importance of their effort in the shaping of personality, the periods of temporary satisfaction, out of which again new incentives arise. Learning and investigation are the same effort to achieve constantly and a new intellectual order within oneself. Observation and criticism, hypotheses and verification, statements and conclusions have the same role in both processes[8].

Taking the point of view of this dynamic, personal conception of investigation and learning, we approach the new conception of teaching. Instead of being the function of transmitting ready knowledge it becomes an activity having as its purpose the deepening of the intellectual culture of man. Therefore the former methods of teaching, applied to the tasks of popularizing information, must be supplanted by methods serving to shape people. They should take into consideration the heterogeneity of human interests and needs and the heterogeneity of intellectual inclinations and possibilities. It is not the kind of information and the way of its transmission which is important. It is important to know to whom and what is necessary.

We cease to value teaching by the quantity of memorized information, and begin to value it by the degree of the intellectual culture of the pupil. Since, however, the acquiring of this intellectual culture can be only due to his own efforts, therefore

---

[8] The first to draw attention to the similarity of investigation and learning and to oppose the former views which decidedly separated them, was F.M. McMurry, who in his book *How to Study and Teaching how to Study* (Boston 1909) analysed the process of learning of children and compared it with the investigative process. The conclusion of McMurry that there is no essential difference between the intellectual work of young people and that of grown-ups, that the process of learning and the process of investigation may and should include the same elements, has become a point of departure for further psychological analyses and didactic recommendations. In the light of these analyses it appeared more and more clearly that human ability for intellectual work should not be divided into separated processes of investigation and learning, for, there is in reality only one mechanism of human thought which may be termed either investigation or learning. We may quote here also the book of J. Dewey *How We Think?* (London 1909). It includes a subtle analysis of the process of thinking. This analysis shows that the deeper forms of learning become investigation and that investigation is, in its very essence, a deeper form of learning.

teaching must assist the individual in his own process of learning. The fruitfulness of teaching ceases to be dependent upon the clarity of presentation and the peremptoriness of the teacher in his executive powers. It is dependent upon the degree in which it is able to awaken and develop the need of learning and the understanding how to learn in the deepest sense of the word. Learning, as we know, is in its deepest sense learning with one's entire self. Such learning involves investigation. Teaching should give incentives and create situations helping deeper learning.

Now we may define from this point of view our relation to the two basic forms of teaching. Teaching may be direct or indirect. In the first case we present to the pupil the store of information which he is to master or we demonstrate to him the kind of efficiency which he should acquire. All this we do especially for him and we try to offer it to him in a form which will be easiest for him. In the second case we perform certain activities because they are important for ourselves, because we want or are obliged to perform them for certain concrete reasons. We do not attempt to teach anyone, we are interested in the thing itself. But our activity is accessible to other people who, either take part in it together with us, or observe it. Our activity teaches them, although its direct purpose was not teaching.

Of these two forms of teaching the second has the strongest influence and gives the pupil an incentive to independent intellectual efforts. For, the situation in which someone especially tries to transmit knowledge to another leads rather to memorizing and repetition than to the deeper forms of learning. He who knows and he who does not know and is to learn what the first has known for a long time differ too much from one another. No matter how cleverly the teacher might model the activity of his pupils on the pattern of a scientific laboratory, he is unable to hide the fact that he knows in advance at what results they will arrive, that the uncertainty of the pupils is not his uncertainty, and that their investigative efforts are not his. The situation is different when teaching is not the intended and direct aim. Then, both the master and the pupils conduct true work. Thanks to this the pupils are more interested in their work and feel more responsible for it than in the first case. They also acquire a considerably greater intellectual independence.

It is easily understood that from the point of view of the theory of knowledge and education, which was called by us objective, the first form of teaching was valued higher. For, according to it, teachers were called upon to reflect ready knowledge, to graft it into the minds of the pupils and by using their master's authority make them retain it there. In the dynamic conception of the process of investigation and learning the transmitting of knowledge loses its raison *d'être* and only the independent acquisition of intellectual culture is important. Teaching understood as a transmitting of knowledge becomes unnecessary and even harmful, since it accustoms the pupil to lower, passively receptive forms of learning.

From this point of view a new definition of the teacher is possible. If teaching is to assist the individual in deepening his intellectual culture, the teacher cannot be only an «objective man» who acts as an intermediary between the producers and consumers of knowledge. He must be a man who constantly educates himself, for only he who is himself learning may encourage others to learn. Most discouraging for pupils is the knowledge that the teacher knows everything in advance, that he is not interested in the knowledge which he possesses and only speaks about it because he wishes to transmit it. This situation implies a sort of unfairness, since everyone understands that the process of learning does not have and should not have any limit. Therefore, the teacher who has not ceased to work for himself reacts more strongly to pupils than the one who is already «made».

But to learn by oneself means to investigate. Therefore, the investigator is the most successful teacher. In order to appreciate this, one must free oneself from superficial didactic prejudices, according to which the success of teaching depends on knowing how to apply dexterously didactic rules and tricks. These prejudices are based on the conception of learning as an isolated intellectual process. According to our conception of learning the teacher may exert a greater and better influence upon the pupils by his own investigative enthusiasm than by his didactic abilities. In this way is explained the fact that - independently of the didactic methods applied - investigators who are strong personalities always find good pupils. Thus teaching is a natural, although a secondary result of investigations. The type of a teacher who is a bored repeater of information is disappearing. The investigator becomes a teacher by the very fact that he investigates and allows others to take part in his work. Such teaching is the highest form of teaching. It is inseparably connected with investigation and learning. A perfect example of such intellectual work being at the same time investigation, teaching and learning was given by Socrates. In his dialogues with his pupils he did not transmit to them ready made solutions, but with their help was seeking for truth. He was an investigator, teacher and pupil at the same time.

Conceiving in this manner the process of investigation, learning and teaching we stand in complete opposition to the former point of view, which treated them as functions of the «objective man», as functions of creating, transplanting and fixing knowledge. Investigation, teaching and learning become actions deeply personal. They are connected with the whole personality of man and are not a result of his purely intellectual abilities. Investigation, teaching and learning require the whole man. Otherwise, they are only the performance of a job.

Intellectual processes and the life situation

The above analyses of the processes of learning, investigation and teaching permit us also to draw certain conclusions concerning the character of intellectual efforts undertaken by men. Intellectual work in each of these three forms is closely connected with the whole personality of man. Neither learning, nor teaching, nor investigation can be lasting and fruitful if it absorbs only the intellectual sphere of the human mind, if it does not also touch the other spheres. The conception of intellectual activities as isolated processes was characteristic of the former «psychology of faculties» to-day completely ignored. With the help of its simplified outlines ideal requirements were defined which the investigator, the teacher and the pupil must obey. In accordance with these requirements they were to become objective people. From the point of view of modern psychology which conceives man as a structural whole, such an isolation of the intellectual processes is absolutely wrong. The activities of learning, investigation, and teaching become the activities of a living person and not functions of the «intellect». The intensity and success of intellectual work are closely connected with the intensity of the whole of our personal life and with the kind of plan of life we have. This may be seen especially clearly in schools of various degrees, and pedagogical psychology has doubtless rendered great service to the analysis of the heterogeneous forms of this dependence. The investigator and the teacher are subject to it to an equal degree.

The adoption of such a conception of intellectual activities has important consequences. If these activities cease to be processes performed in an isolated sphere of the mind, if they are connected with all the other spheres of personal life, then not only factors of that same category, i.e. intellectual factors, influence them, but also all others. Thus, the field of the conceptions of investigation, teaching and learning is broadened. This is most evident in relation to these two last conceptions. The deepest form of teaching is – as we have seen – the creating for the individual of such situations which would awaken his intellectual life and make him participate in those activities of other persons which would require of him independent work and, thanks to this, impel him to undertake the intellectual effort of investigation and learning. From this point of view we must acknowledge the teaching value of many situations in life which have nothing in common with «teaching», conceived, according to former principles, as purposeful transmission of information. All these situations which provoke the individual to intellectual activity, which interrupt his intellectual sleep, destroy the comfortable paths of routine, undermine old and unproved dogmas – all these situations possess in full a teaching value, and their influence is often many times greater than the influence of institutions, founded especially for teaching work, that is, schools. The popular conviction that man learns more and better from «life» than in school, possesses – from this standpoint – its deep meaning. From this point of view, we can also understand how such people as heroes, political leaders, artists and saints can be considered as teachers, although they have nothing in common with teaching understood as a purely intellectual activity, but «teach» others by the strength of their spirit, the greatness of their deeds and their impelling example. The result of such teaching is often a new style of living, an «opening of the mind» to new truths, a breaking with dogmatism and routine.

These considerations make us appreciate the social sources of intellectual phenomena, their connection with the social milieu. This social milieu may awaken or hinder intellectual activity. Therefore, desiring to make this activity as common and as thorough as possible, we must take care to create such conditions which would incite people to intellectual effort and aid them. The requirement of people that they study and educate themselves, that they care more intensively for their own intellectual development, is, however, only a *pium desiderium*, since conditions of life enforce upon them mechanized and unpleasant activities. Thus, whoever cares for the spreading of learning without taking into account social conditions, cannot hope his efforts to be successful. Now, in a milieu where the intensity of intellectual processes is growing, there arise the best conditions for the development of scientific investigations and studies. In this way the closely allied activities of investigation, teaching and learning are connected with the whole of life of human beings, enrooted in a defined social and historical milieu. The intensity of the intellectual processes is a direct expression of the intensity of life.

The exact analysing of the connections between intellectual activities and the structure of the milieu is of great interest for sociology and the history of culture. The realization of this connection, as a result of the analysis of investigation, teaching and learning, allows us to see in a new light many problems concerning science and education.

Science and education as factors in forming of intellectual culture

The treatment of science and education as functions of the enrichment of knowledge and of its diffusion among people, leads, as we have seen, to a conflict between science and education and to a more and more distinct depravation of each of them. Science, impersonal and separated from educational needs, is threatened by the danger of narrow specialization, by the production of insignificant contributions to problems of minor importance and by mere «hunting» after results. On the other hand, education popularizing knowledge breaks its contact with the essential needs of man and becomes an instrument for professional career. On the basis of the dynamic conception of investigation it is, however, possible to conceive the relation of science to education and the mutual relations of investigators, teachers and pupils in an entirely different manner. On the basis of this conception we may do away with a considerable number of the former conflicts and dangers.

First of all, the conception of science is changed. Science does not pretend any more to reflect reality, and gives up exaggerated ambitions of ruling people in conformity with this reality. To-day scholars are more modest, but in their modesty more faithful to themselves. Who knows what is Truth, and before whom will she uncover her face? It is possible that we know more about the world than we knew before, but do we know *better* – who has the right to judge? The knowledge of yesterday does not satisfy us, just as its art, ethics, technique. We build therefore a new knowledge which better answers our questions and better satisfies our sense of fairness. Understanding that the value of science does not consist in reflecting reality but in enabling us to see it in such a way as the changing intellectual needs of man demand, we make of it a means of deepening our intellectual culture. By means of scientific effort as with a plow, we till the unknown soil of reality, gathering constantly an equally laborious, new, and short-lived harvest of understanding.

The overcoming of the old objective attitude towards science is to be seen among investigators more and more distinctly. The more institutional and formalized investigative work became, the more it began to be threatened by the danger of turning into a profession which might be executed without the help of the whole self, – the more strongly began to sound those voices which demanded the treatment of science as a matter of personal importance for the investigator, who should approach the ideal of a wise man and not that of a functionary or a producer.

This is the meaning of the increasing demands made of science for «humanization». They come especially from America, where they are a reaction against the appreciation of everything-scientific investigation included – according to its productiveness.

Both humanists and the specialists in the field of natural sciences make these demands. The eminent historian of the exact sciences, G. Sarton, defines the task in the following manner:

> The only way of humanizing scientific labor is to inject into it a little of the historical spirit... However abstract science may become, it is essentially human in its origin and growth. Each scientific result is a fruit of humanity, a proof of its virtue... Each time that we understand the world a little better we are also able to appreciate more keenly our relationship to it. There are no natural sciences as opposed to humanities; every branch of science or learning is just as natural or as humane as you make it. Show the deep human interest of science, and the study of it becomes the best vehicle of humanism one could devise; exclude that interest, teach scientific knowledge only for the sake of information and professional instruction and the study of it, however valuable from a purely technical point of view, loses all educational value.[9]

The conception of scientific work as an activity personally important, which engages the whole man and shapes, according to his deepest needs, his attitude towards reality is not the only novelty brought by those who fight for the humanization of science. Science being closer to man, is closer not only to the investigator but also to the consumer. It is from this standpoint that the problem of the humanization of science was taken by the American historian J. H. Robinson. In his book *The Humanizing of Knowledge*[10] he attempts to show that science of yesterday has developed thanks to its independence from the outlooks, needs and longings of man; thanks to the constant and often soulless rationalization and mechanization of phenomena and to the overgrowth of specialization in research. Thus, science became «dehumanized». But today it is time to overcome these opinions and to do away with methods which, useful in the era of the early development of science as a safeguard of its objectivism and independence from the older cultural forces, such as religion and metaphysics, become now a hindrance in its further development. Science must find the road to man.

---

[9] George Sarton: *The History of Science and the New Humanism*, New York 1931, 68–69.
[10] James Harvey Robinson: *The Humanizing of Knowledge*, London 1924.

The language of the scholars must be accessible to all. A scientific book cannot be a boring treatise, it must be lively and exciting. It should awaken interest, instruct, incite the intellect to work, and not be like a school textbook to which one never returns and which it is impossible to read to the end without effort. A scientific book should influence the reader, and this influence, and not the display of one's knowledge and learning, should be the aim of the writer. Not only for the common good but also for the good of science itself do we need to-day the new type of investigator. We need an investigator who would be able to see reality, which so often has been hidden behind verbal abstractions, able to stand above personal ambitions and his own theories, able to arrive at a synthesis, avoiding the danger of narrow specialization for the deeper understanding of things[11].

To the changed conceptions of science correspond changed theories of education. Those who learn do not receive from the hands of wise men the miraculous stone of wisdom, the key opening one by one and without difficulty the gates of mysteries. They do not receive the ready treasure of knowledge, nor the rigid pattern of ability. Education requires of the pupils the undertaking of labour. No one can do this work for anybody else. Each must perform it for himself.

Educational activity now gains an entirely new meaning. Its tasks and possibilities are immeasurably widened. The greatest Polish poet Mickiewicz did not exaggerate when he said: "National education is a result of great events in the life of the nation and spreads itself by living transmission. The nation gains enlightenment in religion, in politics, in morals only by the help of great examples. It needs for this great deeds, and what follows – great men."[12] Education in this sense leads to a deepening of social life, a broadening of horizons, the creation of a definite attitude towards life. Education in the more exact meaning of the word, i. e. intellectual education, is closely allied with the former, and cannot be fruitfully applied in just any type of social conditions. It requires a favourable atmosphere. It develops by itself in a stimulating milieu which imposes ideals and impels people to creative activity.

With this dynamic and personal conception of science and education, we see better the possibility of their co-operation. They become co-working factors in the creation of intellectual culture. Lively and deep educational interests not only form the best atmosphere for scientific work, but they unconsciously lead to investigative work. The investigative labour of the scholars has in turn a direct echo in education.

The border line between science and education cannot be now drawn so distinctly as from the point of view which conceived them as two qualitatively different and loosely connected fields of human activity. The opinion which attributes a separate

---

[11] Such syntheses will not be a superficial summing up of the results of investigation in various fields. They will arise from investigations of indivisible complexes of reality. The investigation of parts of these wholes from special points of view, will never grasp their essence. This explains the fact that it is usually difficult to class the best scientific books in one discipline.

[12] Adam Mickiewicz: *Rzecz o literaturze słowiańskiej wykładana w Kolegium Francuzkiem, rok czwarty (1843–1844)*, 2nd ed., Poznań 1851, 90–99: 98.

place to science and to education is false. In reality science and education permeate one another forming an alloy in which each of the ingredients may occur in various quantities. The relations between the investigator, the pupil and the teacher appear now in a new light. According to former principles, it was possible, by analogy to producers, middle-men, and consumers, to know definitively whom we are to call the investigator, whom the teacher and whom the pupil. Appreciating intellectual work by the results achieved, it was possible to divide people into different categories, and to call investigators those, who, thanks to systematic studies, have been able to formulate something new in science, – teachers, those who transmit these new acquisitions to others, and pupils those who acquire and memorize them. From our new point of view, the dividing lines of the former categories disappear. Everyone who learns may be treated as a pupil. But at the same time the pupil is, to a certain extent, an investigator. Influencing his surroundings he also becomes a teacher. Attainment and deepening of one's intellectual culture are common to all these people. The teacher and the pupil cease to look upon the investigator as upon a magician who is master of the forces of nature, or as upon a dictator who builds knowledge. The investigator surpasses others in his eagerness to search for establishing intellectual order, in the sincerity in expressing his thought, in his passion to observe, in his conscientious and responsible attitude towards words, in the humility with which he obeys the demands of truth. He becomes the model for a certain type of life. The teacher ceases to be an intermediary and a salesman of knowledge. He performs a social service in awakening his fellow-men and helping them to educate themselves. The pupil ceases to be a slave loaded with information, a passive receiver of the products of science, and becomes a human being in search for a form of culture which would suit him.

This new conception of scientific and educational problems based on the human need of educating oneself gives a vision of a new social order, of new conditions of existence, of a new organization of professional work, of a new world in which the building of intellectual culture will be accomplished as a direct and natural expression of personal needs. In that new world the falsity of the old world will be overcome, a falsity connected with the fact that the requirements of intellectual work, imposed by the ideal of science and by the schools, are opposed at every step by real life, which hinders the intellectual development of man. Overcome will be the injustice, which admits only a few individuals into the luxury of an intellectual culture.

Social and cultural significance of both theories

The difference between these two points of view, one of which demands from knowledge a faithful description of reality, while the second – the development of men – is not exclusively a doctrinal difference. These points of view are connected with different philosophical opinions and have their cultural and social background. The first point of view attained a decisive prevalence in the second half of the XIXth century. The positivistic theory of knowledge, then in vogue, had a dogmatic character and justified the faith in the objective value of science. Investigation was to obtain results which had to reflect reality. In connection with the positivistic tendency to conceive life on the basis of naturalistic determinism, it was held especially important to make people conscious of those objective natural laws to which they had to submit in their personal and social life. Education consisted in transmitting this information. Spencer, in his studies on education, represents this point of view very distinctly. In his opinion this knowledge was supplied, above all, by biology and psychology.

Such a conception of knowledge as the mirror of reality, and the conception of education as an automatic reflection of this knowledge was also connected with the basic conviction of liberalism, that natural harmony best regulates relationships between men and assures the individual the greatest happiness. Just as every man holds his situation in his own hands and no governmental laws or philanthropic institutions have any right to affect its course, so the spiritual life of man belongs – in the opinion of liberalism – only to him, and none has here any right to interfere. Therefore, all education, that is moulding of man by man is, in principle, lawless. Spencer is in perfect agreement with Rousseau when he argues that moral training consists only in making men understand the natural necessity. Education has only to transmit knowledge which is the reflection of reality, and what the individual will do with this knowledge is his personal affair. The teacher has to inform, not to mould his pupil.

This point of view was widely voiced in the epoch of positivism. «Education» was under suspicion, firstly because of compulsory regulations apparently connected with it, and secondly because of its dealing with a personal and inner process, while the epoch cared first of all for the external fruits of activity and was full of enthusiasm for the growing tempo of production. We must remember that it was the epoch of capitalist industrialism. Just as in the economic field, the ideal was efficiency and the increase of production, without regard for its true utility for the consumers, and without regard for the good of the workers, who were treated as a working power, so, also in the field of culture, in the sphere of intellectual activity the ideal was the increase of scientific production, the enrichment of knowledge and its utilization without regard for the inner relation of man to science and to truth. The definition of the functions of the investigator, the teacher and the pupil was the expression of that industrial and productivist conception of man. Man was an instrument of economic activities and an instrument of the enrichment and conservation of objective knowledge. His value was measured by his efficiency in executing these functions.

These views on science and teaching in the epoch of liberalism and capitalistic industrialism found expression likewise in the evaluation of the social classes. In their hierarchy the first place was in some countries occupied by the «intelligentsia», a class which had the privilege of finding itself closest to «knowledge». Even when this class did not directly create knowledge, it had the easiest approach to it. In this function the intelligentsia saw its social superiority and, conscious of this privilege, felt its duty to fulfil the educational mission. This mission was to popularize knowledge without regard for the real needs of those who learn, without regard for the educational role which it was to play, without regard for its moulding power. In connection with this, education received a marked intellectualistic (in the worst cases – encyclopaedist) and urbanistic stamp.

The change of views on investigation and teaching which we have witnessed in the present epoch, is obviously connected with the transformation of our world-outlook and the cultural and social changes going on. In the theory of knowledge we see the victory of this idealistic point of view which conceives science as something satisfying our needs of knowing. The value of such a science depends on the educational power which it has for man. Adopting this conception, we have thus to give up the ideal of productivism in investigation and the ideal of encyclopaedism in teaching.

This point of view finds support in the growing strength of modern humanism which seeks to free man from the burden of culture which does not contribute to his improvement. We are beginning to understand more and more deeply that not only school, but also professional work should educate man, and not stifle him with the ballast of information and mechanized routine. We are beginning to understand that the spirit of man may be quenched not only when he serves economic demands but also when he serves science, if this service is ill-understood as an activity which does not engage the personality. We are beginning to tend towards a new, uniform culture. Our object is to make the person of the scholar a union of the investigator and the man, and to make the pupil and the teacher live and develop through the satisfaction of their spiritual needs.

This is connected with the conviction that society has the right and duty to care for the education of each individual, a conviction opposed to liberalism, which professed a complete neutrality in this respect. The point of view we plead for demands not only a suitable educational programme but also such reforming of life's conditions as to permit and excite each individual to the deepening of his intellectual culture. This standpoint, according to which, the effort of learning should not be a phenomenon isolated from concrete life and occuring only in the laboratory or school, but should directly grow out of the real and every-day conditions of existence, – this standpoint is also to be found in other fields of culture. The conviction is growing that aesthetic experiences should occur in the common every-day life of men, not in an isolated and artificial atmosphere of museums and theatres[13]. These convictions are the expression of a deepening dissatisfaction at the development of modern culture, in which there occurs a constantly greater gap between every-day life and the so-called spiritual culture. Culture finding its expression in religion, morality, science and art has become accessible in the every-day life only to specialists, while for the masses it

---

[13] I have tried to justify this in my paper: "Wielkość sztuki a odrodzenie kultury" [The Greatness of Art and the Rebirth of Culture], *Życie Sztuki* [The Life of Art], vol. 2 (1935), 1–55.

is still a thing exceptional and rarely attainable. To make culture an object of true and genuine needs of the masses, to make it universal, uniform and harmonious – this is the ideal of the future.

Psychological significance of both theories

The pointing out of connections between views on science and education and social and cultural life does not yet explain everything. The difference between the two standpoints has its source in the duality of the spiritual needs of man. We aim in our activities to attain permanent and objective results. Our thoughts and our dreams, our feelings and our designs, everything that is accomplished within us and is important for us, – all this seems to us incomplete reality, if it does not obtain some objective sanction. But there is in us also another desire. We do not always look obstinately for the reality to approve of the value of our efforts. We are eager to improve ourselves guided by an inner faith that in this effort we will not get lost. We treat reality not as an authority pronouncing sentences, but as an instrument of our own development. When the first of these tendencies prevails, we define intellectual processes in objective categories, conceiving investigation, teaching and learning as functions being in the service of knowledge which faithfully reflects reality. When the second of these tendencies prevails, we are more inclined to subjective definitions, stressing the value of these efforts in the shaping of man. In the first case we place the chief stress on the results whose value may be proved in an objective manner. By these results we control the activity of the individuals and only from them does it derive its sense. In the second case it is the search after truth itself which matters, and its results are appreciated as personal attainments of the individual.

Evaluation of both theories

Since each of these two theories of investigation and teaching has its cultural and social background, since both are conditioned by the epoch and the type of man, it is impossible to pass an impartial judgment concerning the value of either of them. In the course of centuries the scales of value have been over-balanced now on this, now on that side. To-day we are rather inclined to consider the second of these points of view as superior, but to be impartial we must admit that failure threatens on both opposite extremities equally. The dangers which the objective conception of science and education implies, like all errors of yesterday, are only too conspicuous to our sight. It is more difficult for us to see the risks of the humanistic standpoint. Notwithstanding this, we may say that this standpoint is seriously threatened by the danger of subjectivism. In order to face this danger, we must realize that to educate oneself does not mean to care only for oneself. A man educates himself best in deeds which attain results of social or cultural importance. The improvement of one's self is, as it were, a supplementary result of these activities, which have the character of a social service, just as activities not undertaken with the purpose of teaching may prove themselves instructive. But there is no room for the improvement of man's self in two cases: when human life loses its character of service to social and cultural ideals, or when that service takes on the form of slavish and passive adaptation.

The above analysis of the intellectual processes showing their connection with the personality of man, can be instructive only when we define education properly, namely when – discarding the dangers of mechanization and subjectivism – we conceive it as a process of human development by means of activities directed by social requirements and cultural values. This will allow us to define intellectual activities as *personality-shaping activities pursuing impersonal aims* and thus to avoid both dangers: the making of science and education superficial and mechanical phenomena, and the treating them in an impressionistic and subjective manner as well.

The interpretation of the educational rôle of investigation, teaching and learning, stressing the importance of super-personal service, demands the right appreciation of truth in science and the respect of the man in education. The truth which we seek, the truth which we put into words and deeds, the truth which we sometimes give up and to which we return again, forms the ideal measure and aim of scientific investigations. Devoting ourselves to this work, which we pursue systematically, with endurance and with all responsibility, we pass over the bounds of our egotistic selves, we give up the pleasant easiness and comfortable formulas, often we give up many pleasures indeed, and sometimes with sorrow behold how the inevitable result of our investigations destroys our personal habits and demands our renunciation of the world created by our imagination and our desires. But these limitations and renunciations which scientific work requires, affecting only the sphere of our egotistic interests, are, in fact, beneficent, for limitations lead to concentration and renunciations make our inmost part expand.

Similarly as scientific activity is guided by the super-private ideal of truth, so education must be guided not by subjective likes and dislikes but by the super-private ideal of the improvement of man. While teaching or learning we must not adopt a purely subjective, impressionistic, parasitical attitude. We should get lost, if we care only for ourselves, for our own «beautiful soul», for our own «wisdom». This is not a way to attain fullness and strength, for only in disinterested devotion to work may we find within ourselves, almost unexpectedly, the fruits which will enrich us permanently. Only in such effort do our horizons broaden. This truth is proved in many fields. For instance, we will not attain aesthetic training and aesthetic sensibility by the mere fact that we carefully try for it. It develops within us all by itself, as a natural result of complete devotion to works of art, of a disinterested union of ourselves with them. Likewise, we shall not deepen our education if we do not rise to the level of super-private effort, of self-forgetfulness in service. The sense of this service is the awakening and improvement of the human being within us. In order to understand this, it is necessary to realize the fact that our life embraces various spheres of depth. In the most superficial sphere, we egoistically oppose our fellow-men, we struggle for existence and allow ourselves to be lost in trifling cares, we treat knowledge as a power we can use and think of ourselves in categories of our egotistic self. In the deeper sphere we feel our union with fellow-men, we are joined with the world by disinterested bounds, we get ready for contemplation and heroism, we are trying to find our human dignity: In the sphere of superficial life education becomes an instrument for struggle or a servant of egoistic impressionism. Only in the deeper sphere of life does it acquire a deeper meaning. There it awakens, calls forth and unveils for us unexpected possibilities within ourselves. There, forgetting our own selves, we find our true selves; working for our own improvement we work for the improvement of the man. For, it is within us and through us that he tends to attain his expression within defined limits of time and space. And he will attain it, were we only able to overcome our comfortable egotistic care for trifles and to make of the work of education a real service to human improvement. If not, – stifled by the trifles, superficiality and interestedness of our present life – he will lose this opportunity once for all. This very responsibility for the fate of man, which falls on everyone, makes us hope that educational activity will become a service for the improvement of man and be saved from the danger of subjectivism.

Adopting such a conception of science and education as discussed above, we grasp the most valuable – in our opinion – meaning of this viewpoint which stands in opposition to the objective one. But, at the same time, discarding dangers which threaten it – by stressing the importance of super-individual aims and tests such as truth in science and the improvement of man in education – we are able to evaluate more justly the value of the viewpoint characterized by us as «objective». Its intention was to defend science and education from the overflow of subjectivism. The need of such defence is obvious and in this lies the value of the objective point of view. But the methods of this defence no longer suit us. One cannot withhold the flow of subjectivism by making science and education a mirror of reality ruled by the natural laws and independent of man. Human life then breaks into two parts: one which is the sphere of subjective arbitrariness and the other which is the sphere of obedience to the laws of nature[14] and man becomes in turn a slave to his egotism or a slave to things. In certain situations of our present life he must be a machine and an instrument, and this does not favour the development of a strength capable of mastering his subjectivism, of overcoming the bonds of his individual self. The only effective way to bound subjectivism is the inner work of improving the personality by engaging it in the service of ideals and values.

---

[14] The connection, which, in spite of apparent dissimilarities, occurs between scientific naturalism and egotistic sentimentalism, has been analysed thoroughly by I. Babbitt in many of his studies.

The Urgent Need for Mental Education in Poland

The Necessity for a Fundamental Teaching Reform in Middle Schools and New Institutions for Scientific Research in Relation to this Reform

Antoni B. Dobrowolski

ANTONI BOLESŁAW DOBROWOLSKI (6 June 1872, Dworszowice Kościelne, Congress Poland – 27 April 1954, Warsaw, Polish People's Republic) was a geophysicist, meteorologist, and theorist of pedagogy. After initial philosophical studies in Zurich, he attained a degree in biology and geophysics from Liège Univ. (1897), upon which he joined the Belgian Antarctic expedition (1897–1899). Returning to Poland in 1906, D. taught at middle schools and prominently published his Antarctica material, partly funded by the Mianowski Fund. After Poland's independence in 1918, he first joined the new educational ministry and later the national meteorological institute. He was appointed prof. of theoretical pedagogy at the Free Polish Univ. in 1926. After 1945, D. taught pedagogy at Warsaw Univ.

D. was highly involved in the science studies milieu of Interwar Warsaw. He frequently attended the meetings of the Science Studies Circle, published broadly in Nauka Polska and beyond. His text "The Urgent Need for Mental Education in Poland" was originally published in Polish in vol. 1 of Nauka Polska to launch a larger research project. After vol. 1 (1918) had sold out in the meantime, the text was partly reprinted in vol. 6 (136–140) in the first edition of the series "Researching the Genesis and Development of Scientific Creativity" (*W sprawie badania i genezy rozwoju twórczości naukowej*) published in vol. 6 (49–140). The series was continued in Nauka Polska and Organon (▶ **IN THIS VOL.** Białobrzeski 1927, ▶ Dobrowolski 1928, ▶ Borel 1936, ▶ Krogh 1938). While the larger part of D.'s text was translated by Tul'si (Tuesday) Bhambry for this vol., a short passage from the original (498–502) had been translated for vol. 1 of Organon (Dobrowolski 1936, see below 297–300). Typographic changes were indicated by square brackets.

ORIGINAL REFERENCES
Antoni B. Dobrowolski: "O pilnej potrzebie wychowania umysłowego w Polsce: o konieczności zasadniczej reformy nauczania w szkołach średnich oraz stworzenia w związku z ową reformą nowych placówek pracy naukowej," *Nauka Polska*, vol. 1 (1918), 489–502.
— "'Biografja' myśli twórczej," *Nauka Polska*, vol. 6 (1927), 136–140.
— "Letters to the Editor," *Organon*, vol. 1 (1936), 290–294.

FURTHER TEXTS MENTIONED
▶ Czesław Białobrzeski ["C.B."]: "Szkic autobiograficzny i uwagi o twórczości naukowej," *Nauka Polska*, vol. 6 (1927), 49–76.
▶ Emile Borel: "Contribution aux *Documents sur la Psychologie de l'Invention dans le Domaine de la Science*," *Organon*, vol. 1 (1936), 33–42.

- Antoni B. Dobrowolski: "Archiwum materjałów do badania twórczości," *Nauka Polska*, vol. 6 (1927), 140.
- Antoni B. Dobrowolski ["A.B.D."]: "Mój 'życiorys naukowy,'" *Nauka Polska*, vol. 9 (1928), 68–216.
- August Krogh: "Visual Thinking," *Organon*, vol. 2 (1938), 86–94.

The "general" middle school is entrusted with the unique task of ensuring the moral and mental education of the future *intelligentsia*, the nation's brain. Here I will only speak about *the education of the mind*.[1]

This education, usually referred to as *"general education,"* can only be provided by the middle school. Elementary school is insufficient for that purpose, while secondary school is too specialized.

"General education," understood as the attainment of a certain amount of knowledge, a certain entirety of knowledge about the world, tolerates no criticism. The knowledge that middle school can provide is not real knowledge – scientific, specialized, ready for use, but rather pseudo-knowledge – vague, popular, and of not much practical use. Middle school provides a certain minimum of information and technical skill indispensable for studying at secondary school – a minimum that is relative, as it depends on the middle school's agreement with the secondary school. It initiates pupils into the sacred books of national history, though here we already enter the realm of moral or religious education. Besides this, however, what is taught at middle school is neither indispensable nor sufficient for life or for actual scholarship, i.e. independent research. It can only serve as a medium, as indispensable material for a more serious goal: to *develop the pupils' minds,* to *train their thinking.*

This is the indispensable and sufficient goal, and not any other, that should be the guiding light of middle school teaching. Those who are to enter the nation's intellectual elite must learn to think, to develop their abilities. It is not a disgrace if an "intellectual" does not know about some thing unrelated to his thinking or his work, but it is a disgrace if he is unable to think, if he has not cultivated his abilities. But if he has nourished his abilities, if he has learned to put his brain to use, he will be able to acquire the knowledge he needs independently.

"Teach your pupils to think – and only that, nothing more, and the rest, that is to say knowledge, will come to them through natural growth." This is the new princi-

---

[1] The first, strictly pedagogical part of this presentation is a schematic summary of a lengthier work that will soon be published under the title "Teaching Reform in Middle Schools". There is no space in this presentation for presenting a proper argument; I include only premises and the resulting practical conclusions. Both are adequately expressed but undeveloped. I believe, however, that the ideas alone, without detailed evidence, will suffice for the moment – at least for those who are able not only to oppose thoughts but also to see eye to eye. I shall add that the plan for a fundamental reform, sketched out in this presentation, does not just or simply result from a critical deconstruction of contemporary teaching practice, but is chiefly the fruit of seven years of a teacher's conscious efforts.

ple – the only one that is reasonable. The old principle, in comparison, is: "Teach your pupils 'the subject,' after all, that itself will force them to think, so by that alone they will learn to think." But how? Blindly; their teacher of thinking will be chance, and chance will not always be a good teacher. You do not learn to fence by getting into fights…

The conscious, planned and systematic learning of various "subjects" in middle school should be transformed into a conscious, planned and systematic learning of thinking against the backdrop of those "subjects".

The subject teacher should be transformed into a teacher of thinking, into a cultivator of the mind.

This is how he will know *what* to teach, *how* to teach, how to *assess* and control the results of his teaching.[2]

1. The teacher of thinking must have a detailed and concrete knowledge of *what* he is really supposed to teach. He must *know the art of thinking*: he must know the catalogue of *the faculties of the mind*; he must know their applications, both separately and in various combinations depending on concrete circumstances, just like a fencer must know all parrying and offensive techniques separately and in combination depending on circumstances. The teacher of thinking must be familiar with the mind's tendencies, its age-related development, its errors and faults. It will not do to be versed in logic, epistemology and general psychology of the mind. The teacher of thinking must be a *practical* psychologist of the mind, he must know the *grammar of actual thinking*.

We already know a little of that grammar, which means that we have a provisional base for the education of thinking. However, that base must be developed earnestly in order to continue to improve that education.

By "faculty" [of the mind] I mean simply the ability to perform a task. In order for that performance to be reliable, efficient and dexterous, all knowledge must be awakened, developed and formed – it must be *exercised*. It is possible for that exercise to happen unaided, somehow or other, by chance, in passing, as a side product of the individual's life. But in that case the result is simply fortuitous, and as such it is usually not the best, especially when it comes to such varied and complex activities as the activities of the mind, which are so often exposed to clumsiness and error, so often

---

[2] The notion that the sense and goal of teaching in middle school should be a *conscious, planned and systematic* education of thinking based on academic subjects leads us to postulate the practical *appreciation* of the faculties of the mind, the *means* of training and developing them, and finally the means of *assessing* the results of that education of thinking. To be sure, these concepts have little in common with the older theory of disciplinary teaching, which was but a shapeless general theory that did nothing to indicate reliably *what and how*, that did not even seriously pose those questions. But these concepts are also distinct from today's general trend towards heuristics and the "work method," which are seen as a panacea for the education of the mind even though no one has defined precise *what* heuristics and *what* "work method," and *what and how much* it is actually able to develop or train.

affected by retardation, distortion and stagnation due to insufficient or inappropriate training. The faculties of the mind have a much greater need for *conscious*, planned and systematic exercise, as well as for continuous and skilful adjustment and readjustment, than the faculties of the body.

Nonetheless, it is the conscious, planned and systematic exercise of various physical skills, which have been practiced since time immemorial, which continues to develop and diversify, while *the conscious education of thinking* has not yet been put into practice. Those who teach fencing, gymnastics or drill *know* in detail, not only in general terms, what actions constitute their art, and what simpler motions constitute any action. If a pupil performs some complex movement poorly, the teacher *knows* well what part of that movement was not up to scratch, and will now pay attention to that part in all kinds of complex sequences – he will try to give it the appropriate training. *He is familiar*, at least in practice, with the catalogue of necessary physical skills along with their connections and coordination. This is his didactic foundation. He knows *what* to teach.

What teachers of academic subjects are missing – even those teachers who really wish to use their subject, their lesson, to "develop their pupils' minds" and to become teachers of thinking – is precisely that foundational knowledge, be it even a practical knowledge, about the concrete components of the process of thinking, their connections and coordination. Teachers content themselves with their rather too *general* familiarity with thinking; they simply believe or have an overall impression that this or that subject or this or that method – be it a "heuristic" method or a "work method" – will in and of itself "develop the mind". They do not examine in detail, in each single case, *which* faculties are trained and to what extent really; they do not prepare and arrange their lesson plan, and do not teach their subject from *that* point of view. They do not draw on their pupils' errors and clumsiness to find out which faculties of the mind are particularly unsound and in need of special training. They have no clear vision of the pedagogical goal; they omit to critically assess the formative value of each of their lessons, of each moment in their lessons. They are oblivious to what goes on in their pupils' brains during class. This is why the time has come to finally *find out what we are actually doing with the brains of our youth in school*. We will gain that knowledge once the education of thinking becomes connected to its indispensable foundations, i.e. our understanding, at least in broad terms, of the mind's ideal activities and of the grammar of actual thinking. This can be a rough sketch, as long as it is not too vague.

It would already be an achievement if teachers became aware that the ability to classify, for instance, is a complex one, and that it depends on reliable separation and connection, two faculties that are based on systematic comparison, which in turn hinges on the ability to perceive similarities as well as differences. Further, teachers ought to realize that each of these faculties comprises different types, orders, varieties and special cases, and that classification can mean, for example, either creating classes or attributing a given thing to one of the already existing classes, or identifying and extracting those parts of a given complex that belong to an already existing class. The first kind of classification, to give another example, can be either natural or artificial, in which case the task would be to define the limits of arbitrariness. Classification can be academic or practical; it can be suitable for future elaboration or it can be merely provisional. Teachers should also recognize that the principles of classification depend on its purpose and so on. Concerning differences, the task is to identify real differences rather than apparent ones, to identify not just differences that are due to chance or superficial but deeper ones; not just the obvious ones but also subtle variations, not only visible disparities but also hidden ones; to distinguish unconditional from conditional variations, permanent from changeable ones, primary from secondary differences; to identify differences that are characteristic, essential. It is important to be able to judge the relative value of differences from a given point of view (lesser, more significant, and most important differences), and so on. Finally, teachers must understand that we have to *teach* our pupils all these individual abilities – along with many, many other ones – through our work with various materials (things, phenomena, relationships; material and spiritual ones; innate, human), in other words, through different "subjects" of the school curriculum. We should not only use each and every occasion to train those faculties, but consciously create such occasions by planning and structuring our lessons and even the entire subject course accordingly, especially when it comes to faculties of the mind that are important but more difficult to develop.

2. Teachers must know *how* to teach thinking. That is only possible through a system that allows them to create suitable ad hoc *exercises*, each of which would train a specific intellectual faculty or complex thereof in a concrete situation. The exercises must be based mainly on the pupils' *individual work*, which should be as *independent* as possible – just as in martial art learners perform exercises with their own hands and weapons. Thus the exercises should take the form of *problems* that pupils solve with as little help from their teacher as possible. The *material* – and only the material – is supplied by various "subjects".

"Subject courses" ought to be *transformed* point by point into series of such exercises. Where elements of courses or even entire courses are not suitable for such exercises, they should be eliminated or substituted by more appropriate ones. The "subject," that is to say knowledge, must be strictly subordinated to the education of thinking. (This does not mean that the subject must become a series of unrelated fragments, or that we should arbitrarily overturn the logical order of the course. That would thwart the acquisition of knowledge indispensable as material and background for those exercises. What is more, a lack of continuity and logical consistency in consecutive lessons would render them unnatural, coming out of the blue, with no relationship to the preceding lesson. This would certainly cause the pupils to lose interest. It would also entail a chaotic absorption of material instead of a consistent arrangement of concepts. Finally, it would prevent the involuntary surfacing of questions, preventing pupils from formulating them independently. In other words, disrupting the continuity of subject courses would stand in the way of the planned and systematic education of thinking, thus going plainly against our proposition). Even courses or parts of courses that are indispensable for future higher study (reading and self-expression in the broadest sense of the term, elements of mathematics, geography, foreign languages) should as far as that is possible be substituted by such exercises, which surely can be done. The results of such a radical transformation should be worked into a *handbook for teachers*, which would contain clear and sufficient lesson-by-lesson guidelines.

We already have the beginnings of this arduous project. We must only develop them – bring them to a conclusion – and employ them in schools. We will certainly find a handful of teachers who are willing and able to introduce those reforms into their practice. Motivated by their example, their good results, and especially by the ideal that lifts their guild's flag up high, less talented teachers will follow in their tracks.

3. Teachers must know how to *assess and control* the results of the new teaching method; they must be able to examine the development of their pupils' minds, their pupils' progress along the course of their training in thinking. The process of teaching itself will be a continuous examination. But besides this testing in the group, it is indispensable to examine each student individually at regular intervals – each quarter, each year. The differences will become the measure of quarterly and yearly progress. For that we need a system of suitable *surveys*. For younger children we already have surveys of a similar kind – the "tests" devised by pedologists; they only need to be improved for our purpose, and above all, to be linked to the content of various "subjects". For adolescents we must yet create a system of tests. Attempts are already underway.

This is how the fundamental reform of teaching in middle schools presents itself, in a schematic and condensed form. It is necessary – we must make middle school meaningful, once and for all! It is possible – individual efforts have already initiated it. It is inexorably linked to the nation's interest: the better we nurture young minds, the better they will learn to think, and the more the nation's general creativity will increase in quantity and especially in quality.

Work on the great reform can be divided into two areas:

First area: Putting the reform into effect *in the nearest future*. To arrange what we already have into a temporary grammar of thinking and to draw on this in order to transform "courses" into series of exercises in using the mind; to prepare pertinent guidelines for teachers, to put the reform into practice.

Second area: Perfecting the reform, which is to say working to *refine the grammar of actual thinking*. This is a challenging and large-scale task that demands steady work, which is why it must be initiated immediately. *Research on the human mind must be made to tackle precisely those areas that practical necessity indicates.*

This task will hardly be accomplished through the private initiative of either individuals or organizations. But even the state, even with its best intentions, with the best possible ministry, will not suffice on its own, will not be equal to the task. This is not a formal or average matter; it is a question of renewing middle school from scratch. What we need are creative forces rather than capable administrators. Only intense collaboration between competent and creative individuals persuaded by the greatness of the task, along with the committed assistance of social institutions conscious of its importance, will guarantee its success.

When it comes to the fundamental task of the reform's foundations and refining – our ever broader and deeper understanding of the mind's normal functions in various real-life situations, in academic settings as much as when knowledge is applied in practical life, in a word, when it comes to the continuing development of that grammar of actual thinking – the collaboration of all who are capable of advancing this fundamental task becomes indispensable.

The Mianowski Fund could play a key role in this project, not through sporadic support but through planned and systematic contributions.

The Fund would engage in small-scale efforts such as organizing a competition for "testing" adolescents and publishing teachers' handbooks appropriate to the new system, etc. Besides these, however, the Fund's activities would be large-scale efforts focused exclusively on this fundamental task: to support and organize research on the intellect in the above-mentioned *practical area* in order to elaborate an ever fuller and more detailed answer to the question *"how do we really think?"*.

Several *sources* can help us understand concrete forms and instances of thought. We must draw on each source, making it a centre for our planned efforts, and disseminating our results through suitable publication series.

First source: *Experimentation.* Two series of works:
1) To make use of the results of current experimental psychology by extracting, collecting and systematizing the concrete data we need for our task.
2) To make experimental psychologists aware of the practical importance of the question "how do we really think?," thus motivating them to direct their research towards this field; in particular, to encourage them to focus their pedological research on this question and to expand it to include *adolescents*.

These two series of works would correlate with two series of the Fund's publishers, carried out either by permanent employees hired especially for this purpose, or by inviting entries for competitions.

Second source: A phylogenesis of intellectual achievements in science and technology; identifying the earliest embryos of those achievements, the steep and windy roads and paths of development, regression, errors and mistakes.
Two series of works:

1) Applying to our purpose the works of previous generations on the development of individual scientific truths and inventions; deducing from those works the concrete data we need and systematizing it.
2) Encouraging research on the phylogenesis of various greater and lesser intellectual achievements. By asking "how do we really think?," this research would not content itself with a general grasp of thinking, as previous generations have done, but would seek as detailed and exhaustive an answer as possible.
These two series of works would correlate – as in the category above – with two series of the Fund's publishers.

Third source: Analysing particularly interesting *research and lawsuits*, both of which supply a great variety of intellectual strategies in different circumstances. This would furnish conclusions pertinent to real-life situations rather than to science. Work in this direction should be encouraged e.g. by sponsoring a pertinent competition. A corresponding series with the Fund's publishers would combine individual contributions and systematic studies and syntheses.

Fourth source: An analysis of practical or conceptual *projects* that individuals or organizations executed with particular skill in difficult circumstances. This venture would aim to account for the success of those projects, especially where it depended on intellectual virtues.

An incentive to engage in such work would be provided e.g. by advertising an appropriate competition.

A corresponding series with the Fund's publishers would combine individual contributions with systematic studies and syntheses.

Fifth Source: Systematic daily behavioural observation of children's minds during appropriately designed lessons that force them to perform continuous independent work.

Encouraging the more capable teachers to work in this direction through appropriate promotion. A corresponding series with the Fund's publishers would combine individual contributions (e.g. publishing teachers' observational journals) with systematic studies and syntheses.

Sixth source – the most important: ontogenesis of the attainments of the intellect. Identifying the birth, development and adolescence of the thought of creative individuals: discoverers and inventors in all areas of human life.

Considering the content of a creative element, human activity can be separated into three categories: the first – repetition of the ancient, reproduction, but not actual creation, routine; the second – systematization of the attainments up to now; the third at last – doing and making something new: actual creativity. All three are, without a doubt, important for life; alas, the latter one is not only the most important and most difficult, but also needs particular circumstances and stimuli, as it is apt to impairment, and even to complete decline, as there have been very long periods in peoples' histories in which any creation of something new stopped, at least in many branches (e.g. in China). It is an unbelievably important matter for ensuring the progress of mankind to reveal, partially at least, the secrets of creative thought, the mechanism and the living sources of discoveries and inventions.[a]

For, there are two possible ways of investigating the creative thought of an individual worker:

1) One of them, indirect in its character, proceeds from without: the successive stages of thought, as reflected in their outward upshots, are being ordered, compared and connected with one another. Thus the internal processes of thought are being traced and interpreted by means of their external symptoms.

2) The other way, direct in its character, proceeds from within: here, again, it is the outspoken confessions of the research worker that are taken into consideration: it is on the ground of his own utterances that we ascertain how his thought originated, developed and matured, whilst all the external circumstances of this process are duly taken into account, too.

It is evident that the first way cannot be regarded as sufficient: it does not lead to any decisive conclusions; for it offers presumptive evidence only, which must remain unverified.

It is only a straightforward confession of the author (in the form of genuine and sincere notes), which should, as far [as] possible, be made on the spur of the moment, that can afford a reliable first-hand material.

Now, we are much in need of such material. Generally speaking, authors are not in the habit of recording the course and the fate of their ideas, and the circumstances and stimuli under which they arise. Such confessions as *e.g.* E. A. Poe's narrative of the origin of his "Raven," or Poincaré's account of how he developed Fuchs' functions, are of very rare occurrence. Biographers of great discoverers and inventors do their utmost to lay bare all the details of their lives; they investigate their correspond-

---

[a] The following paragraphs (until „[...] their willing response to the present appeal." on p. 300) had been translated in 1936 to appear in vol. I of the Organon as a "letter to the editor" (290–294). They were introduced as follows: "The Editor of Nauka Polska (a Polish periodical which has been founded with the object of fostering science, investigating its needs and the conditions of its development, especially the psychological aspects of the creative worker and of creative work), has received an interesting letter, written by Professor A.B. Dobrowolski, the well-known geophysicist. The ideas contained therein should, in our opinion, be diffused to such an extent as to stimulate other men of science, and gain their responsiveness in the form of pertinent contributions. With this end in view, we take the liberty of publishing Professor Dobrowolski's appeal, in the hope it will prove both interesting and provocative.

Professor Dobrowolski writes: 'To get insight into the mysterious workings of a creative mind, which result in a scientific discovery or a technical invention, we have to avail ourselves of what is rightly deemed one of the most important clues to it (maybe, it is the most important of all): these are the personal confessions of the creative worker himself, revealing, partially at least, the secrets of creative thought, the mechanism and the living sources of discoveries and inventions.'"

ence, their conversations etc.: in this way, they strive to find an explanation of the course the author's thoughts took, and to get a glimpse of the springs which prompted them. It is all in vain: the deficiencies are far too great, because of reticence and reserve on the inventors' and discoverers' part. So, all the efforts (and they have been innumerable) to establish an "ontogenesis" of discoveries and inventions have failed to yield any positive results; all that we possess in this department are mere surmises, based on a very frail ground. The various "theories" of the creative faculty of the mind are sheer speculations, more or less ingenious.[3]

So we see that the first indispensable condition of successfully investigating the creative power of individual workers is to secure adequate *first-hand material*. To attain this it is essential that authors themselves should tell us „how they do it".

Authors should be induced to undertake this.[b]

On having the question placed before their minds and represented in all its importance, authors should be appealed to for goodwill, and requested to tread in E. A. Poe's and Poincaré's steps, *i. e.* to speak out without reserve, and tell the story of their creative ideas.

They who transform, improve, discover, invent – in one word, they who do original work in the field of science, art, technics – shall initiate the world not only into the results of their creative thought, but also into its "biography," this last consisting of notes, arranged in chronological succession, and giving details about the origin of their thought, its development and its coming to maturity, and about everything that has, somehow or other, contributed to the creative process.

Such an appeal is bound to be made one day; it is already in the air. People's concern about the mysteries of creative work grows more and more intense. Volumes are being written about creative workers in order to trace the psychogenesis of their pro-

---

[3] The theories of creative faculty in the technical department are somewhat better, as they approach the subject from a specific and matter-of-fact point of view; speaking more strictly, we have to emphasize the theories of patent obtaining, which are being formed by "heurologists": Hertig, Kohler, Engelmeyer, Schanze, Rossman and others. They all strive to single out the class of inventions "proper" i. e. such as could be good subject-matter for a patent; so they are compelled to look out for essential features of inventive work. ||| The footnote on theories of creative faculty was only added for the partial 1936 translation. It was supplemented by the Organon editors, presumably shortly before the volume was printed: "Professor Dobrowolski has informed us that the very latest, to the best of his knowledge, contribution in this field is 'The Psychology of the Inventor: A Study of the Patentee' – by J. Rossman, Patent Examiner in the U.S.A. Patent Office, and Editor of the Patent Office Society; in this last we occasionally come across articles dealing with theory as well. Still, even the theories of 'heurologists,' superior as they are to others with regard to their methods, suffer, in Professor Dobrowolski's opinion, from the same deficiency as literary and philosophical and psychological theories it is reliable first-hand material that they lack. – *The Editor's Note*." See Joseph Rossmann: The Psychology of the Inventor: A Study of the Patentee, Washington, D.C. 1931.

[b] In the 1936 translation, the Organon editors added the following footnote referring to ▶ Dobrowolski 1928: "With regard to Professor Dobrowolski's efforts to secure reliable material for research work upon the mechanism of inventions and discoveries, cf. Nauka Polska, vol. 9. My Scientific Biography (by A.B.D.) – *The Editor's Note*."

ductions. Questionnaires are being addressed to them with a view to obtaining details about their creative work. Still, all this is but an expression of literary curiosity or philosophical speculation, whereas it should be put forward as a novel question of weighty importance. But, there is no doubt that this question will come into prominence ere long.

Why should the Poles not make a start?

Such an appeal would be heard most widely if uttered by a circle of experts. They could assemble around an *international medium* devoted to the questions of creativity. The medium could serve as a head office, and creators could send their reports of the actual order and circumstances of their thoughts in particular discoveries and inventions. People with different expertise, and deeply interested in creativity, should work in the editorial office.

Why should the Mianowski Fund not start this beautiful international endeavour, assemble a thoughtfully chosen group of Polish colleagues and be the publisher of such a medium?

When authors have realized the importance of such introspective individual work, they may feel disposed to confide to the outside world the mysteries of the inner course and process of their thoughts, when making discoveries and inventions. Their communications, however scarce and rather inaccurate at first, would gradually increase in number, and improve in their character; little by little, a suitable system of making notes would be worked out – which is a matter of experience: and an adequate discipline in thinking over the course of their work would be introduced – which is a matter of habit. The notes sent in by less important authors would be, form the scientific point of view, as significant as those sent in by great authors (like the phenomena which we note in natural science: all of them, regardless of their greatness, are equally valuable). In a short time, we should be in possession of a considerable amount of raw material which would be tabulated, generalized, linked together and explained – to the best of our ability –, let it be even in an empirical way: there would be nothing objectionable in it: even the subtlest branches of human science originated in rough empirical methods.

All these "Biographies of ideas" are bound, even in case they are perfectly sincere, to suffer from one very serious deficiency: they will not be complete. No author will ever be able to become aware of all the stages, constituents and factors of his thought: apart from all those details which will have been concealed on purpose, through self-conceit and undue ambition, his notes are sure to lack many others items too, more or less important, more or less numerous.

Still, "creative-thought science" will not be made impossible by this. The abovementioned deficiencies will not be systematic in their nature; they will be accidental. If a sufficient supply is handy, then a given note, however deficient in itself, is sure to come in touch with another one, which will be quite similar, or with a whole group of other similar notes; these will be deficient too, but gaps will occur in them in other places. So, the gaps occurring in one document will be filled in by the corresponding complete spots in other similar documents. Such cases are very common in scientific practice, e.g. in paleontological research work.

Another deficiency of "biographies of ideas" will proceed form inaccuracies, or even forgeries, conscious or unconscious (caused by self-conceit, by the sense of priority, and other "very human" motives). But, science does possess many a means of detecting such blemishes. Moreover, these blemishes will not be systematic in their nature, either; they will be accidental. Besides, whereas there is one Truth only, deviations from it are many, and as different as human individuals themselves. As a consequence, in an abundant collection of documents, those which are similar in their nature will manifest striking, and obvious in their anomaly, differences in those places which have, consciously or unconsciously, been forged. Thus, the forgeries will be disclosed.

When man carefully collects all possible vestiges of some creative work, in his endeavour to trace and detect the real source and the real course of the discoverer's and inventors' thought, he may call out: "If the author had put in here but one word of explanation, I should understand it all". This can be seen e.g. in the fine research work of S. Gelblum, to whom a single off-hand sketch by Watt, in a letter to his partner, instantly disclosed the origin of the famous "parallelogram" in the primitive steam-engine.[4] This can be seen, too, in Le Châtelier's research work upon Lavoisier[c]: many a department of this great man's scientific activity would be unintelligible, if it were not for a handful of accidental information about some, seemingly unimportant details. On the ground of these facts, we make bold to anticipate a rapid advancement of research work upon inventors and discoverers, provided the authors themselves convey the intelligence of which we stand in need.

It is not on the stage only that we want to see creative workers with their treasure-trove before our eyes; we want to be admitted behind the scenes, into the very laboratory of their thought. We expect them to offer us not only their trophies, but also the secrets of their victories, and to present us with the whole Truth that is in their possession: and we expectantly look forward to their willing response to the present appeal.

---

[4] In his [Gelblum's] work "Par l'historique à la science des idées," which was printed at Liège in 1906 [sic!]. Unfortunately it has been left in the form of proof-sheets, as a consequence of the author's death. Four years later, Peter Engelmeyer, the well-known technicist and heurologist raised this very question of the origin of the "parallelogram." But his solution is not satisfactory: he was not aware of the difficulties which had been pointed out by Reuleaux (Cf. Franz Reuleaux: *Cinématique. Principes fondamenteaux d'une théorie générale des machines*, Paris 1877.). ||| In 1918, the footnote contained only the first two sentences. See also ▶ Dobrowolski 1927, 318 and for Gelblum ▶ Dobrowolski 1928, 322. In the 1936 English translation, the Organon editors referenced Reuleaux' text here, too (cf. editorial note b on p. 298). The Engelmeyer remark was only added in the 1936 English translation by Dobrowolski himself, probably referring to Peter Engelmeyer: *Der Dreiakt als Lehre von der Technik und der Erfindung*, Berlin 1910 (cf. ▶ Dobrowolski 1928, 324).

[c] Henry Le Châtelier (ed): *L'air et l'eau. Mémoires de Lavoisier* (= Les classiques de la Science, vol. 8), Paris 1923.

This is how we envisage the great project of teaching reform, of introducing a new kind of education: the education of thinking.

This is how we envisage the possible and desired contribution of the Mianowski Fund.

This project, undertaken for a *practical* reason, will undoubtedly contribute to advance various *theoretical* aspects: the development of the psychology of the intellect. Aspects of this project that will contribute significantly to the fundaments of the *theory* of thinking are: an ever more detailed understanding of the norms of intellectual activity in different real-life situations; an augmented catalogue of concrete faculties of the mind; an ever improving state of research on the mind's natural tendencies, commonly referred to in terms of "ability," as well as its errors and faults. This project will also provide plenty of material to determine and classify various *types* of intellect (perhaps it will allow us to understand, for instance, the nature of the Polish mind): the primitive attempts we have seen in this domain to date will evolve into reliable knowledge. In this country, an attempt of this kind was made in the mid-nineteenth century by Michał Wiszniewski in his work "Charaktery rozumów ludzkich" (which the author published in 1853 in London as *Sketches and characters, or: The Natural History of the Human Intellects* under the pseudonym "James William Whitecross").[d]

The great teaching reform will thoroughly transform the role of the teacher, elevating it high above the common and uninteresting part it has played until now. A new era will begin: the era of the conscious cultivation of minds. A new teacher will appear: the teacher of thinking, the educator of the mind in the literal sense of the term.

The great teaching reform will equally well take place without us. Schooling is moving in that direction, though in zigzags, the world over. Other developments that will take place with or without us are the evolution of the new field of research, i.e. of the systematic study of inquisitive and inventive thought, and the new habit by which creators publicly describe the sources and development of their creations. Such developments are already making themselves felt.

Let us not shun from actively taking part in these great issues of general human concern. Let us have the courage to initiate work in this direction *ourselves*!

---

[d] Cf. Michał Wiszniewski: *Charaktery rozumów ludzkich* [Characters of Human Intellects], Warszawa 1876, i.e. James William Whitecross: *Sketches and Characters, or: The Natural History of the Human Intellects*, London 1853.

An Autobiographical Sketch and Remarks on Scientific Creativity

Czesław Białobrzeski as 'C. B.'

CZESŁAW BIAŁOBRZESKI (31 August 1878, Poshekhonye near Yaroslavl, Russian Empire – 12 October 1953, Warsaw, Polish People's Republic), theoretical physicist, astrophysicist, and philosopher of science. B. studied (1896–1901) and later worked as an assistant at the Univ. of Kiev. After two years of studying with Paul Langevin at Collège de France (1908–1910) B. became a docent and later prof. (1914) at Kiev Univ. He also taught physics in higher education programmes for women and engaged in the Polish Univ. College (1917–1919). In 1919, B. moved to independent Poland for chairs at Cracow and later Warsaw Univ. (1921).

Starting in experimental physics, B. became a central figure in theoretical physics holding posts in all important institutions and societies. He also presented at the Science Studies Circle (► **IN THIS VOL.** List of Sessions) and published in Nauka Polska. His text *An Autobiographical Sketch and Remarks on Scientific Creativity* (signed "C.B.") was included in vol. 6 and opened the first edition of the series "Researching the Genesis and Development of Scientific Creativity" (*W sprawie badania i genezy rozwoju twórczości naukowej*). Next to Białobrzeski's text (A), the section's first part "Autobiographical Material" contained the report "Ways of my Intellectual Development" (77–136) by economic and social historian Franciszek Bujak, who had signed as "F.B." (B). Both texts were introduced by the following note: "For the purpose of this study we venture to include from the editorial collection two autobiographies kindly submitted by scientific researchers" (49).

The second part (136–140) consisted of several pages from an earlier article by Antoni Bolesław Dobrowolski (► Dobrowolski 1918, 1927a), while the final part demanded founding an archive for biographic research on creativity (► Dobrowolski 1927b). The series was continued in Nauka Polska and Organon (► Dobrowolski 1928, ► Borel 1936, ► Krogh 1938). This translation leaves out some passages from the original text, in which B. describes his early childhood and education until becoming a docent in Kiev (53–56 in the orig.), his time in Paris and the return to Kiev (59–60), and some of his work on radiation (64–68). Translation by Tul'si (Tuesday) Bhambry.

ORIGINAL REFERENCE
Czesław Białobrzeski ["C.B."]: "Szkic autobiograficzny i uwagi o twórczości naukowej," *Nauka Polska*, vol. 6 (1927), 49–76.

FURTHER TEXTS MENTIONED
► Emile Borel: "Contribution aux *Documents sur la Psychologie de l'Invention dans le Domaine de la Science*," *Organon*, vol. 1 (1936), 33–42.
Franciszek Bujak ["F.B."]: "Drogi mojego rozwoju umysłowego," *Nauka Polska*, vol. 6 (1927), 77–136.

- Antoni B. Dobrowolski: "O pilnej potrzebie wychowania umysłowego w Polsce: o konieczności zasadniczej reformy nauczania w szkołach średnich oraz stworzenia w związku z ową reformą nowych placówek pracy naukowej," *Nauka Polska*, vol. 1 (1918), 489–502.
— ["A.B.D."]: "'Biografja' myśli twórczej," *Nauka Polska*, vol. 6 (1927), 136–140.
- — "Archiwum materjałów do badania twórczości," *Nauka Polska*, vol. 6 (1927), 140.
- August Krogh: "Visual Thinking," *Organon*, vol. 2 (1938), 86–94.

I received the Editors' invitation to write a brief response to the problem indicated above. If I have decided to do their bidding, it was only after careful consideration and long hesitation, even though I deeply appreciate the trust placed in my abilities.

The task before me is not only rife with difficulties, but it is also a delicate and thankless one. By writing this I must, as it were, exhibit my intellectual individuality for public viewing; I am sure to meet with a great variety of accusations, be it of conceit or of false modesty and so forth. But I undertake to produce this autobiographical sketch about my creativity in the conviction that if other researchers, more worthy than I, follow in my footsteps, this will unquestionably be of use for a great cause, namely the effort to understand the development of scientific creativity – the foundation of modern civilization.

In particular, young adepts, who after a long and rocky road are striving towards the summits of knowledge, might find guidance and encouragement in this type of sketch.

But does the question of scientific creativity really deserve more careful examination? If we gauged its value by the interest it has met with so far, an interest that is expressed in the quantity of relevant literature, the answer would certainly be negative. Discussions on this subject are few and rather improvisational.

In this respect, poetic and philosophical creativity has fared better. Thorough and tireless critics continue to elucidate the genesis of literary masterpieces and the conditions for the development of fine arts in general. Great thinkers' sources of creativity have also been studied in depth. It would seem that the creative process is principally the same everywhere, and that tackling creativity with respect to individual sciences should yield no significant results.

Certainly, some characteristics are common to all manifestations of creativity, which is the highest form of human mental activity. A particularly close relationship should exist between scientific and philosophical creativity. And yet the differences in thought content and method across the sciences are immense, and the diversity of the creative processes at work in them is evident from the variety of skills required for their successful practice. What is more, within each science we can single out a handful of distinct types of creative ability on whose concurrence the rapid expansion of knowledge is contingent.

Now, considering that in contemporary civilization science has become the most powerful agent of progress and the leading force of social life, the usefulness of studying scientific creativity cannot be denied.

We can expect that in due course this research will provide guidelines on how to detect and to form creative abilities.

I will focus my attention on the problem of creativity in the exact sciences, especially in the field of physics, which I practice myself.

The creativity of a physicist is unique in that it is doubly constrained. First, the requirement for consistency with the experiment limits the soaring imagination, and second, the results must be presented according to an exact, usually mathematical, form, whose indispensable condition is relentless exactitude and precision of thought.

This is not to say that the imagination plays a secondary role here. On the contrary, searching for fundamental explanations of the natural world, the physicist has created the supranaturalistic world of atoms, electrons, and invisible radiation. The imagination, curbed by mathematics, discovers the relationships at work in this world and uses them to understand phenomena that are available to the senses. The power of this imagination makes itself felt to all who use electromagnetic waves in their room in Warsaw to listen to a concert performed, say, in London.

Before proceeding to outline autobiographical particulars concerning creativity, we must answer one serious and apparent doubt.

"Granted," they will say, "it may be worthwhile to study the secrets of the creativity of great scholars whose bright and fecund thoughts have shown the way to generations of scientists and illuminated entire realms of being. But is it worth the effort to accumulate material about the creativity of that multitude of scholars who have followed in the masters' footsteps, making only modest contributions to the repository of knowledge?" This complaint is unjust for several reasons. Above all, science has become second nature to humanity. A great number of people work on its development and on the transmission of scientific accomplishments from one generation to the next. The fruits of this labour are colossal and the group effort serves to bolster and to provide material for the geniuses who produce the great syntheses.

Furthermore, it is not as if ingenious scientists resembled lonely peaks rising high above the plain. There is an intangible gradation of talent. Each individual, endowed with abilities in this or that domain, possesses a certain amount of creative intuition. Epoch-making discoveries, at least the experimental ones, have often occurred to people whose mentality had none of the characteristics of greatness. Given how scientific work is organized today, it would be no easy task to determine if science owes more to ingenious minds or to the shared labour of talented or simply capable people. What I just said does not imply that scientific work is accessible to all. There is a certain level to which one must rise, and even among people working in the sciences only some are able to attain it. Finally, there is no doubt that to some extent creative abilities can be cultivated.

Being naturally endowed with no exceptional abilities, one can still cultivate and prepare one's mind for creative work, and one can skillfully put to use the potential for intuition supplied by nature. Counting myself in this category, and drawing on my own experience, I believe I am justified in my conviction that the initiative of the editors of *Nauka* merits the scientific community's active support in every respect. The broader the framework for accumulating documents on creativity, the more interesting and consequential will be the results, both for the psychology of creativity and for the task of cultivating the mind's highest faculty.

Beginning now to sketch out my experiences and resultant conclusions, I become aware of the difficulties that this endeavour presents. Above all, the task can be understood more or less broadly. Will it suffice to point out in what way a certain manifestation of scientific creativity had been prepared and carried out, or ought I to reach deeper and attempt to outline the development and formation of these creative faculties?

This broader treatment would demand great powers of introspection that perhaps only few people possess. The first attempts might therefore turn out somewhat awkward. But gradually, through a comparative approach, the most appropriate way of solving the problem proposed here will emerge and the difficulties will be reduced to a minimum.

I will try to carry out my task from a broader perspective, that is to say by attempting to outline the type of mentality I represent. Every type possesses traits that are characteristic of the individual, and therefore I must stress them, too. I will, however, disregard unduly minute details that might put to the test the reader's patience, and whose existence only seems justified when portraying an individual of exceptional calibre.

Aside from the difficulties related to introspection, the task as defined above demands great impartiality of judgement. At any rate one must, as it were, step out of oneself and observe one's psyche like an object to be studied scientifically.

It is possible that in the great majority of cases an astute outside observer is much better placed to satisfy this demand for impartiality than the subject studying himself.

And yet, accounts of personal experience have a unique value that is not called into question even by a certain partiality or self-love, or by the lack of perspective that results from too close a proximity to the object under investigation. But it is time now to leave these preliminaries and get to the point.

I will try to limit to a minimum the external details of my life. But we cannot ignore the backdrop against which the individual develops.

[...]

Now I will do my best to define the typical attributes of my own mentality along with my more exceptional individual qualities.

There have been varied attempts to classify minds. In my opinion all people for whom scientific work is a life goal are marked by one of two intellectual tendencies. These tendencies correspond to the division into 'philosophical' and 'expert' minds.

Of course, here, too, we will encounter varied gradations of the attributes, but in their pure manifestations both tendencies are clearly distinct. The expert mind is characterized by exceptional aptitude in a given field and shows little interest beyond what seems to be an innate calling. Particular abilities usually go hand in hand with an interest in and fondness for one's field of expertise as well as a drive toward independent work. There are born mathematicians just as there are born naturalists and philologists.

The philosophical mind possesses no sharply defined abilities at all. Its curiosity is open to the whole world. The designation is based on the fact that philosophers usually possess such a universal inquisitiveness. This is not to question that creative work in philosophy requires certain special intellectual qualities. To be sure, even the discipline of philosophy is currently experiencing increased specialization, a narrowing of scope, but universality still characterizes most philosophers.

The wealth of aptitudes and interests of an expert mind is focused on a given conceptual field, while for a philosophical mind it is dispersed across a number of fields that can be even be distant from each other.

A few chosen individuals combine the attributes of both types: on the substrate of their many-sided interests and abilities there grows a powerful talent, which is then directed to a given field of knowledge. Thinking of the more distant past, there comes to mind the names of Pascal, Descartes and Leibniz. Amongst nineteenth-century physicists, Helmholtz can serve as an example of such a richly gifted individual.

Among minds who chiefly belong to the first category, a combination of outstanding abilities in two separate but related branches of science can be uncommonly beneficial for scientific creativity, for instance when in addition to a talent for physics there is a talent for mathematics. The unsurpassed model for such double genius is Newton, followed closely by two other great Britons, Lord Kelvin and Maxwell. But let us come down from these summits and tackle our actual subject.

I unquestionably belong to the second category, the philosophical mind. This is evidenced by my keen interest in fields outside of physics and related subjects such as mathematics and chemistry; as I already mentioned above, my interest spans the entirety of knowledge, and philosophy in particular.

I have always had a predilection for physics, but for quite a long time I also toyed with the idea of devising a philosophical system after having explored and done some independent work in this science. I saw physics as the most appropriate preparation for a synthesis on account of its accuracy and its ability to yield fundamental concepts about external reality. At that time I believed somewhat naively that thanks to science the eternal secrets attending human life would not just be unveiled one day, but that they were being unveiled already. I have since abandoned this project, indeterminate though it was, as forever surpassing my abilities; nonetheless I believe that each generation ought to produce such a synthesis for itself and their immediate successors, and that individuals with a calling for that will always be found.

Given how the different abilities have been characterized above, the question arises if such minds as I describe as philosophical are able to work effectively in the specialized sciences at all. Of course there is no doubt concerning the expert minds. The diffusion and indeterminacy of a philosophical mind's abilities apparently condemns it to an unproductive encyclopaedism. I wish to say a few words in defence of this type of mentality.

Creative faculties and various abilities are not isolated and independent but support one another. The broad mental culture that the philosophical mind longs for will always have a favourable impact on creative performance in the specialized field. Acquired mental qualities strengthen one's thinking in that field, once it has been chosen, and come together in it.

In my view a youth who has no particularly strong leanings towards some branch of science should not stifle his broader interests, if they exist. That would make him a useful worker, not a creator. A general mental culture will strengthen his abilities in all disciplines, and in particular in the favoured field. Naturally, one must beware of falling into dilettantism, setting out at once on a course of independent work.

Here the objection will be raised that in this way the best years will be lost for creativity. There is quite a widespread belief, formulated by Ostwald, that after turning thirty talented individuals basically lose their greatest creative abilities and continue to draw on ideas conceived in their youth. This opinion strikes me as glaringly unjust and applicable at most to a specific type of creative person. Minds of the philosophical type seem particularly capable of long-term development. I can say from my own experience that my mental development was complete no earlier than around the age of 35. Of course, the deepening and broadening of a working person's professional knowledge and general outlook on the world continue beyond all boundaries of age. I only wish to affirm that in my case the efficiency of my thought processes and creativity continued to develop more or less until the age indicated above.

Until that age I was distinctly aware that my thinking was not only becoming more and more precise, but also that the ideas that came to me had increasing scientific value.

I used to write down ideas for my research work; comparing them I see that they were becoming gradually more mature and that they rose to higher standards by degrees. This material was of immense value to me, but unfortunately almost all of it was lost as a result of my hasty departure from Kiev in 1919. I was able to reconstruct only an insignificant part of it from memory.

[...]

Now let us turn to the most important problem – characterizing the creative process based on experiences observed through self-analysis. I do not intend to describe how each of my scholarly works came into being: that would be a pretentious abuse of the reader's patience, and besides I would no longer be able always to reconstruct with sufficient precision the processes at play. Once the goal has been achieved, the road that leads to it ceases to be of interest and slowly fades from memory. This is why I would like to point out to all scientific workers who recognize the importance of the problem of creativity that creative moments must be caught in the act and preserved through appropriate notes.

I will discuss in detail the genesis of only two of my relatively more important theoretical works. These are: my work on the role of radiation in the internal stellar equilibrium, as well as my work on the internal dispersion and absorptivity of radiation energy.

I only wish to make a few rather general observations on research in physics, beginning with the experimental kind.

Some researchers, especially early in their careers, tend to make discoveries by testing whether any physical factor might influence a given phenomenon, e. g. if a magnetic field affects fluorescence, or if X-rays affect the spectrum of live steam.

It happens sometimes that experiments of this kind lead to new observations. As a rule, however, one should beware of such simple ideas, unless they are based on well-considered theoretical foundations that augur a positive result. This is why good experimental research can happen in two ways only. Research of the first type builds on new ideas; its success is determined by their value, which in turn depends not only on creative abilities but also on a perfect grasp of the phenomena under investigation. Therein lies the necessity of an excellent command of the theory for creative experimental work. It happens more and more often that experimenters execute ideas indicated by theorists. This reciprocal dependence of theory and experimentation – the dependence of the experiment on theory and vice versa – will no doubt continue to intensify.

Scholars who perform the second type of experimental work do not pursue new paths at all; they try to achieve the most perfect exactitude within the domain of research opened up by scholars of the first type. Here, creativity tends to manifest itself in ideas for new laboratory equipment that would facilitate more precise measurements. Physicists who work in this way can be seen as actual experimenters who do not require such an excellent command of the theory.

Sometimes it happens that the abilities proper to both types are combined, and this is what makes the most perfect kind of experimental physicist. Works that are unsound either in terms of precision or in terms of the value of the ideas are admittedly the most common, but their scientific worth is negligible. Experimenters should keep their eyes open to all details of the phenomena they study; nature often reveals quite unexpected things, and we must be well prepared in order not to overlook them. Great discoveries have often happened by chance; Roentgen, for example, was among these lucky ones. But not everyone is able to recognize that unveiling of nature's secret and to make use of it.

Experimental work that is born of a novel idea rarely yields results that correspond exactly to what was predicted. Therein lies the incompatibility between nature and our minds. Oftentimes in the course of carrying out a project we are diverted onto entirely new tracks. I can cite an example from my own experience. In 1913, I began an experimental study on the influence of powerful ionization on absorption spectra.[a] The first results were quite promising but vague. But then, carefully examining photographs of absorption spectra in gaseous benzene, I observed a shifting of one of the ultraviolet emission lines of the mercury arc, which I had used as a source of radiation. This phenomenon is quite puzzling, since it had never been observed under similar conditions and was not easy to account for: after all, it is only in very rare and exceptional cases that the frequency of light oscillations changes.

The outbreak of the European war forced me to interrupt my work, and the photographs were lost during the turmoil of the Bolshevik Revolution. I have not had the opportunity to return to those tests, but it seems that no one has detected that mysterious shifting yet – most likely because it only occurs under narrowly defined conditions.

Theoretical research is clearly dominated by one of the two tendencies, which correspond to the two types of scholars theorizing physical phenomena. Most scholars work closely with experimentation, either by trying to account for experimental data, or by submitting theoretical conclusions for experimental verification. There exist theorists of a different kind, those who undertake to improve existing theories, mainly in terms of mathematical precision, or to solve the more difficult problems posed by those theories. Mathematicians who tackle physics problems belong to this type. Creative work in the natural sciences tends to be done by physicists of the first type. In any event, excellent mathematical skills are indispensable to all physicists aspiring to theoretical research, while more and more branches of mathematics are finding applications in physics, thus requiring broad and arduous studies.

The preparation and skills needed for creative and genuinely valuable work in theoretical physics are generally much more considerable than in experimental physics. For the experienced physicist, possessing sufficient erudition and keeping up diligently with scientific literature facilitates the formation of ideas for experimental research. For this reason it is natural and common for a physicist, even one with a passion for theory, to start out with experimental work; it is helpful to have an immediate experience of the phenomena that we want to theorize.

Most physicists never leave this path of experimental research. For this reason good schools preparing them for creative experimental work are indispensable for physics to flourish in any given country. Unfortunately, experience at home and abroad suggests that we cannot take for granted that universities will accomplish this task. It must be added that these institutions are overburdened with teaching duties, and doctoral theses often fail to rise above the level of school projects. Herein lies the rationale for an independent research institute whose scientific output would be subject to monitoring by an independent Council; the structure of the university does not actually generate a professor's responsibility, at least where scientific work is concerned.

---

[a] I wish to thank Professor Hugon J. Karwowski (University of North Carolina at Chapel Hill) for correcting my translation of the sections dealing with physics. Any remaining mistakes are, of course, my own. [T.B.]

Turning again to my above-mentioned research, I shall at first make a few remarks on the search for new ideas that invite scientific development. It seems that most researchers try to find these ideas by studying the latest scientific literature. Poring over works within his field of interest, the researcher looks out for new topics that their authors had not yet considered. My way of getting down to business was different. Reading scientific papers, usually only minor ideas would come to me. If they are good researchers, authors present well-considered issues that have been investigated as comprehensively as possible, so that following their train of thought we will find it difficult to find a way out of it. My search for problems ready to be solved happened in a different way. Considering a broader set of phenomena I tried to find those that had not yet been explained. Of course, this method requires a good command of the subject, better even than when we identify research problems by examining the literature. Furthermore, the horizon of the unknown broadens and deepens in the mind's eye as the researcher gains a better understanding of the field. The hardest part is to select from the realm of the unknown those problems that allow for connections with what has already been studied, and that lend themselves to existing methods of experimental and mathematical research.

[...]

While the fundamental idea in my work on the role of radiation in the stellar equilibrium was, as I said, relatively easy to develop, the theory of absorption required me to work by the sweat of my brow. Before the right road towards explaining and developing a novel idea in physics is found, one's thoughts blunder and go astray to a greater or lesser degree. So it was with me. To extricate oneself from these difficulties, one must deduce the broadest possible conclusions from the assumptions that come to mind; in this way it becomes clear if those assumptions correspond to reality. This requires a sharp critical sense that refuses to accept vague concepts, as well as an aptitude for correct deduction. But these mental qualities can only suffice to reject flawed concepts and to formulate a theory if we have found true assumptions, that is to say ones that lead to conclusions faithfully representing the phenomena under investigation. A system of true assumptions cannot be discovered by way of deduction. Here repeated acts of creative intuition are indispensable, and these always occur unexpectedly and after the mind has had a certain period of rest. It seems to me that the difference between the scientific work of a mathematician and that of a physicist lies in the fact that the part of deduction predominates for the former, while the latter's work is governed by intuition, the part related to searching for the right concepts and assumptions.

I shall not tire the reader with accounts of my struggle against the difficulties that I had to overcome in order to understand how the pressure of radiation transmitted by particles produces heat, that is to say how it increases the energy of the particles' random movement. I shall only indicate the decisive moment. At one point the idea occurred to me that a regular light wave, before it is transformed into the irregular thermal motion of particles, must first itself change into irregular light waves, that is to say it must be dissipated. When I proved later that the pressure of dissipated waves always has a positive impact on particles, that is to say that that impact causes a loss of light energy and a gain in thermal energy (I called this theorem the principle of fluctuation impact), the theory was generally established. Thus the first stage in the process of transforming radiation energy (of light or invisible rays) into heat is a gradual dissipation of regular waves by the matter into which they enter; the second stage is a transformation of energy waves dissipated into heat through the pressures that the waves exert onto the particles.

Besides having explained absorption, I consider it significant that I paid attention to the difference between waves dissipated within matter and dissipated radiation that is dispersed in the surrounding space; until then only the latter had been studied.

Before I conclude this topic, I should call attention to one more rather characteristic detail. It happens that highly sensitive people like myself are so deeply affected by some difficulty or contradiction that their mind, instead of taking them to fight, surrenders with resignation.

When I had found the ultimate mathematical formula for the quantity of heat produced by the transformation of radiation energy, that formula turned out to conflict with the fundamental and experimental law of absorption (Lambert and Bouguer). This presented itself to my mind as an indisputable fact, and I came to the conclusion that my theory of absorption, built on the classical theory of light, collapsed as a result of this incompatibility with the experiment. I did not quite lay down my arms and started looking for solutions with the help of quantum theory, which, as is well known, has proved indispensable to account for a range of phenomena in radiation, but until now it has not been reconciled with the classical wave theory of light.

I did manage to find this solution, and the incompatibility I mentioned was resolved. I considered the problem successfully solved and stopped thinking about it. After a while, however, I returned to this issue, intending to incorporate the results into a longer disquisition. Now almost immediately the realization struck me that the incompatibility with classical theory was specious, since that formula of which I mentioned above is not directly related to the law of absorption. I was left to wonder why this simple idea had eluded me for several months.

That erring was beneficial to me, since it forced me to work out a theory based not only on classical but also on quantum laws, and to demonstrate, with application to absorption, the compatibility of these two points of view – a rare case in present-day theoretical research on radiation.

What I presented here suggests that when we are elaborating a more far-reaching argument, the creative process often involves a rise and fall of ideas, quiet triumphs and painful disappointments, as well as nerve-racking assiduity. But the victorious conquest of nature's secrets leads to great joy. Initially it explodes in flames, but, like great love, it does not go out completely but leaves in the soul a glimmer that brightens the grey routine of life.

Let me throw in a few more remarks from my personal experience. As for the method of my work, I should point out that it resembles a mosaic. Once there is a leading idea, a centre to crystalize concepts, I begin the preparatory stage during which, poring over the subject at hand, I jot down everything that enters my head and seems valuable. Then, if the work seems promising, these loose ideas, lines of reasoning or calculations turn into a whole that warrants elaboration. In this case I create a more or less detailed plan and with this solid preparation, I usually write the work according to the plan directly in its final form.

In this context I should note that one of the most important mental qualities for those who work in science, and probably also in literature, is an ability to focus on a subject intensively and for a lengthy period of time. This demands tedious effort and willpower. There are gifted people who dislike this kind of effort, which makes them unproductive researchers. They can fully state their scientific interests by reading research publications; their motivation to do creative work is not strong enough to overcome this kind of mental sluggishness. But those who possess the virtues outlined here along with a mastery of the subject, they should boldly launch themselves into creative work. In this we ought to satisfy ourselves with modest goals for our work and not overestimate our abilities, but at the same time we must also have the courage to be passionate about the big things even while maintaining critical distance towards our own ideas.

It is a good thing sometimes to have a broadly defined research goal: with skilful development, some result, though perhaps a much narrower one, will be achieved.

But, the reader will tell me, all this is of no avail without the appropriate abilities. Without abilities, however, it is impossible to really master a subject as difficult as physics in the first place, and mastery precedes scientific research as a necessary condition. One could still ask what characterizes a physicist's aptitude specifically. This question was in fact already answered at the beginning of this sketch. A particular kind of imagination is at play in the formation of ideas in physics, and to foster fruitful scientific creativity in this field of learning it is crucial to possess and to develop it. This imagination must grasp causal relationships in the course of nature with maximum precision. Here the contemporary physicist will be aided by the extrasensory world of electric atoms that influence one another through the network of electromagnetic forces created in the space around them. The imagination tries to reproduce various phenomena based on these elements of reality, thereby discovering and embodying powerful instruments that do not exist in nature, at least not in an isolated form or aiming at a particular goal. Radio-telegraphy and telephony can serve as examples. It is clear how sharply and precisely the imagination must work in order not to get lost in the chaos of innumerable relationships. When one possesses it to a sufficient degree, and when the conditions outlined above are satisfied, success is a sure thing. The theoretical physicist must then also be able translate the mechanism drawn up in the imagination into the language of mathematics.

They will ask, what about creative intuition, an indispensable and perhaps the most important mental factor. Unfortunately, this is also the most elusive one. The hierarchy of scientific talents probably depends most on the resources of that intuition. Nonetheless I believe that, conventionally speaking, a person who shows the mental qualities listed here will possess it to a greater or lesser degree.

If we work skilfully, we should pay no heed to failures, or shrink from difficulties that for the time cannot be solved. Let us think patiently about the problem that interests us. If the mind is momentarily discouraged and weary of the topic, it helps to let it be for some time and to return to it later with redoubled energy, if we have a feeling that the chosen path promises success. In this case it is likely that our intuition will not fail us: at some auspicious moment the solution will flow from the subconscious depths or we will see clearly the road that leads to it.[1]

The formation of solutions or intuitive ideas cannot be represented in a clear light; that would require looking into the secrets of the subconscious psyche.

I can only throw in a couple of remarks on the external circumstances accompanying that process.

Freshness of the mind is crucial, as well as its full conscious activity. Intuitive revelation happens neither in dreams nor before sleep, when the mind is tired, even if it is in a state of excitement.

Since to explain or solve our problems requires tranquility and prolonged reflection, we creative scientists must be allowed to remain alone with our thoughts for some time. A rural surrounding can be stimulating; far from the hustle and bustle of the city we can devote ourselves and our thoughts to contemplating nature. Impressions of beauty create a mood that rouses the soul's creative faculties from inactivity. I am particularly sensitive to light and colours. Among the genres of painting I am particularly fond of landscapes. Perhaps it is no coincidence that as a sun worshipper my thoughts have constantly turned to problems related to light.

While solitude is sometimes necessary for the scientific researcher, it can be both influential and beneficial to entertain relations and to enter into discussion with people working on the same or similar problems. One ought to make particular use of opportunities to establish closer ties with leading personalities. A lack of this influence was a great handicap in the formation of my talent for creative work.

---

[1] In his letter to Olbers (1805), Gauss emphasizes the revelational character of intuition that brings about a solution: "You will remember... perhaps my complaints about a theorem that... had thwarted all my efforts to find sufficient proof...; all search had been in vain, and each time I was forced to put down my quill in sadness. Finally a couple of days ago it worked out – but the success, I almost want to say, springs not from my laborious quest but from God's Grace. Just as lightning strikes, so the riddle was solved: I myself would not be able to trace the thread connecting what I had known before, what I had done in my recent experiments, to that through which it worked out." ||| Białobrzeski does not give the detailed reference. See, however, *Wilhelm Olbers. Sein Leben und seine Werke* (vol. 2: Briefwechsel zwischen Olbers und Gauss, Erste Abteilung), ed. by C. Schilling, Berlin 1900, 268–269.

If in a given environment no atmosphere of learning is created, a young scholar will not find that nutritious fluid that sustains his concentration from without and motivates him to work: under these unfavourable conditions even a great talent will get discouraged and decline.

Besides my modest contribution on the mechanism of scientific creativity, this sketch, as I have already said, aims to outline the author's mental profile. That being so, I cannot pass over his attitude towards those very general questions where the author's attempted solutions constitute his philosophical worldview. Failing to take this subject into account, this attempt at characterizing the philosophical mind would be incomplete, especially as minds of this type aim to understand the entirety of our experience as far as possible at any given moment in the history of the human spirit. Worldviews include metaphysical concepts among others. There is a widespread belief that metaphysics is a purely subjective creation, the poetry of concepts. I do not share this view, but believe that metaphysical statements can be justified to a satisfactory degree if we curtail excessive ambitions in this arena. There is no doubt that the subjective element plays a more significant role here than it does in the exact sciences, since the general worldview includes, besides the totality of studies and scientific interests, the individual mind's deepest inclinations. This is not the place to elaborate on this point, so I shall content myself with recording my views on the most important philosophical questions, leaving aside their formation and transformations.

Within the theory of knowledge I hold a realistic position that recognizes a reality that is independent of and different from human consciousness. I believe that science is slowly gaining an understanding of that reality, an essential understanding, not one that is limited to mere symbols concealing an unknowable being. In my opinion the critical investigation of the conditions and methods of knowledge, which is the subject of the theory of knowledge, is an important discipline that fosters competence in every domain. I have a lively interest in the methodology of physics, which has been studied little and is so unique on account of the strange fusion of empirical with mathematical, that is to say axiomatic, methods. It is striking that sensory elements are almost completely eclipsed from fundamental concepts in physics, that the discipline should be founded on extrasensory entities.

Let us turn to the question of metaphysics. With regards to the essence of being, I favour the spiritualistic current, which recognizes that an underlying cause, a being akin to our soul, is hidden beyond natural phenomena; we can ascribe to it the qualities of a substance or we can simply view it as a set of mental symptoms developing through time. I see no reason to reject the concept of the soul as a substance.

In the conflict between determinism and indeterminism, I stand on the side of the latter, and therefore I presume that in the course of nature there exists an immanent purposefulness, which is noticeably present in living organisms. This does not contradict the existence of unshakeable natural laws: mechanism and theology can complement one another; they are not mutually exclusive.

From among the three most common positions regarding notions of God, the theistic, pantheistic and atheistic, I hold with the first, which best matches my metaphysical views as outlined above. Assuming that God exists above the perceptible universe, I tend, like many naturalists, to tinge my theism with a pantheism that implies a closer connection between God and nature.

Asked to name a more or less contemporary philosopher whose worldview overlaps significantly with mine, I should mention Lotze. But I must note that until recently I knew little about Lotz and I have not read his works.

Coming, as promised, to the deepest fundamentals of the individual soul, I wish finally to summarize in a few words my relationship to religion. After an initial youthful phase, once I recognized the untruth of the claims of science that had lead to widespread antireligious views (at the end of the nineteenth and the beginning of the twentieth century), I tried to comprehend, mostly through philosophy, the problem of religion and of religious life.

Now, trying to come to an end as soon as possible, I shall summarize without further ado my views on religion in three sets of statements:

1) Religion is an inextinguishable necessity for most human souls. Even among declared opponents of religion it is often easy to detect a religious temper. This results above all from the human being's position in the universe. We are aware of our weakness and dependence on some higher order, one that transcends our complete understanding but still makes itself felt to the human mind in that we can sense its greatness. This is why I believe that a religious temper suits the scholar better than anyone else. People who spend their lives studying nature and looking into its precipitous depths without erecting in their souls a temple to the unknown God, they have a different mental constitution than I and those who are like me.

2) A person's inner worth is determined by the moral principles that guide his or her general behaviour and individual deeds.

The ethical question is the single most important problem in human life. It can, of course, be treated separately from religion. But I believe that the ultimate and immovable justification of the concept of good, of ethical responsibility and spiritual value in general, can be found only in religion, in the faith in God as the highest good.

3) Doubtless, religion is in a sense the most personal affair of individual conscience. And yet, as a social creature the human being aspires to spiritual fellowship with others and wants to create shared spiritual goods. This tendency is reflected in religion. And so from all the religions that humanity has managed to create throughout its history, the most perfect, in my opinion, is the Christian religion.

Here I shall conclude; I hope that, in accordance with the wishes of the editor of *Nauka Polska*, this sketch has acted as an incentive for other ones to be written, and I am sure that among them will be found documents of the highest value.

C. B.

An Archive of Materials for Research on Creativity

Antoni B. Dobrowolski

ANTONI BOLESŁAW DOBROWOLSKI (6 June 1872, Dworszowice Kościelne, Congress Poland – 27 April 1954, Warsaw, Polish People's Republic) was a geophysicist, meteorologist, and theorist of pedagogy. For further information see the biographical note to
► IN THIS VOL. Dobrowolski 1918.

D. was highly involved in the science studies milieu of Interwar Warsaw, frequently attended the meetings of the Science Studies Circle, and published broadly in Nauka Polska. His call to establish "An Archive of Materials for Research on Creativity" closed the first edition of the Nauka Polska series "Researching the Genesis and Development of Scientific Creativity" (*W sprawie badania i genezy rozwoju twórczości naukowej*) in vol. 6 (49–140). The first of its three parts ("Autobiographical Material") included an autobiographical sketch by theoretical physicist Czesław Białobrzeski (signed "C.B.", ► Białobrzeski 1927), and the report "Ways of my Intellectual Development" (77–136) by economic and social historian Franciszek Bujak (signed "F.B."). The second part (136–140) was a reprint of several pages from D.'s 1918 article in vol. 1 of Nauka Polska (► Dobrowolski 1918), which, according to a short editorial note (136) was reprinted since the vol. 1 had been sold out in the meantime. The series was continued in Nauka Polska and Organon (► Dobrowolski 1928, ► Borel 1936, ► Krogh 1938). Translation by Tul'si (Tuesday) Bhambry.

ORIGINAL REFERENCE
Antoni B. Dobrowolski: "Archiwum materjałów do badania twórczości," *Nauka Polska*, vol. 6 (1927), 140.

FURTHER TEXTS MENTIONED
► Czesław Białobrzeski ["C.B."]: "Szkic autobiograficzny i uwagi o twórczości naukowej," *Nauka Polska*, vol. 6 (1927), 49–76.
► Emile Borel: "Contribution aux *Documents sur la Psychologie de l'Invention dans le Domaine de la Science*," *Organon*, vol. 1 (1936), 33–42.
Franciszek Bujak ["F.B."]: "Drogi mojego rozwoju umysłowego," *Nauka Polska*, vol. 6 (1927), 77–136.
► Antoni B. Dobrowolski: "O pilnej potrzebie wychowania umysłowego w Polsce: o konieczności zasadniczej reformy nauczania w szkołach średnich oraz stworzenia w związku z ową reformą nowych placówek pracy naukowej," *Nauka Polska*, vol. 1 (1918), 489–502.
— "'Biografja' myśli twórczej," *Nauka Polska*, vol. 6 (1927), 136–140.
► — ["A.B.D."] "Mój 'życiorys naukowy,'" *Nauka Polska*, vol. 9 (1928), 68–216.
► August Krogh: "Visual Thinking," *Organon*, vol. 2 (1938), 86–94.

A concrete way of realizing A.B. Dobrowolski's initiative can be found in the above-mentioned brochure entitled "Par l'histoire à la science des idées,"[a] whose creation was influenced by Dobrowolski. It contains the proposition to accompany each published piece of original research with an addendum containing a "biography" of the creative scientific thought of that work, its origin and development.

Thus an archive would emerge with materials for the research we are concerned with here. Autobiographies do not usually include these materials, as they describe the course of intellectual development as seen from a certain distance in time; these descriptions are, of necessity, abbreviated, and they omit some of the developmental connections that could have been recorded if one's creative thoughts were observed almost *in statu nascendi*.

---

[a] This refers to the second part in the opening edition of the Nauka Polska series "Researching the Genesis and Development of Scientific Creativity" (see the editorial introduction above), which contained a partial reprint of ▶ Dobrowolski 1918. There (300), as well as in ▶ Dobrowolski 1928, 324 Dobrowolski refers to the text as *Par l'historique à la science des idées*.

My "Scientific Biography"

Antoni B. Dobrowolski as 'A.B.D.'

ANTONI BOLESŁAW DOBROWOLSKI (6 June 1872, Dworszowice Kościelne, Congress Poland – 27 April 1954, Warsaw, Polish People's Republic) was a geophysicist, meteorologist, and theorist of pedagogy. For further information see the biographical note to ► **IN THIS VOL.** Dobrowolski 1918.

D. was highly involved in the science studies milieu of Interwar Warsaw. He frequently attended the meetings of the Science Studies Circle, published broadly in Nauka Polska and beyond. His text "My 'Scientific Biography'" (*Mój 'życiorys naukowy'*) was originally published in vol. 9 of Nauka Polska in 1928 (signed "A.B.D."). It opened the second edition in the series "Researching the Genesis and Development of Scientific Creativity" (*W sprawie badania i genezy rozwoju twórczości naukowej*), which had been initiated in vol. 6 (► Białobrzeski 1927, ► Dobrowolski 1927). Apart from D.'s text, the 1928 section contained two further reports: a so-called "Second Biography" (signed "X.Z."), and "Memories about Ways into Scientific Work" (signed "J.Z."). The series was continued in Nauka Polska and Organon (► Borel 1936, ► Krogh 1938).

In the text, D. describes his scientific life from his days in grammar school until the early 1920s, when he worked at the National Meteorological Institute. This translation concentrates on the passages, in which D. describes the history of the initiative to collect „biographies of scientific thoughts" (Dobrowolski 1928, 68–70, 154–158). An extended version of the text was published posthumously (Dobrowolski 1958). Translation by Tul'si (Tuesday) Bhambry.

ORIGINAL REFERENCE
Antoni B. Dobrowolski ["A.B.D."]: "Mój 'życiorys naukowy'," *Nauka Polska,* vol. 9 (1928), 68–216.

FURTHER TEXTS MENTIONED
► Emile Borel: "Contribution aux *Documents sur la Psychologie de l'Invention dans le Domaine de la Science*," Organon, vol. 1 (1936), 33–42.
► Antoni B. Dobrowolski: "O pilnej potrzebie wychowania umysłowego w Polsce: o konieczności zasadniczej reformy nauczania w szkołach średnich oraz stworzenia w związku z ową reformą nowych placówek pracy naukowej," *Nauka Polska,* vol. 1 (1918), 489–502.
► — "Archiwum materjałów do badania twórczości," *Nauka Polska,* vol. 6 (1927), 140.
— *Mój życiorys naukowy*, Wrocław (1958).
► Czesław Białobrzeski ["C.B."]: "Szkic autobiograficzny i uwagi o twórczości naukowej," *Nauka Polska,* vol. 6 (1927), 49–76.

▸ August Krogh: "Visual Thinking," *Organon*, vol. 2 (1938), 86–94.
anonymous ["J. Z."]: "*Wspomnienia o drogach do pracy naukowej*," *Nauka Polska*, vol. 9 (1928), 246–259.
anonymous ["X. Z."]: "*Drugi życiorys*," *Nauka Polska*, vol. 9 (1928), 217–245.

CONTENTS: 1. Introduction. 2. Research on inventions and discoveries.

1. Introduction

I was among those whom the editors of *Nauka Polska* asked for a "scientific biography". A great honour, but I wouldn't say that I accepted the proposition with enthusiasm. Above all precisely because it seems too great an honour: I do not think that highly of my scientific activity; after all, whatever I might have achieved are but small, insignificant things. What's more, I am not even thinking about my archives at this moment: there will be time for that once I stop being able to do anything better; in the meantime I'm all about the future, I hardly ever think about the past, I have neither the desire nor the time for that. So it was with reluctance that I accepted the invitation. But I accepted it. Those who know the editor will know how hard it is to turn him down.

Besides, I have certain reservations. I am not convinced that "scientific biographies" of this kind will actually prove useful as material for *scientific* research on creativity. They might be "interesting" to those who are interested in the human mind and soul; they might even be "instructive" for young researchers or teachers, by providing certain guidelines, based on personal experience, about what ought to be done *perhaps* and what one should *perhaps* avoid. But collecting such descriptions is unlikely to yield anything accurate, because the descriptions are neither accurate nor complete or systematic; they lack *immediate* observations and contain too many interpretations, distortions and stylisations, since facts are reflected in the waves of memory. In my opinion the only method that might yield more serious results is not the "scientific biography" of a given person, but the biography of a given *work*: the researcher's immediate confession about the birth of a given idea, its development and maturation, as well as of the circumstances that influenced the research process in one way or another. This kind of declaration ought to be made in real time: fresh, full, undistorted[1].

Thus I have no pretensions to "scientificity" in the following account, where I will attempt to describe the stimuli, factors and circumstances of my scientific career, the distinctive features of its course as a whole, and finally the genesis of the more inter-

---

[1] Cf. Antoni B. Dobrowolski: "'Biografja' myśli twórczej" ["Biography" of Creative Thought], *Nauka Polska*, vol. 6 (1927), 136–140 and ▸ Dobrowolski 1918, 296–301. ||| The first text referenced here was a partial reprint of ▸ Dobrowolski 1918 in the second part of the series "Researching the Genesis and Development of Scientific Creativity" (see editors' note above). The same section was translated for vol. 1 of the Organon in 1936 (see editors' note in ▸ Dobrowolski 1918).

esting among my scientific works, with particular attention to the methodological aspect. Of course I will focus only on things that might be of "general" significance, that might be "instructive," or at least "interesting". Though I will not curtail myself too much precisely for this reason: I must give the reader an *image*, not a dry scheme, of my scientist's life, and *explain* that life along with its creations. I will try to treat myself as a specific *example* of a life that could be interesting because it is common to many: it is the life of an *Intellectual*. Here and there I will also use my observations and experiences to deduce certain general comments on Science, its issues, the conditions for its existence. Thus I will dissolve my "person," a less important matter, in the magma of things that through their general significance are more important than the mere "person."

2. Research on inventions and discoveries

I have never forgotten that my first and most important motivation to work in Science is to find out, personally and from my own point of view, what Science is, what it really is. When my Antarctic works were ready for print, and even before that, I felt the need to review and synthesize what I had learned about Science until then. This is when I started thinking about *the mechanism of discoveries and inventions*, essential factors for the progress of Science, and about the moments, equally essential for the research process, when ideas appear most clearly, most powerfully, most intensely. It became apparent immediately that this question went beyond the realm of Science alone, that it also spanned Technology and Art, and that it was a fundamental question concerning the general problem of creativity.

I began with the field of Science – and this beginning is where I ended. I do not like to surrender to difficulties, but here, after almost two years of auxiliary study and various trials, I was forced to surrender all the way down the line, or rather, I had to lay down my arms – indefinitely. The phylogenesis of the achievements of the mind as contained in the works of Science involves detecting the earliest embryos of these achievements, detecting the steep and windy roads and paths of development, detecting regression, errors and mistakes. All this would be very "tempting," very "interesting" and "instructive" – but nothing more. It could only amount to auxiliary labour with respect to the fundamental task, which is to study the *ontogenesis* of inventions and discoveries by tracing the birth, development and maturation of the thought processes of creative individuals – discoverers and inventors. But here the commonly applied *indirect* external method of arranging, comparing and connecting a given idea's successive stages, *in as far as these stages happened to acquire some outward expression*, appeared completely inadequate. It is impossible to decipher in this way the actual course of the thought processes that lead to discovery or invention. At the most we might come away with some more or less amusing suppositions, which might appear more or less probable but can in no way be verified. This is because the data is random; there are too many gaps and there is not enough good *first-hand material*! Such material could only be obtained from *the creator's immediate confession* about the birth of a given creative idea, its development and maturation, the external circumstances accompanying the process – a confession given "in real time" if possible, or at any rate after not too long an interval. We have insufficient material of this kind. Creators do not generally confess their ideas or the stimuli, conditions and outcomes of those ideas, or how they were actually realized. Mostly they remain silent. This is why so far none of the many attempts to describe the ontogenesis of various inventions and discoveries have gone beyond literary-philosophical feuilletons of no scientific value; a fortiori, the same must be said of the various "theories" of creativity, scientific and otherwise, and about the works of the German "heurologists" and others: they are all theories without material. In my opinion we had to "introduce new habits" so that creators would begin to tell us "how they do it," so that they would record and communicate with us everything they know about their inventions and discoveries; in one word, so that every original work – ingenious or modest – would be accompanied with *its* history, *its* "biography". So it was necessary to make the appropriate appeal – and to wait until the desired material would begin to emerge. How to exploit this material, how to compensate for the grievous shortcoming of incompleteness and the even more grievous one of inaccuracy and falsification in such biographies of works and ideas? I wrote about this parenthetically in the first volume of *Nauka Polska*, and the editors saw fit to reprint this parenthetical remark in volume 6.

I resolved it would perhaps be best to begin with the easiest field – Technics. Here the inventor's thought process is crystalized in its successive stages: it is embodied in sketches, in models. Different phases of the development become more distinctly separate, and the course of the thought process appears less intense than in works of Science or Art. But since I myself was not confident in that field, I decided to look around for a suitable colleague – an engineer-inventor with solid theoretical knowledge and with "the mind of a philosopher". In Liège I found such a partner in the person of Szymon Gelblum.[b] This engineer-designer and recluse philosopher had invented the tank ten years before the war – not for destructive purposes but for peaceful crawling on uneven terrain, especially rocky ground where vehicles on rubber are simply useless. I had seen a model in Gelblum's room – absolutely the model of a tank; only instead of the articulated track it had a thick leather belt with distinctive incisions. At the same time, given the frequent accidents in mine shafts, he was racking his brains for a sensible mechanism assuring the safety of elevators there. In vain he kept applying to the Ministry for a locomotive and carriage that would allow him to perform on-track tests with vehicles being set in motion and brought to a halt, and to determine empirically the relationship of friction to speed. Similarly, he was

---

[b] Dobrowolski referred to Gelblum at several occasions, cf. ▶ Dobrowolski 1918, 300 and ▶ Dobrowolski 1927, 322–324.

never able to persuade an industrialist to invest in tests with his "crawler". When we met, I immediately got him interested in the construction of a microscope. The stand of my old "machine" had become rickety. Besides, for my research on frost I needed a special microscope whose tube could be focused quickly on any point on the glass pane used for the frost to form on – a microscope slide one meter in diameter. My friend made his inventions in an extraordinarily "elegant" manner. His ideas were usually based on "localizing" certain general methodological concepts. For instance, while researching better lift mechanisms, his point of departure was the idea that opposite processes, e.g. setting a vehicle in motion and bringing it to a halt, are principally identical. His "crawler" applied the simple notion that on bad roads a vehicle should make its own good road, that is to say that it should "lay out its track in front of itself". He generated his ideas synthetically, but then he worked them out analytically. Here, he would usually pose a more general question and either provide a general solution through accurate and skillful application of mechanics models, in which case he treated his own idea as a special case, or he would exhaust all possibilities, all possible solutions, all ideas, step by step and through systematic discussion, in order to choose the most appropriate among them. This is how he made two innovations in microscope construction. He made them before my eyes – we were in touch constantly – and he presented them in *Zeitschrift für wissenschaftliche Mikroskopie* in 1903 as general reflections on the problem of automatically stopping the tube at a desired height, and as general reflections on designing a sensible connection between the stand and the micrometer screw.

Gelblum got deeply involved in the difficulty I faced. What is more, he promptly understood the necessity of introducing a "fashion" for recording the biographies of technical inventions. Deciding to give his own example at once, he attached to the manuscript of the second article mentioned above his work's "historique" with a motivational statement to the purpose. The editors printed the article, but they returned "l'historique" to the author, "since there had never been a precedent"... So he tried to interest the circle of Belgian inventors, and at a meeting of the *Association des Ingénieurs sortis de l'École de Liège* in 1905 he gave a very nice presentation, logically constructed and based on engineering concepts. The effect was bewildering: apparently no one recalled a meeting that was similarly passionate and stormy. That those valiant engineers understood nothing, literally nothing, is perhaps not surprising, given that they have no intellectual culture or interests, given their narrow-mindedness, to speak plainly. But that the presenter should have been received with cries of outrage and all but shaking fists – "You want me to take my ideas to confession?!" – that made no sense to me at all. Finally, Van Beneden and Spring, to whom I described this affair, were able to shed some light on it: "The shouters were thieves and potential thieves of the ideas of others!"... The rabidness was so great that numerous appeals were made to stop the "scandalous" presentation from being printed in the Bulletin of the "Association" – an extraordinary demand that apparently had never been voiced in the organization's history. Only the "placating" stance of Dwelshauwers, a professor of industrial chemistry, convinced them to withdraw their petition, and the presentation did get published.

Following this unexpected fiasco we devised a more serious plan of action: to address our well founded appeal not to the congregation but to the world, and to all, irrespective of their order – academics, artists, engineers. Some people would respond; perhaps only a small percentage, but they would surely be the best. Our goal was to start the ball rolling with a small brochure – a probe. My friend took on the task of writing it, drawing on his technical background but phrasing it in general terms; I was supposed to follow suit with "something bigger". To get the readers' direct interest, Gelblum decided to present a concrete example conveying the importance of the creator's confession, showing how hard it is to understand anything without that, how sometimes one random note, one word, can explain a lot, and therefore how quickly research on creativity will advance once creators begin to speak up. For this purpose he used his beautiful inquiries into Watt's invention of the steam engine, in particular the utterly unfathomable origin of his famous "parallelogram," which the German engineer-philosopher Reuleaux had in vain tried to explain. One involuntary "wordless confession" – a random hand-drawn sketch in a letter to a partner – immediately threw light on this, allowing at least to decipher the logical scaffolding of the idea.[2]

This brochure, printed in Liège in 1906 with the title "Par l'historique à la science des idées" was poorly distributed because the author, not quite satisfied, was planning to republish it in "better" shape.

But that never happened.

Like me, this man was undermined by existential questions – this is what connected us so deeply. He, too, though undermined, was able to work in hope of "experiencing illumination"... But that hope had not grown at all since our encounter. For me, this was the beginning of the period when my eyes were the most "wide-open"; it was the period of my greatest doubts. Only the shadow of hope remained. That shadow of hope allowed me to live and to work, but it was so faint and so hidden somewhere in my soul that my friend probably never saw it. A great spiritual crisis soon befell him: sleepless nights of brooding destroyed his last glimmer of hope. He could not live, seeing no meaning in life and having no hope of ever finding it – he, crystalline like Spinoza and consistent like him. So he cut his life short. There was no other way out.

I lost my first and last "philosophical friend". I never wrote that "something bigger" we had planned...

[...]

---

[2] Four years later the well-known engineer-heurologist Peter Engelmeyer took on the same task in his short work *Der Dreiakt als Lehre von der Technik und der Erfindung* (Berlin, 1910). But his solution was only superficial: he remains unaware of the difficulties that Reuleaux already pointed out in his *Kinematics* (see Franz Reuleaux: *Cinématique. Principes fondementeaux d'une théorie générale des machines*, Paris 1877.). ||| In 1936, the editors of *Organon* added this reference to the English translation of ▶ Dobrowolski 1918 in vol. 1 of Organon (cf. footnote 4 on p. 300).

Contribution aux Documents sur la Psychologie de l'Invention dans le Domaine de la Science

Emile Borel

EMILE BOREL (7 January 1871, Saint-Affrique, France – 3 February 1956, Paris, France), mathematician, best known for his work in probability and measure theory and also strong political engagement. After graduating from École normale supérieure, B. lectured at Lille Univ. before becoming prof. at the Faculté des sciences de Paris. B. was a member of parliament and briefly served as naval minister. He was an active organiser of science – for example, he presided the Société mathématique de France (1905), founded the first statistical institute in France in 1922 (Institut de statistique de l'université de Paris, today de Sorbonne Université), and in 1934 was president of the Académie des sciences – and as such he was interesting for science studies contexts in Warsaw.

B.'s text was published in vol. 1 of Organon as a contribution to the collection of "documents on the psychology of the invention in science" in the eponymous section. According to the editorial comment therein, it should commence a series of autobiographical documents on creative action, which, in fact, continued the series "Researching the Genesis and Development of Scientific Creativity" (*W sprawie badania i genezy rozwoju twórczości naukowej*) from Nauka Polska (► **IN THIS VOL.** Białobrzeski 1927, ► Dobrowolski 1918, 1927, 1928, ► Borel, ► Krogh 1938). The contact between Borel and the journal was made by publicist, historian of literature, and Polish consul general in Paris Zygmunt Lubicz-Zaleski, who helped with international contacts and distribution. Minor typographical errors were corrected without special indication.

ORIGINAL REFERENCE
Emile Borel: "Contribution aux *Documents sur la Psychologie de l'Invention dans le Domaine de la Science*," Organon, vol. 1 (1936), 33–42.

FURTHER TEXTS MENTIONED
► Czesław Białobrzeski ["C.B."]: "Szkic autobiograficzny i uwagi o twórczości naukowej," *Nauka Polska*, vol. 6 (1927), 49–76.
► Antoni B. Dobrowolski: "O pilnej potrzebie wychowania umysłowego w Polsce: o konieczności zasadniczej reformy nauczania w szkołach średnich oraz stworzenia w związku z ową reformą nowych placówek pracy naukowej," *Nauka Polska*, vol. 1 (1918), 489–502.
► —"Archiwum materjałów do badania twórczości," *Nauka Polska*, vol. 6 (1927), 140.
► — ["A.B.D."]: "Mój 'życiorys naukowy,'" *Nauka Polska*, vol. 9 (1928), 68–216.
► August Krogh: "Visual Thinking," *Organon*, vol. 2 (1938), 86–94.

Nous avons l'intention de présenter à nos lecteurs une série de documents autobiographiques sur l'activité créatrice dans le domaine de la science. Nous nous promettons de publier ainsi un certain nombre de confidences personnelles de grands savants et chercheurs.

Cette série va commencer par un exposé qu'a bien voulu nous faire parvenir M. Emile Borel, dont l'œuvre scientifique de si haute portée se complète efficacement par une activité d'ordre social et politique. Auteur de travaux mathématiques universellement connus, professeur et savant, Président de l'Académie des Sciences de Paris, docteur honoris causa de plusieurs universités, entre autres de celle de Varsovie, M. Borel a exercé en même temps les fonctions les plus élevées comme homme d'État. Député, deux fois Ministre de la Marine, Vice-Président de la Commission des Affaires Etrangères, Vice-Président de la Commission des Finances, Président de la Fédération Française des Associations pour la Société des Nations, Président du Comité Français de la Coopération Européenne – M. Emile Borel personnifie ainsi l'idéal d'intégration des forces créatrices scientifiques dans la vie sociale.

La Rédaction

L'exposé de M. Emile Borel commence par la lettre suivante:

Monsieur le Directeur,

L'enquête que vous entreprenez sur le développement de la pensée créatrice dans les sciences me paraît fort intéressante et fort utile. Je suis en effet de ceux qui pensent que la découverte scientifique est l'élément le plus important du progrès de l'humanité et que son importance ne cessera de croître dans les siècles qui vont venir. Je consens donc bien volontiers à apporter ma modeste contribution à votre œuvre, et j'essaierai de suivre de mon mieux le plan que vous avez bien voulu m'indiquer, car l'étude comparative des réponses diverses que vous publierez sera certainement facilitée si tous les auteurs se conforment à un même plan.

– I –

Il me semble que l'éveil de la curiosité scientifique se produit à peu près de la même manière chez tous ceux qui, dès leur jeune âge, ont du goût pour les mathématiques. La science mathématique a en effet ceci de particulier qu'elle est beaucoup plus ancienne que toutes les autres sciences et que ses progrès ont été très considérables dans de nombreuses directions depuis déjà bien longtemps. Par suite, il faut de très longues études pour atteindre les limites de la science actuelle, limites au delà desquelles peuvent être faites de véritables découvertes; mais, par contre, l'esprit de découverte et de recherche peut s'exercer dès l'étude des éléments, et il s'exerce effectivement chez tous les jeunes élèves qui ont l'occasion d'étudier avec de bons livres et de bons professeurs. C'est peut-être à l'occasion des problèmes de géométrie élémentaire que cet esprit de découverte peut s'exercer le plus facilement, mais il trouve également l'occasion de s'exercer dans la résolution de nombreux problèmes d'arithmétique et d'algèbre. C'est souvent par la liaison des méthodes différentes de l'algèbre, de l'arithmétique, de la géométrie, que se produit la découverte.

Si l'on examine les problèmes qui sont proposés aux candidats dans les examens élémentaires de mathématiques, on peut les diviser en deux catégories: ceux qui peuvent être résolus sans esprit d'invention, d'une manière en quelque sorte mécanique, par tous les élèves qui ont consciencieusement étudié leur cours, et ceux qui exigent pour leur solution une part d'ingéniosité et d'invention. Les bons examinateurs savent d'ailleurs mélanger ces deux catégories de questions d'une manière appropriée au développement intellectuel des candidats qu'ils ont à examiner. Il faut en effet observer que, même chez ceux qui étudient les mathématiques uniquement en vue d'applications pratiques, il n'est pas inutile de développer l'esprit de découverte. Dans bien des problèmes fort simples qui se posent dans la pratique, la solution se trouve obtenue beaucoup plus vite par telle remarque ingénieuse que par l'application de méthodes classiques.

Prenons par exemple cette question simple que l'on pose parfois aux enfants: Combien font le tiers et le demi tiers de 100? Celui qui prend à la lettre cet énoncé calculera plus ou moins péniblement le tiers de 100 = 33,333, puis en prendra la moitié et fera l'addition. L'élève qui a quelques notions d'arithmétique remarquera que le tiers et le demi tiers font en tout $^3/_6$, c'est-a-dire $^1/_2$, et que, par conséquent, il doit prendre simplement la moitié de 100. Celui enfin qui a l'habitude de l'intuition géométrique apercevra immédiatement une longueur OA divisée en 3 parties égales par deux points B et C et verra intuitivement que le milieu M de BC est également le milieu de OA:

Il aura ainsi la solution de la question par une vue directe et immédiate sans aucun calcul.

Je crois donc que l'histoire du développement du goût pour les mathématiques et pour l'invention mathématique est à peu près la même chez tous ceux qui ont eu ce goût dès leur jeunesse et que les exercices scolaires, la préparation des examens, avec les différences et les analogies qu'ils présentent dans les divers pays, contribuent essentiellement à ce développement. Il faudrait une longue étude qui sortirait de notre cadre pour rechercher par quels moyens ce goût de l'invention est le mieux développé et dans quel sens devraient être modifiés les programmes d'enseignement pour faciliter l'éclosion des vocations mathématiques. J'insiste à nouveau sur le fait qu'en développant l'esprit d'invention mathématique par l'enseignement, on ne contribue pas seulement à la formation d'un petit nombre de spécialistes que sont les mathématiciens, mais on rend également service à tous ceux pour lesquels les mathématiques sont seulement étudiées en vue de leur application pratique.

– II et III –

Il est tout naturel que la curiosité scientifique et l'esprit d'invention, après s'être exercés dans des problèmes posés par des professeurs qui en connaissaient eux-mêmes la solution, cherchent à s'exercer dans des domaines jusqu'alors inexplorés. Pour les raisons que j'ai indiquées tout à l'heure, ces régions inexplorées, en ce qui concerne les mathématiques, s'éloignent de plus en plus des éléments et il faut d'assez longues études pour les atteindre. La longueur de ces études nécessaires pour atteindre les limites de la science augmente avec une rapidité à certains égards inquiétante et il est extrêmement désirable que certains mathématiciens appliquent leurs efforts à simplifier et à faciliter l'accès de ces régions où peut s'exercer la découverte. C'est aussi une invention et une découverte que de permettre l'acquisition plus aisée d'un bagage important de connaissances et l'exploration plus facile d'un vaste domaine dans la science.

À l'époque où j'ai terminé mes études à l'École Normale Supérieure de Paris, en 1892, la théorie des fonctions analytiques était un des domaines de la science dans lesquels d'importantes découvertes avaient été faites pendant les décades précédentes, mais où il restait certainement encore beaucoup à faire. Je fus attiré par ce domaine, et notamment par l'étude de l'influence des points singuliers sur les propriétés des fonctions. La représentation géométrique de la variable imaginaire par un point du plan était classique depuis longtemps; les problèmes à étudier se posaient ainsi sous une forme à la fois géométrique et algébrique, et ce mélange, dans la recherche, des méthodes de la géométrie et de l'algèbre me plaisait beaucoup. Pour essayer de donner, même à ceux qui ne sont pas familiers avec les spéculations mathématiques, une idée de la nature des problèmes qui se posaient à moi, je vais les simplifier en me bornant à la considération des ensembles de points, dans un plan.

Imaginons un mètre en bois ou en métal sur lequel on a porté les divisions décimales; on a ainsi figuré les décimètres, les centimètres, les millimètres. Pratiquement, il n'est guère possible d'aller plus loin et il faudrait des appareils fort délicats pour arriver à tracer des traits suffisamment distincts, espacés d'un dixième de millimètre. Toutefois, le mathématicien n'a pas l'habitude de s'embarrasser de ces contingences pratiques, et c'est un procédé naturel chez lui que de généraliser une méthode dont il a pu faire les premières applications. On peut ainsi concevoir que sans pouvoir les tracer effectivement, on imagine après les divisions en *cm*, en *mm*, en dixièmes de mm, des divisions encore plus fines, en centièmes *de mm*, en millièmes de *mm*, etc, etc. Si nous admettons que nous puissions avoir des microscopes à grossissement indéfini, si nous admettons également que la règle conserve le même aspect sous ces grossissements considérables, rien ne nous empêche de prolonger indéfiniment par la pensée ces divisions. Nous avons ainsi marqué sur notre règle une infinité de points dont chacun correspond à une fraction décimale simple; par exemple la fraction 0,3241732 correspond à un des dix millions de points qui marquent la division du mètre en dix-millièmes de *mm*. Dans le langage de la théorie des ensembles on dit que tous les points qui sont ainsi marqués sont *denses partout* sur la droite. Il n'y a pas en effet une portion, si petite qu'elle soit de la longueur du mètre, sur laquelle tous ces points décimaux ne soient pressés les uns contre les autres.

Il y a d'autre part sui la droite des points qui ne sont pas des points décimaux, ce sont tous les points qui sont représentés par une fraction décimale illimitée, soit une fraction périodique comme 0,333…, soit une fraction irrégulière comme celle des décimales du nombre π: 0,14159265…

Sur toute portion de la droite, si petite qu'elle soit, se trouvent à la fois des points décimaux et des points non décimaux. Si donc, on veut comparer l'ensemble des points décimaux avec l'ensemble des points non décimaux par la méthode naturelle et classique qui consiste à diviser la droite en intervalles de plus en plus petits, on n'aboutira à aucun résultat; quelque petit que soit l'intervalle, on constatera que dans cet intervalle il y a à la fois des points décimaux et des points non décimaux. Il semble donc qu'il ne soit pas possible de décomposer la droite en intervalles qui renferment tous les points décimaux sans renfermer en même temps tous les points non décimaux.

Cette impossibilité n'avait peut-être pas été énoncée explicitement, mais elle était implicitement admise par tous les mathématiciens. Ceux-ci cependant, pour diverses raisons tirées de la théorie de la puissance des ensembles de Georges Cantor, tirées aussi de considérations de calcul des probabilités plus ou moins confuses, que nous préciserons tout à l'heure, savaient que les points décimaux devaient être regardés comme plus rares que les points non décimaux. Si l'on tire au sort les chiffres d'une fraction décimale, pour que cette fraction soit un nombre décimal limité; il faut qu'à partir d'un certain rang tous les chiffres soient égaux à zéro, et c'est la une éventualité qui doit être regardée comme très peu probable. N'y avait-il donc pas un moyen de distinguer par la considération d'intervalles suffisamment petits les renfermant tous, les nombres décimaux des nombres non décimaux?

C'est en réfléchissant sur ce problème et en essayant de me représenter les traits eux-mêmes par lesquels l'infinité de nombres décimaux peut être marquée sur une droite, que j'eus l'idée fort simple suivante: si ces traits sont suffisamment fins, leur largeur totale pourra être rendue très petite et inférieure à la longueur totale de la droite, et dans ces conditions il y a bien des chances pour que certains points de la droite ne soient pas recouverts par ces traits, car il serait paradoxal qu'on puisse recouvrir la droite entière par des traits dont la largeur totale est inférieure à sa longueur. Cette simple réflexion mettait sur le chemin de la découverte et il ne fallait plus qu'un peu d'attention et un peu de patience pour arriver à la formuler.

Si l'on reprend l'image du mètre sur lequel sont marquées des divisions et si l'on donne aux divisions centimétriques une largeur d'un *mm*, l'ensemble de ces 100 divisions centimétriques couvrira 10 *cm*; si l'on donne ensuite aux divisions millimétriques une largeur d'un centième de *mm*, l'ensemble de ces divisions millimétriques occupera moins d'un *cm*; on peut ainsi continuer et s'arranger pour que l'ensemble des divisions marquant les dixièmes de *mm* occupent en tout moins d'un *mm*, et ainsi de suite. Dans ces conditions, lorsqu'on sera arrivé jusqu'au bout, c'est-à-dire lorsque l'on aura marqué tous les nombres décimaux limités, même ceux qui ont un très grand nombre de décimales, l'ensemble des traits tracés occupera seulement une fraction de la longueur totale de la droite. On pourra même s'arranger pour que cette fraction soit inférieure à un nombre très petit donné à l'avance.

On arrive ainsi, en choisissant convenablement les intervalles, et en définissant ces intervalles à *partir* des points décimaux même que l'on veut étudier, à renfermer tous ces points décimaux dans un ensemble d'intervalles dont la longueur sera par exemple inférieure à 1 *mm*, alors que ces points sont infiniment serrés sur toute la droite qui a un mètre de longueur. C'est là un résultat très simple qui aurait dû être connu depuis fort longtemps, mais qui cependant est apparu comme très nouveau et très paradoxal.

Notre imagination géométrique se représente en effet très difficilement ces intervalles qui renferment tous les points décimaux et qui cependant, ne constituant qu'une très faible fraction de la droite entière, laissent en dehors d'eux bien plus de points qu'ils n'en renferment a leur intérieur. L'arithmétique a beau nous donner l'assurance que la longueur totale de ces intervalles étant extrêmement faible, il n'est pas possible qu'ils renferment à leur intérieur tous les points de la droite, l'intuition géométrique de ce résultat ne nous est pas naturelle.

Ce n'est pas ici le lieu de développer les conséquences qu'a eues la méthode ainsi créée, méthode consistant essentiellement à construire les intervalles à partir des points qu'on étudie, au lieu de chercher à étudier la répartition de ces points dans des intervalles formés à l'avance d'après une règle fixe. Il me suffira de rappeler les résultats qu'a donnés cette méthode pour l'étude des fonctions analytiques dans certains domaines singuliers, et de rappeler également que tout le développement de la théorie de la mesure des ensembles et de la théorie célèbre de l'intégration de M. Lebesgue se rattache directement à cette méthode.

– IV –

Comme je viens de l'indiquer, mes premières recherches scientifiques ont porté sur la théorie des fonctions et des ensembles, et je pense que les premiers travaux sur la mesure des ensembles dont je viens de rappeler le principe ont été ceux dont la répercussion sur le développement des mathématiques contemporaines a été la plus considérable. Si l'on me demandait de caractériser par un trait commun la méthode que j'ai appliquée dans ces travaux ainsi que dans mes travaux sur les fonctions entières, sur les séries divergentes, sur la croissance des fonctions, je crois que le trait commun aux méthodes diverses que j'ai utilisées est un souci constant d'étudier les êtres mathématiques en eux-mêmes, comme le biologiste étudie les êtres vivants, de me familiariser avec eux, et de ne pas me laisser influencer dans cette étude intrinsèque des individus par les préjugés et les traditions.

Cette méthode de travail m'a conduit à une conception réaliste des mathématiques, qui distingue les êtres mathématiques pouvant être effectivement définis, de ceux dont l'existence est purement hypothétique, mais ce n'est point ici le lieu de reprendre une discussion sur les définitions où intervient l'infini, discussion qui ne sera sans doute jamais close.

J'ai indiqué tout à l'heure quel lien pouvait exister entre la théorie des probabilités et mes premières recherches sur certains ensembles de points. J'ai été ainsi conduit à étudier, à côté des probabilités finies et des probabilités continues qui avaient, seules, été considérées par les mathématiciens, les probabilités que j'ai appelées probabilités dénombrables. Leur étude a permis de préciser de nombreux résultats de la théorie générale des probabilités, notamment en ce qui concerne la loi des grands nombres. Il serait trop long d'indiquer ici en détail les raisons de l'importance que le calcul des probabilités a prise depuis un demi siècle dans les sciences les plus diverses. En m'y intéressant d'une manière particulière depuis près de vingt ans, je ne me suis pas éloigné autant qu'on pourrait le croire de mes premiers travaux sur la théorie des fonctions, car il y a bien des questions, notamment la théorie générale de la mesure des ensembles, qui se rattachent à la fois à la théorie des fonctions et à la théorie des probabilités.

– V –

Il est très malaisé d'indiquer quels sont les domaines de l'activité intellectuelle qui intéressent en dehors de la spécialité à laquelle on s'est consacré, car on risquerait de donner l'impression d'une richesse excessive si l'on mentionnait tout ce à quoi on a pu prendre quelque intérêt, ou d'une pauvreté trop grande si l'on s'astreignait à ne désigner que des domaines que l'on peut avoir la prétention d'avoir explorés d'une manière véritablement sérieuse et approfondie.

Il me semble donc que l'on doit se contenter d'indiquer une tendance générale. Si je m'examine, il me semble que je m'intéresse aux diverses branches de l'activité intellectuelle, surtout dans la mesure où j'ai pu y acquérir davantage la connaissance des hommes.

On peut être tenté de considérer ce fait comme une réaction contre le caractère abstrait des mathématiques. Ceux qui ne sont pas mathématiciens sont en effet portés à considérer les mathématiques comme une science essentiellement inhumaine dans laquelle la personnalité des hommes qui s'y consacrent ne joue aucun rôle.

Une telle conception des mathématiques me parait complètement inexacte et s'il m'est permis, étant donné la nature de cette enquête, de revenir encore sur un sujet personnel, je crois discerner que la sympathie et les oppositions de caractères entre les hommes jouent un rôle important dans la recherche et la découverte scientifique.

Il n'est pas douteux que mes premiers travaux dont j'ai essayé de dire quelques mots tout à l'heure et qui ont orienté pendant de nombreuses années toutes mes recherches ont été singulièrement influencés dans des sens opposés par deux mathématiciens allemands de caractères fort différents.

J'ai été extrêmement séduit, dès l'âge de 20 ans, par la lecture des travaux de Georg Cantor dont je n'eus le plaisir de faire la connaissance que quelques années plus tard au Congrès de Zurich en 1897. Georg Cantor a apporté dans l'étude des mathématiques cet esprit romantique qui est l'un des côtés les plus séduisants de l'âme allemande. Weierstrass, au contraire, a essayé de soumettre la théorie des fonctions de variables imaginaires découverte par Cauchy à cette discipline stricte et sévère qui est aussi un des côtés du caractère allemand. Il y a eu certainement pour moi un très grand plaisir intellectuel à montrer, grâce à la théorie des ensembles de Cantor, que les fonctions ne se laissaient pas aussi facilement discipliner qu'avaient pu le croire Weierstrass et ses disciples. Par ailleurs, en prouvant qu'un élément de fonction analytique au sens de Weierstrass, lorsque les coefficients étaient choisis au hasard, n'était susceptible d'aucune prolongation et représentait par suite une fonction d'une nature extrêmement particulière, j'ai montré combien était artificielle la méthode par laquelle Weierstrass voulait ramener l'étude de toutes les fonctions à une même méthode, tandis que l'étude directe des propriétés des fonctions, en se plaçant au cœur même de leurs singularités est autrement instructive et autrement profitable.

Mes travaux sur les séries divergentes, en montrant combien un certain formalisme que Cauchy n'avait adopté qu'à regret et à titre provisoire était excessif, se rattachent aussi à une tournure d'esprit romantique analogue à celle de Georg Cantor.

S'il m'était permis d'ajouter quelques mots de conclusion à ces réflexions dont je prie qu'on excuse le décousu, je voudrais dire qu'à mon avis, l'unité de l'esprit humain est plus importante que sa diversité et que les méthodes d'invention, d'imagination, de découverte, sont beaucoup plus analogues qu'on ne le croit généralement dans les divers domaines où s'exerce l'activité de l'esprit.

Emile Borel

Visual Thinking: An Autobiographical Note

August Krogh

AUGUST KROGH (15 November 1874, Grena, Denmark – 13 September 1949, Kopenhagen, Denmark), physiologist and zoologist. K. initially studied medicine, yet finally graduated in zoology (1893–1899) from Kopenhagen Univ. He became full prof. in 1916 after lecturing for some years already and was awarded the Nobel Prize in Physiology or Medicine (1920) for his research in *capillary circulation*.

K.'s text was published in vol. 2 of Organon in 1938. An editorial note introduced it as a further contribution to the series "Documents Concerning the Psychology of Creativeness in Science and Scholarship", which, according to the note, had been initiated by an autobiographical account of Emile Borel in the preceding issue (▶ **IN THIS VOL.** Borel 1936). However, the editorial note does not mention Antoni Bolesław Dobrowolski's letter to the editor in the same vol. (▶ Dobrowolski 1918, biographical note), which, in fact, was a translated passage from the author's initial call to collect the material in question (▶ Dobrowolski 1918). Also unmentioned goes the equivalent collection of (auto)biographical material that had been published in vol.s VI and IX of Nauka Polska (▶ Białobrzeski 1927, ▶ Dobrowolski 1918, 1927, 1928). Krogh's text was translated into Polish for vol. 24 of Nauka Polska (1939).

ORIGINAL REFERENCES
August Krogh: "Visual Thinking," *Organon*, vol. 2 (1938), 86–94.
— "Myślenie wzrokowe," *Nauka Polska*, vol. 14 (1939), 35–42.

FURTHER TEXTS MENTIONED
▶ Czesław Białobrzeski ["C.B."]: "Szkic autobiograficzny i uwagi o twórczości naukowej," *Nauka Polska*, vol. 6 (1927), 49–76.
▶ Emile Borel: "Contribution aux *Documents sur la Psychologie de l'Invention dans le Domaine de la Science*," *Organon*, vol. 1 (1936), 33–42.
▶ Antoni B. Dobrowolski: "O pilnej potrzebie wychowania umysłowego w Polsce: o konieczności zasadniczej reformy nauczania w szkołach średnich oraz stworzenia w związku z ową reformą nowych placówek pracy naukowej," *Nauka Polska*, vol. 1 (1918), 489–502.
▶ —"Archiwum materjałów do badania twórczości," *Nauka Polska*, vol. 6 (1927), 140.
▶ — ["A.B.D."]: "Mój 'życiorys naukowy,'" *Nauka Polska*, vol. 9 (1928), 68–216.

CONTENTS: Visual Thinking. An Autobiographical note by August Krogh. Sessions of the Science Studies Circle (1928–1938).

Personal confessions of scientists and scholars constitute the most important material for the study of creative thought. These confessions may relate to their intellectual development, from childhood to maturity, or they may trace, step by step, the evolution of one creative idea only, so giving, as it were, a biography of this idea; they may also tend to characterize the type of their mental abilities. Every secret of creative work so disclosed by vigilant introspection constitutes material of great importance for the psychology of creative work.

The first volume of ORGANON initiated the publication of documents of this kind with the autobiographical note of Professor E. Borel. In this volume we are happy to present to our readers an autobiographical note kindly sent us by Professor August Krogh.

August Krogh, Professor of Zoophysiology at the University of Copenhagen, recipient of the Nobel prize in physiology and medicine, Fellow of the Royal Society of London and member of several other scientific societies, doctor honoris causa of the Universities of Edinburgh, Budapest and Lund, is one of the most eminent physiologists of to-day. His discoveries are particularly important for two branches of physiology, where they mark a new epoch. They are: the study of respiration and the study of capillary circulation. By carrying on the great work of Christian Bohr, the founder of the Scandinavian physiological school, Professor Krogh contributed largely to the formulation, in terms of physics, of the problem of gaseous metabolism of animals. His investigations concerning the circulation of the blood revealed the part played by the capillaries in these phenomena. They brought about a revolution in the classical opinions on this subject and still constitute a source of new ideas for research in the physiology and pathology of the circulation of the blood.

Professor A. Krogh's pre-eminent position among the founders of modern physiology is due to his exceptional creative abilities and to his gift of expressing his ideas in a simple, clear and precise form.

The Editor

Visual Thinking. An Autobiographical note by August Krogh

I shall begin this autobiographical note by stating my firm belief that the essential traits in the intellectual make-up of a person are inherited and can be modified only to a comparatively slight extent by environmental influences.

Of my intellectual heritage I know but little. 300 years back my known ancestors were small independent landholders and farmers in North Slesvig. My great grandfather showed initiative in migrating to the capital and became a well-to-do brewer. Through my paternal grandmother I am related to a dispensing chemist (Hoffmann) who showed some inventive ability. My maternal grandfather who was a government official (Dreckmann) was very possibly of a gipsy strain. My father and mother were cousins. I was born as their eldest son in 1874 in Grenaa in Jutland. My mother, who was well educated in literature and languages, took much interest in the teaching of her children.

The memory of my early childhood reveals two essential traits which must be inherited. One is the power and desire to visualize, to form mental pictures. This, in my case, is not a pronounced visual memory. I do not remember pages in books and my memory for faces is distinctly inferior, but I noticed as a small child that I could at will «see pictures on the ceiling», pictures in which something happened and in which I might or might not take part myself. These pictures could be very clear and I could see any detail that I wanted. This power has been developed and found very useful as mentioned more in detail below.

The other trait was a very lively interest in animals, especially in small animals, and their behaviour. My impressions of some peculiar insects which I have seen between the age of three and six years like a peculiar *Sphinx* larva and a *Cerambyx* beetle were so vivid that I could recognize the species many years later, and I remember from the age of seven onwards that I spent many hours watching the ways of spiders.

I have found no clue to any of these traits in what is known about my ancestors.

My father, who did not get much in the way of intellectual education, but taught himself not a little and had considerable technical ability, became as a young man interested in physics and got some semi-popular works on physics[1] and on the history of invention[2] and he also performed for himself some of the simpler experiments in electricity. These works became a treasure to me and constitute without doubt from an intellectual point of view the most important environmental influence to which I have been exposed as a boy.

Under the spell of these books I attempted many things in the way of construction and experimentation, most often without success. In the small provincial town resources were severely limited, and I believe that this helped to develop a habit of always being on the look-out for things which would serve my purposes, a habit which has in later life proved very useful. A further habit of economizing with regard to apparatus and paper and thereby wasting time has been quite difficult to suppress when many years later it was no longer demanded by circumstances. This trait was conspicuous in my father and was quite probably inherited.

The teaching at school was of very little help to me. The dull textbooks in natural history which I had to peruse from the age of eight onwards almost quenched my interest in that subject, which had during the two first years at school been greatly stimulated by an excellent textbook for beginners (Zahlerts), now completely forgotten. Later on one of the teachers helped me a great deal by giving moral support to my pursuits, by talk and by lending of books.

During two years at high school and the two first years at the University (1891–95) I was, as far as I remember now, a voracious reader interested in many things, but

---

[1] Adam Paulsen: *Naturkræfterne, deres Love og vigtigste Anvendelser. En almenfattelig Fremstilling* (3 vols.), København 1874–1879.

[2] André Lütken: *Opfindelsernes Bog* (7 vols.), København 1877–1883.

without any clear idea of what I wanted to do. Beginning the advanced study of zoology in 1895 I was told, not at the University, but by an old and wise friend of mine, the zoologist William Sørensen, that to become a scientific zoologist I would have to acquire a working knowledge of anatomy and physiology, subjects which were not at that time taught to students of zoology, but only in the medical school. I had scarcely any idea what physiology might mean, but listening to the very first lecture by Christian Bohr, dealing with the methods for determining the quantity of blood in the body, it struck me that this kind of work appealed to me more than anything else.

In 1896 I began in my small student's room serious experimentation on the hydrostatics of *Corethra* larvae. As a part of this study I developed methods for gas analysis under the microscope, but nothing was published until many years later. It was, I believe, in the course of this research that I became conscious of my own experimental and constructive methods. A considerable part of my work was done in bed during the night when I would try to visualize the processes studied and the experiments to be carried out. I found that I could visualize fairly complicated apparatus and all details of their working. The constructive ideas would come, apparently, out of nowhere, but the visionary examination of them was a conscious and rational affair. I never made, and even now never make, drawings, not even rough sketches, until the construction of an apparatus was complete, because I found that a drawing would hamper the free flow of ideas and bind me down to that particular solution of the problem. I believe that the work was often going on unconsciously and during sleep, because I would wake up with an apparently new idea, which was then immediately worked up and put to all kinds of imaginary tests. It goes without saying that the final practical test often showed that something had been overlooked and that matter failed to behave as I had imagined. Because the imagination had been quite clear and definite each such failure taught me a lesson which stuck.

In suitable cases the physiological processes studied were visualized in essentially the same fashion, but such visualization was of course always hypothetical and served as a basis for the designation of suitable experiments to put the hypothesis to the test. A trait which I believe to be essential to success is to be easily dissatisfied with the working of apparatus or with the course of experiments and to make the defects the starting point for fresh constructive thought.

In Christian Bohr's laboratory where I worked for nearly two years as a student and became an assistant after taking my master's degree in Zoology in 1899, I was taught quantitative work especially with gases and not only did the attainment of precision soon come to have a great fascination for me, but I became definitely conscious of the quantitative point of view on studying physiological mechanisms, so as to be satisfied with an explanation or a theory only when it could be shown to be quantitatively sufficient.

I would like to illustrate my points by reference to some of my papers and first to my study of the possible excretion of free nitrogen from organisms undertaken more than 30 years ago. In this case, as in several others, the research was initiated by an outside stimulus, the Seegen prize offered by the Vienna Academy of Science. The problem fascinated me because it would require experimental work of the utmost precision, and I spent a long time before attempting or drawing anything, trying to establish in my mind the conditions required to obtain such precision and the best ways to avoid or reduce the sources of error. The main point, arrived at early and by a logical train of thought, was that a respiration apparatus of the closed type, initiated by Regnault, to be very accurate must be kept completely immersed in a constant temperature bath and must for this and other reasons be small. I succeeded in reducing the size to 300 cm^3 and could accommodate small cold-blooded and warm-blooded animals. On an average a nitrogen production was observed amounting to about 0,1 % by volume of the oxygen absorbed, and this for a long time I thought to be real, but I had overlooked the fact that in a closed circuit apparatus the oxygen which is admitted all the time remains as compounds taking up space and will not leave the gas volume unaltered. A slight increase in volume of the CO_2 absorbing fluid was noted in an experiment of long duration and when this trivial observation was followed up and allowed for as the volume of the combined oxygen, it reduced the apparent production of gaseous nitrogen to the significant figure of about 0,01 % of the oxygen, well within the limits of error.

Also in another case a seemingly trivial observation provided an important clue. I had to contend with a high mortality of the eggs subjected to experimentation. A corresponding mortality in closed circuit apparatus had been noted before and was ascribed by Seegen to organic poisons given off by the organism. Seegen had found that by taking the circulating air through a combustion tube it could be purified so as to become innocuous. I had therefore introduced into my apparatus a glass vessel containing a platinum spiral kept at a red heat during experiments. It did not help much, but when, late in the series of experiments, a new tube of slightly altered dimension was put in, a faint yellow precipitate was seen to be formed on the wall during experiments. This turned out to be mercuric oxide and the mortality was due to mercury vapour from minute drops of mercury left over in the gas sampling tubes. I was able to show that mercury vapour was also responsible for the symptoms observed by Seegen, but in this case the combustion tube appears to have been efficient.

I think it is an important principle, too often disregarded, to study carefully the experimental evidence which appears to be at variance with one's own and to detect, if possible, the reasons for the differences, and for this task the habit and power of visualization is very helpful, but I have to admit that too often the descriptions given are insufficient. In the present instance I had to discuss the famous paper by Regnault and Reiset "Recherches chimiques sur la respiration..." from 1849[3], in which a variable production of nitrogen up to 1,5 % of the oxygen absorbed was generally observed,

---

[3] Victor Regnault and Jules Reiset: "Recherches chimiques sur la respiration des animaux des diverses classes," *Annales de Chimie et de Physique* (Série 3), vol. 25 (1849), 299–519.

with occasional absorption of similar quantities. Thanks to the extremely careful descriptions given in the paper I found it possible to show that the main source of error lay in the temperature determinations made by thermometers in the water jacket surrounding the animal chamber. The air temperature in a somewhat similar container could be shown to differ systematically from that of the water jacket and the rise observed *during* an experiment would explain satisfactorily the apparent nitrogen production recorded by Regnault and Reiset on the assumption of an unaltered temperature.

The cases of apparent absorption were investigated in detail. In almost all of them the animals were under inanition or came to suffer in some way, resulting in a fall in temperature which could explain the result. I succeeded in explaining qualitatively, by means of the data given by Regnault und Reiset, all the experiments except three. In one of these, and in this one only, the dates were given (March 27th–30th 1848, coinciding with the outbreak of revolution) and from the meteorological tables a change in temperature could be deduced which would also in this case make the result conform to expectations.

The principle by which I have been guided in most of my physiological researches is frankly teleological. How do the organisms solve their problems? or rather: How is the particular problem which (more or less accidentally) has caught my attention being solved?

Preparing lectures on respiration in the animal kingdom I had to consider the difficulties of insect respiration in forms which are known not to make respiratory movements. The explanations given in the literature, mainly to the effect that the heart-beat and body movements would cause some renewal of the air in the tracheae, could not be visualized as producing any significant ventilation and it occurred to me to try and figure out the amount of ventilation which could be produced by simple gas diffusion in the tracheal tubes. A priori I did not expect this to accomplish much, because I had in mind the microscopic picture of the tracheae as very long and narrow tubes and failed to appreciate their aggregate cross-section and absolute lengths which in an insect cannot be more than a few millimetres.

The measurements and calculations made showed that in the majority of insects diffusion in the tracheae is sufficient to cover their respiratory needs, and it turned out further that this would hold only up to a certain size of animals, diffusion being decreasingly adequate with increasing dimensions of the body.

I have not found it possible really to visualize microscopic structures and processes in their true dimensions, and this is a very serious drawback, because functional possibilities depend very largely upon the actual dimensions. A combination of visualization and more abstract reasoning is often necessary.

My work on the regulation of capillary circulation began when I was writing a monograph on Respiratory Exchange and had to consider the mechanism of internal respiration, especially in muscles where the demand for oxygen and material for combustion can be increased to more than 20 times the resting figure. The conception of a fixed system of capillaries, through which the blood was flowing slowly when the muscle was at rest, and at a tremendous rate when the muscle was working near the limit of its capacity, seemed very unsatisfactory from a teleological point of view and I came to visualize a functional arrangement in which the majority of capillaries would remain closed during rest while the open ones were equally spaced on the transverse section of the muscle. This, again through teleological reasoning, led to the visualization of capillaries opening and closing in a sort of rotation providing a uniform supply of oxygen, when considered over a certain length of time, to all parts of the tissue. These general conceptions were substantially verified by observation, but in order to arrive at a real understanding elaborate measurements and determinations of the rate of diffusion of oxygen in animal tissues were necessary.

When I am to describe in more general terms my mental characteristics I must put the power of working by mental pictures, produced by an easy flow of associations, first, as underlying almost all other abilities which I may possess. It is closely bound up with a fair knowledge of elementary physics and with a selective memory for all kinds of apparatus and materials and also for animals and plants. It has often puzzled me that my memory for words, verses, quotations and foreign languages which is, I believe, above the average is not visual in character, and that the visual memory fails in a number of cases mainly, it is true, such to which I attach no special interest. My memory for chemical facts and formulas is decidedly poor.

The problems which fascinate me in biological science are mainly those which give a free play to the faculties mentioned, but the attainment of a high degree of precision in measurements never fails to appeal to me and I have been a fairly skilled manipulative worker.

I have no special aptitude for mathematics, and have mastered only the elements including those concerned with probability.

I look upon myself, perhaps with no good reason, as a cautious observer. At any rate I am easily made to doubt my own results as well as those of others, and the conclusion that anything in biology is «endgültig bewiesen» [definitely proven] rarely fails to arouse my deep suspicion.

Sessions of the Science Studies Circle
(1928–1938)

as described in the consecutive *Reports on the Activities of the Science Studies Circle* (*Koło Naukoznawcze*)

number and date of meeting/ attendees, discussants	Presenter, title (original Polish title) – References to relevant report sections and to – *Nauka Polska* and *Organon* for published papers
I 14.06.1928 11, 5	Bogdan Suchodolski on Max Scheler's *Die Wissenschaftsformen und die Gesellschaft (Probleme einer Soziologie des Wissens)*, Leipzig 1926. – Sprawozdanie (1929), 353. – NP 11 (1929), 379–381.
II 10.12.1928 14, 7	Franciszek Bujak, The Man of Action and the Student (Działacz i badacz) – ▶ SEE THIS VOL. Sprawozdanie (1929), 353–354. – NP 11 (1929), 11–23, ▶ Organon 1 (1936), 20–32.
III 07.02.1929 22, 7	Paweł Rybicki, Science and the Forms of Social Life: Issues at the Intersection of Sociology and Theory of Science (Nauka a formy życia społecznego. Kilka zagadnień z pogranicza socjologji i teorji nauki) – ▶ Sprawozdanie (1929), 354. – ▶ NP 11 (1929), 24–64.
IV 07.04.1929 20, 8	Tadeusz Kotarbiński, On the Skills of a Researcher (O zdolnościach cechujących badacza) – ▶ Sprawozdanie (1929), 354–355. – NP 11 (1929), 1–10.
V 27.04.1929 22, 10	Wojciech Przybyłowicz, Remarks on the Relation of State and Science (Uwagi o stosunku państwa do nauki) – Sprawozdanie (1929), 355. – NP 11 (1929), 65–91.
not counted 25.11.1929 12, 6	Stanisław Ossowski, The Problematics of the Science of Science (Problematyka naukoznawcza) – ▶ Sprawozdanie drugie (1930), 167. – ▶ NP 20 (1935), 1–12 and Organon 2 (1936), 1–12.
VI 12.12.1929 33, 7	Czesław Białobrzeski, Science and Religion (Nauka i Religja) – Sprawozdanie drugie (1930), 167–168. – NP 13 (1930), 1–15.

number and date of meeting/ attendees, discussants	Presenter, title (original Polish title) – References to relevant report sections and to – *Nauka Polska* and *Organon* for published papers
VII 07.04.1930 17, 6	Stanisław Rychliński, The Influence of Organic Work Ideals on the Development of Science in the Kingdom of Poland after 1863 (Wpływ ideałów pracy organicznej w Król. Polskiem po 1863 r. na rozwój nauki – Sprawozdanie drugie (1930), 168–169. – not published
VIII 20.10.1930 22, 6	Zygmunt Szweykowski, The Creation of the Mianowski Fund (Powstanie Kasy im. Mianowskiego) – Sprawozdanie trzecie (1931), 274. – not published
IX 03.11.1930 34, 6	Witold Doroszewski, Light and Shadow of Scientific Thought (Światła i cienie myślenia naukowego) – Sprawozdanie trzecie (1931), 274. – not published
X 11.02.1931 25, 8	Marian Heitzman, Science and the Scholar and their Relation to Society in Historical Perspective (Nauka i uczony a społeczeństwo na tle historycznem) – Sprawozdanie trzecie (1931), 273–274. – NP 14 (1931), 1–20.
XI 19.12.1931 19, 9	Franciszek Bujak, The development of Polish Science and Letters from 1800 to 1880 (Rozwój nauki polskiej w latach 1800 do 1880) – Sprawozdanie czwarte (1932), 350–351. – NP 15 (1932), 203–239.
XII 26.01.1932 12, 5	Tadeusz Makowiecki, On Establishing a Research Program in the Field of Culture (W sprawie ustalenia programu badań w dziedzinie kultury) – Sprawozdanie czwarte (1932), 351. – NP 15 (1932), 273–277.
[XIII] 12.04.1932 26, 7	At this session, Czesław Białobrzeski's talk New Paths of Modern Natural Sciences (Nowe drogi współczesnego przyrodoznawstwa) was discussed, which B. had given at the 40th anniversary of Kasa Mianowskiego on 28 Feb. 1932 – Sprawozdanie piąte (1932), 91–97. – NP 16 (1932), 1–14.
XIV 21.06.1932 22, 3	Kazimierz Dobrowolski, Studies on Scientific Culture in Poland Before the Late XVI Century (Studja nad kulturą naukową w Polsce do schyłku XVI stulecia) – Sprawozdanie szóste (1933), 231–232. – NP 17 (1933), 17–148.
XV 03.11.1932 27, 8	Tadeusz Kotarbiński, Bacon on the Future of Science and Letters (Bacon o przyszłości nauki) – Sprawozdanie szóste (1933), 232–235. – NP 17 (1933), 1–16.

Sessions of the Science Studies Circle (1928-1938) 343

number and date of meeting/ attendees, discussants	Presenter, title (original Polish title) – References to relevant report sections and to – *Nauka Polska* and *Organon* for published papers
XVI 28.01.1933 28, 9	Czesław Białobrzeski on William McDougall's *World Chaos*. *The Responsibility of Science*, London 1931. – Sprawozdanie szóste (1933), 235–239. – NP 17 (1933), 312–317.
XVII 11.03.1933 30, 3	Jan Łukasiewicz, On the History of Scientific Method (Z dziejów metody naukowej) – Sprawozdanie siódme (1934), 404–407. – not published
XVIII 23.05.1933 19, 3	Bogdan Suchodolski, The Notion Science and its Role in the Development of Stanisław Brzozowski's Thought (Rola pojęcia nauki w rozwoju myśli Stanisława Brzozowskiego) – Sprawozdanie siódme (1934), 407–410. – not published
XIX 28.10.1933 30, 7	Kazimierz Dobrowolski, National Characteristics of Scientific Creativity (Cechy narodowe twórczości naukowej) – Sprawozdanie siódme (1934), 411–415. – not published
XX 16.12.1933 32, 5	Tadeusz Zieliński, Pure and Applied Science in Antiquity (Nauka czysta i nauka stosowana w świecie starożytnym) – Sprawozdanie siódme (1934), 411–415. – not published
XXI 07.02.1934 24, 5	Bohdan Kieszkowski, Science and Governmental Aspirations in Italy (Nauka a dążenia państwowe we Włoszech) – Sprawozdanie siódme (1934), 418–420. – NP 19 (1934), 185–199.
XXII 03.10.1934 34, 3	Marjan Zdziechowski, Sketches on a Philosophy of German History (Próba filozofji historji Niemiec) – Sprawozdanie ósme (1935), 247–250. – not published
XXIII 13.03.1935 54, 6	Jan Rozwadowski, Report on the Project "The Truth of Life" (odczyt, będący sprawozdaniem z pracy p.t. „Prawda życia") – Sprawozdanie ósme (1935), 250–256. – partly printed in Kwartalnik Filozoficzny (vol. I, 1923) and Przegląd Współczesny (vol. XX, 1927).
XXIV 16.10.1935 43, 6	Jan Rutkowski, Creative Scientific Work and Universities (Twórcza praca naukowa a uniwersytety) – Sprawozdanie ósme (1935), 250–256. – partly printed in Kwartalnik Filozoficzny (vol. I, 1923) and Przegląd Współczesny (vol. XX, 1927).
XXV 20.11.1935 39, 6	Bogdan Suchodolski, Investigation and Teaching (Badanie i nauczanie) – Sprawozdanie dziewiąte (1936), 311–316. – NP 21 (1936), 1–44 and in ▶ Organon 1, 43–78.

number and date of meeting/ attendees, discussants	Presenter, title (original Polish title) – References to relevant report sections and to – *Nauka Polska* and *Organon* for published papers
XXVI 12.02.1936 59, 6	Czesław Białobrzeski, Modern Physics and their General Consequences for Science and Letters (Ogólno-naukowe konsekwencje współczesnej fizyki) – Sprawozdanie dziewiąte (1936), 311–316. – NP 21 (1936), 1–44 and in ▶ Organon 1, 43–78.
XXVII 20.10.1936 87, 9	Minister of Religious Denominations and Public Enlightenment Wojciech Świętosławski, On the Organisation of Creative and Inventive Work (O organizacji pracy twórczej i wynalazczej) – Sprawozdanie dziesiąte (1937), 191–203. – not published
XXVIII 18.11.1936 33, 4	Stanisław Ossowski, The Humanities and Social Ideologies (Nauki humanistyczne a ideologia społeczna) – Sprawozdanie dziesiąte (1937), 204–208. – NP 22 (1937), 1–24.
XXIX 09.12.1936 43, 5	Czesław Białobrzeski, Science and Culture (Nauka a kultura) – Sprawozdanie dziesiąte (1937), 208–218. – not published
XXX 24.02.1937 34, 10	Zygmunt Zawirski, Science and Philosophy (Nauka a filozofia) – Sprawozdanie jedenaste (1938), 207–216. – not published, see however Bohdan Kieszkowski's comments in the discussion and his eponymous text in NP 23 (1938), 1–24.
XXXI 20.10.1937 31, 5	Jan Rutkowski, Aspects of the Level of Creativity in Our Science Policy (Zagadnienie poziomu twórczości w naszej polityce naukowej) – Sprawozdanie dwunaste (1939), 187–191. – NP 22 (1937), 37–57.
XXXII 03.11.1937 28, 5	Henryk Jakubanis, The Birth of the Notion of Science. Facts and Perspectives (Narodziny pojęcia nauki. Fakty i perspektywy) – Sprawozdanie dwunaste (1939), 191–199. – not published
XXXIII 24.11.1937 29, 5	Paweł Rybicki, Science and the Irrational Element (Nauka a element irracjonalny) – Sprawozdanie dwunaste (1939), 199–203. – NP 24 (1939), 1–26
XXXIV 19.01.1938 33, 8	Tadeusz Makowiecki, The State and the Humanities (Państwo a nauki humanistyczne) – Sprawozdanie dwunaste (1939), 203–215. – not published
XXXV 26.01.1938 36, 8	Jan Łukasiewicz, The Genesis of Three-valued Logic (Geneza logiki trójwartościowej) – Sprawozdanie dwunaste (1939), 215–223. – not published

number and date of meeting/ attendees, discussants	Presenter, title (original Polish title) – References to relevant report sections and to – *Nauka Polska* and *Organon* for published papers
XXXVI 23.02.1938 30, 9	Kazimierz Dobrowolski, Studies on Primitive Thought (Studia nad myśleniem pierwotnym) – Sprawozdanie dwunaste (1939), 223–230 – not published
XXXVII 27.04.1938 30, 4	Antoni Bolesław Dobrowolski, Introductory Considerations on Higher Schools, and Especially Academic Schools (theory, questions, demands) (Wstępne rozważania o szkołach wyższych, w szczególności akademickich [teoria, zagadnienia, postulaty]) – Sprawozdanie dwunaste (1939), 230–242. – not published
XXXVIII date unknown	Jan Łukasiewicz, On the Tasks of Universities (O zadaniach uniwersytetów) – *Sprawozdanie 57 z Działalności Kasy im. Mianowskiego*, Warszawa 1938: 32.

References for the [First], Second, Third ... and Twelfth Report on the Activities of the Science Studies Circle

- ▶ Sprawozdanie z działalności Koła Naukoznawczego, *Nauka Polska*, vol. 11 (1929), 353–355.
- ▶ Sprawozdanie drugie z działalności Koła Naukoznawczego, *Nauka Polska*, vol. 13 (1930), 166–169.
- Sprawozdanie trzecie z działalności Koła Naukoznawczego, *Nauka Polska*, vol. 14 (1931), 272–274.
- Sprawozdanie czwarte z działalności Koła Naukoznawczego, *Nauka Polska*, vol. 15 (1932), 350–351.
- Sprawozdanie piąte z działalności Koła Naukoznawczego, *Nauka Polska*, vol. 16 (1932), 91–97.
- Sprawozdanie szóste z działalności Koła Naukoznawczego, *Nauka Polska*, vol. 17 (1933), 231–239.
- Sprawozdanie siódme z działalności Koła Naukoznawczego, *Nauka Polska*, vol. 19 (1934), 404–420.
- Sprawozdanie ósme z działalności Koła Naukoznawczego, *Nauka Polska*, vol. 20 (1935), 247–256.
- Sprawozdanie dziewiąte z działalności Koła Naukoznawczego, *Nauka Polska*, vol. 21 (1936), 300–324.
- Sprawozdanie dziesiąte z działalności Koła Naukoznawczego, *Nauka Polska*, vol. 22 (1937), 191–218.
- Sprawozdanie jedenaste z działalności Koła Naukoznawczego, *Nauka Polska*, vol. 23 (1938), 207–216.
- Sprawozdanie dwunaste z działalności Koła Naukoznawczego, *Nauka Polska*, vol. 24 (1939), 187–242.

Part III:
Commentaries

Science Studies in Poland before the Second World War: Institutional Frames

Jan Piskurewicz and Leszek Zasztowt

> *The scientist is the man who invented that term!*
> *[Naukowiec to taki, co wymyślił ten wyraz!]*
> Hugo Steinhaus[1]

Polish Positivism

During the late decades of the nineteenth century, when the Polish territories were still partitioned among the Austrian, Prussian and Russian empires, and after a number of insurrections had not succeeded, a new chapter in the history of Polish thought was to be written. Especially in Warsaw, where the January Uprising of 1863/64 against imperial Russia had been defeated, not only scholars, but much broader sections of the local *intelligentsia* left aside romanticist ideas of insurrection as a means to turn over foreign rule. Instead, they advocated to develop Polish society under the given circumstances and – depending on a writers' individual political stand – eventually gain autonomy or independence. Next to *praca organiczna*, i.e. meticulous 'organic work' on the social body, *praca na podstaw* – 'work at the basis' – focussed especially on lower social strata. Further ideas also concerned social reformation, for example women's emancipation as well as religious and national equality.

These Polish discussions were no exception. Since about the middle of the century, a peculiar cultural revolution had been changing practically all European countries and beyond.[2] While Positivists discussed their ideas not only in political pamphlets, but also in literature, and especially novels, they also took a strong interest in science as a basis for social organisation. At the same time, Catholicism forfeited its uncontested stand in Poland, especially among the educated circles. Science promised new horizons and gradually replaced religion as a framework in political and social discussions.[3] The new common sense heavily relied on rational, research-based ap-

---

[1] Hugo Steinhaus: *Słownik racjonalny*, Wrocław 1993, 58.
[2] Johannes Feichtinger, Franz L. Fillafer, Jan Surman (eds): *The Worlds of Positivism. A Global Intellectual History, 1770–1930*, Cham 2018.
[3] Stefan Amsterdamski: *Between history and method: disputes about the rationality of science*, Dordrecht, Boston 1992.

proaches of scientific experience to improve social life. In many areas, scientism overruled traditional Catholicism.

Science was held in high esteem, it was perceived as the highest-ranking activity – the one and only activity that could ensure civilizational progress, which at that time was synonymous with the advancement of humankind in general. And thus, it was also a matter of time until epistemological reflection of the new civilizational tool increased.

In Poland, and especially in Warsaw, the first generation of Positivists consisted of academics and publicists that were mostly associated with Warsaw's Main School (*Szkoła Główna Warszawska*), which existed between 1862 and 1869 until it was included into the newly founded Imperial University of Warsaw (*Imperatorskij Varshavskij Universitet*). This so-called Main School Generation[4] included for example historian and philosopher Aleksander Świętochowski, known as 'the Pope of Polish Positivism,' sociologist Ludwik Krzywicki, philosopher Władysław-Mieczysław Kozłowski, and geographer Wacław Nałkowski. A not less important figure was Aleksander Głowacki alias Bolesław Prus, an extremely popular fiction writer and journalist at the time.

After 1880, Warsaw Positivism was rivalling with other ideas, for example Socialism and most importantly the literary current of *Młoda Polska* (Young Poland), while also Romanticism kept a stand. An important event in the awakening of this so called second phase was the foundation of the Józef Mianowski Fund (*Kasa im. Józefa Mianowskiego*) in 1881, which was named after the physician and activist (1804–1879), who had been the rector of the Main School. In no time the Fund became an important institution to subsidize research and publications by Polish scholars. Next to a more professional funding, the institutionalisation also entailed a specialisation of interest. Until the outbreak of the First World War, Positivists who were connected to the Mianowski Fund took special interest in investigating the elements of scientific cognition. With stark optimism and a high degree of self-assurance they assumed 'pure' scientific cognition, 'untainted' by any psychological, social, or outlook-related premises. This conviction was a common thread in writings of most authors from Prus to Świętochowski, from Krzywicki to Wacław Nałkowski.[5] However, with some time passing, the great optimism about 'pure' cognition was abandoned for the concession that the scholar's mind is subject to a variety of processes and influences, as it is to so-called common-sense thinking.

Next to the study of psychological detail, also holistic approaches became noticeable. As new autonomous disciplines emerged from Positivist philosophical currents, such as sociology, psychology and anthropology, including social and cultural anthropology, the principal functions of science attracted interest. As was mentioned

---

[4] Stanisław Fita: *Pokolenie Szkoły Głównej,* Warszawa 1980.
[5] Anna Hochfeldowa, Barbara Skarga (eds): *Z historii filozofii pozytywistycznej w Polsce: ciągłość i przemiany,* Wrocław 1972.

above, the Polish context was not separated from international discussions. The works of Émile Durkheim, Auguste Comte, Herbert Spencer, Thomas Buckle and others were widely perceived and these authors and their new disciplines heavily influenced reflections on science of Polish thinkers. The most important and characteristic aspect of their search was the assumption that theoretical considerations on science might be conducted in an empirical fashion – as is the case in mathematics and natural sciences. As Adam Mahrburg wrote in the first volume of the Philosophical Review (*Przegląd Filozoficzny*) in 1897: "Our objective is merely to realize what science is all about, in principle, and to identify the means or measures it uses to achieve its goal; what are the human needs science is meant or designed to satisfy, and how it actually relates to the other areas of our creative effort."[6]

By the time Poland regained independence in 1918, Przegląd Filozoficzny, which was the journal of the Polish Philosophical Society (*Polskie Towarzystwo Filozoficzne*) had become the major forum for Polish scholars to tackle scientific developments and their social determinants and broader conditions. It was in this periodical that the young Tadeusz Kotarbiński published his first essay in the field of science studies as early as 1915.[7] A few other scholars addressed similar issues at that time, mainly in the field of new theories of science, since no proper sociology of science did yet exist. Among others there was the well-known zoologist Józef Nusbaum-Hilarowicz, author of the book *Uczeni i uczniowie* (Scholars and Students, 1910), and embryologist Jan Tur, who published his book *Nauka i uczony* (Science and the Scholar) in 1917.[8]

The strong theoretical development of Positivist epistemological thought was fostered by the ongoing organizational integration of a scholarly milieu. In the Russian partition, it became much easier when the authorities made amends after the Russian Revolution in 1905. After Warsaw had witnessed a long school strike, educational associations could be officially formed and discussions be led more systematically and publicly. The Mianowski Fund and particularly its Academic Section (*Dział Naukowy*), which was led by Stanisław Michalski (1865–1949) was one of the central institutions.

The Mianowski Fund: A hub for Science Studies in Interwar Poland

After the First World War, when Poland re-gained independence, the Academic Section of the Mianowski Fund with director Michalski at its center organized a large convention to assess the state of Polish sciences and letters and to determine the most urgent needs of Polish scholarship. In April 1920, 533 scholars and invited guests

---

[6] Adam Mahrburg: "Co to jest nauka?," *Przegląd Filozoficzny*, vol. 1/1 (1897), 9–29: 9.
[7] Tadeusz Kotarbiński: "Dążności rozkładowe postępu wiedzy," *Przegląd Filozoficzny*, vol. 18/1–2 (1915), 1–19.
[8] Józef Nusbaum-Hilarowicz: *Uczeni i uczniowie*, Lwów 1910; Jan Tur: *Nauka i uczony*, Warszawa e. a. 1917.

from all over the country attended this *I. Congress designated to Questions of the Organisation and Development of Polish Science and Letters* (*I. Zjazd poświęcony sprawom organizacji i rozwoju Nauki Polskiej*) to discuss the status and prospects of science in the new Polish state. The debates were based on two volumes of papers that Michalski and his team had collected in advance. As early as 1918, they had sent out a questionnaire to eminent scholars from all disciplines to ask about their fields' needs and expectations. In fact, this collection of conference papers materialized as the first two volumes of the later journal *Nauka Polska, jej potrzeby, organizacja i rozwój* (Polish Science: Its Requirements, Organization, and Development), where Michalski acted as editor in chief. Besides questions of specific scientific disciplines, the journal addressed general issues pertinent to science as a whole. In his memoirs Michalski wrote:

[...] once practical activities related to supporting science had started to expand within the Fund, the need arose to theorize such activity: the need for a science of science, ["nauki o nauce"] devised to focus on researching the life of science, its sociology, the psychology of scientific creativity, the organization of science, the history of science, the present-day status of science, plus a series of related issues – which together made up the subject matter of a new branch of science, without a dedicated [publishing] organ in foreign literature at that time, to which position *Nauka Polska* was soon successfully promoted [...].[9]

In the fourth volume of Nauka Polska (1923), the editorial postulated the need for integrated research on science to lay down the theoretical foundations for those supporters, organizers and practitioners of science and letters. This innovative programme included research on science as a part of culture as well as the proposition to direct research in such a way that it would be applicable to the practical organization of research itself.[10] It is worth noting that this concept built upon Adam Mahrburg's aforementioned postulate. After all, the journal grew out of long lines of tradition just after the war.

This programme of the Academic Section was supported by the annual publication of Nauka Polska. The editorial team with Michalski at its centre motivated scholars of various disciplines to contribute, but it was mostly sociologists and philosophers who tackled the theoretical problems, which eventually condensed into the idea of a distinct discipline, a "science of science". One of the most interesting initiatives, in which Michalski closely cooperated with Antoni Bolesław Dobrowolski, was a series of especially commissioned autobiographies by outstanding scholars. These texts should offer valuable material for research on the psychology of creative scientific activity.[11]

---

[9] Muzeum Ziemi, Warsaw: Materials of Janina Małkowska: Excerpt from an Unpublished Autobiographical Fragment of Stanisław Michalski, 213–214.

[10] ▶ IN THIS VOL. "Editorial Introduction" (1923), 104, i.e. "Wstęp redakcyjny," *Nauka Polska*, vol. 4 (1923), VII–IX.

[11] See ▶ Antoni Bolesław Dobrowolski: "The Urgent Need for Mental Education in Poland" (1918), i.e. "O pilnej potrzebie wychowania umysłowego w Polsce: o konieczności zasadniczej re-

The first systematic take on a disciplinary project was Florian Znaniecki's 1925 article *The Subject and the Tasks of Science of Knowledge*[12], in which the author proposed an empirical theory of knowledge as a separate science with its own tasks and methodology. This attempt was rooted in sociology and the psychology of cognition – two fields that had recently seen intense development, not least in Znaniecki's cooperation with William I. Thomas.[13] Methodologically, Znaniecki conceived his new *Science of Knowledge* as theoretical generalization based upon empirical facts. It would embrace issues as the influence of practical activities on knowledge (and thus on science), the individual education in theoretical thinking, the social determinants of scholarship as well as the intellectual life of communities, etc. One fundamental issue were the psychological conditions of any creative effort. Znaniecki explicitly sketched the new discipline as an instrument for organizing and supporting science and kept coming back to his treatise in later works.[14]

Most of the authors who contributed to Nauka Polska presented their papers at the Science Studies Circle (*Koło Naukoznawcze*) beforehand. The Circle was founded in 1928 on the initiative of director Michalski to supplement the periodical's mission of developing theoretical foundations for the organization and support of science.[15] Meetings were held at the Staszic Palace in Warsaw, which had been the seat of the Society of Friends of Learning (*Towarzystwo Przyjaciół Nauk*) since 1824 and since 1907 of the Warsaw Scientific Society (*Towarzystwo Naukowe Warszawskie*). Apart from scholars, many other people joined discussions, for example government clerks with an interest in science and its methodology. Apart from initiating this exchange, the organizers sought to develop specific research areas, for example the sociology of creativity or the psychology of scientific discoveries. The papers that were presented at the Circle's meetings addressed philosophical, and also the role and importance of science and its position among other cultural values in society.[16] Until the outbreak of the Second World War in 1939, about 40 meetings of the Science Studies Circle were documented, and Michalski even continued to chair interdisciplinary meetings under the Nazi occupation.[17]

---

formy nauczania w szkołach średnich oraz stworzenia w związku z ową reformą nowych placówek pracy naukowej," *Nauka Polska*, vol. 1 (1918), 489–502..

[12] ► Florian Znaniecki "The Subject Matter and Tasks of the Science of Knowledge" (1925), i.e. "Przedmiot i zadania nauki o wiedzy," *Nauka Polska*, vol. 5 (1925), 1–78..

[13] William I. Thomas, Florian Znaniecki: *The Polish Peasant in Europe and America*, 5 vol.s, Chicago 1918–1920.

[14] Florian Znaniecki: *The Social Role of the Man of Knowledge*, New York 1940, id., *Społeczne role uczonych*, ed. by Jerzy Szacki, Warszawa 1984.

[15] Jan Piskurewicz: *W służbie nauki i oświaty: Stanisław Michalski (1865–1949)*, Warszawa 1993; Stefan Zamecki: *Problematyka naukoznawcza na łamach periodyku "Nauka Polska. Jej Potrzeby, Organizacja i Rozwój." Studium historyczno-metodologiczne, Lata 1918–1947*, Warszawa 2016. See also: ► List of Sessions.

[16] For an overview on presentations and papers published thereupon see ► List of Sessions.

[17] Jan Piskurewicz: "Stanisław Michalski," in: *Kasa Mianowskiego 1881–2011*, ed. by Leszek Zasztowt, Warszawa 2011, 175–232: 219–220.

A central issue for the Circle had always been the emerging vision of studies on science, its objectives and shape. As the report of the third meeting concluded Paweł Rybicki's earlier talk:

> [...] we are aware today of the birth of a new discipline, the science of science ["naukoznawstwa"], which aims to uncover the truths governing science as a phenomenon. A vocation for research in this field is most common among people who do scientific work; in order to make foundations of the science of science as solid as possible, we ought to begin by examining the most concrete issues within the field of sociology of knowledge, such as recruiting of workers or the influence of social conditions on the researcher's work.[18]

Efforts were made to implement this postulate at subsequent meetings, besides analysing issues such as the scholar's personality and conditions of creative work. Contributors included philosophers Tadeusz Kotarbiński and Jan Łukasiewicz, historian Franciszek Bujak, as well as Wojciech Świętosławski, a chemist and biophysicist. The latter, who at that time was Minister of Religious Affairs and Public Education, focused on the organization and funding of research.[19]

Debating the position of science among the other cultural values as well as many other philosophical and methodological questions, the discussions within the Circle did not stand behind comparable international theoretical and programmatic discussions on science (or science studies). The circle had become an important hub for these topics. Moreover, as diverse as the stands in the discussions were at times, attempts for disciplinary integration were introduced, too. In late 1929, Stanisław Ossowski delivered a paper on Issues of Science Studies (*Problematyka naukoznawcza*), which he expanded in the article The Science of Science (*Nauka o nauce*) together with his wife Maria Ossowska.[20]

The Ossowskis understood science as a complex product of cognitive and organizational actions and intended to study it from an epistemological and anthropological point of view. Approaching science from different perspectives, they proposed to classify questions into the following categories: philosophy of science, methodology of science, psychology of science, sociology of science, and history of science, as well as practical and organizational issues. Pointing to the practical purposes of their "sci-

---

[18] ▸ Report on the activities of the Science Studies Circle (1929), 108, i.e. "Sprawozdanie z działalności Koła Naukoznawczego," *Nauka Polska*, vol. 11 (1929), 353–355..

[19] ▸ Tadeusz Kotarbiński: "The Skills of a Researcher" (1929), i.e. "O zdolnościach cechujących badacza," *Nauka Polska*, vol. 11 (1929), 1–10; Jan Łukasiewicz, On the History of Scientific Method as reported in "Sprawozdanie siódme z działalności Koła Naukoznawczego," *Nauka Polska*, vol. 19 (1934), 404–420: 404–407; ▸ Franciszek Bujak, "The Man of Action and the Student" (1936), i.e. "Działacz i badacz," *Nauka Polska*, vol. 11 (1929), 11–23; Wojciech Świętosławski: "Notes on the Organization of Creativity and Scientific Research Work in Poland," *Nauka Polska*, vol. 10 (1929), 1–13.

[20] See ▸ Second Report on the activities of the *Science Studies Circle* (1930) for an overview of Ossowski's presentation on 25 November 1929, i.e. "Sprawozdanie drugie z działalności Koła Naukoznawczego," *Nauka Polska*, vol. 13 (1930), 166–169, and ▸ Maria Ossowska and Stanisław Ossowski: "The Science of Science" (1936), i.e. "Nauka o nauce," *Nauka Polska*, vol. 20 (1935), 1–12 for the article based on the presentation.

ence of science," the Ossowskis concluded that the organization of science "cannot to-day do without studies just as specialised and complicated as those which are required for the construction of large industrial establishments". moreover, "teaching what Science is, contributing to form in the minds of scientific workers this or that conception of Science, [these studies] at the same time influence their further creativeness".[21] This programmatic article was to be a blueprint for the creation of a large, internationally acclaimed Science Studies institute, which should be based on structures like the Science Studies Circle and the Mianowski Fund's Academic Section. However, mostly due to a lack of funding, this project had to be postponed in the early 1930s and actually could never be realized.

The broad programme presented by the Ossowskis rested on a discussion of similar other concepts from Poland and beyond. Next to Znaniecki's proposal for a science of knowledge, it strongly engaged with German ideas of a sociological take on science, for example Werner Schingnitz's *Scientiologie*.[22] Discussing international work in this context was part of a general strategy of Michalski and his colleagues to put Poland on the map of international science studies.

One important step in this direction was the publication of the international review *Organon* – a science studies periodical for foreign readers. Starting in 1936, Michalski invited scholars from the Circle and beyond to contribute to the two volumes that would be published until the outbreak of the Second World War. Thus, the Maria Ossowska's and her husband's programmatic article was translated into English to open the first volume, which also included contributions by the physicist Czesław Białobrzeski, historian Franciszek Bujak, science historian Aleksander Birkenmajer, and pedagogue, philosopher and science historian Bogdan Suchodolski. Apart from further articles, the second volume of Organon presented the replies to a ques-

---

[21] ► Ossowska and Ossowski: "The Science of Science" (footnote 20), 182.
[22] Werner Schingnitz: "Scientiologie", *Minerva-Zeitschrift* vol. 7/5–6 (1931), 65–75 and id., "Scientiologie (Schluss)", Minerva-Zeitschrift vol 7/7–8 (1931), 110–114. It is worth mentioning that discussions in the Science Studies Circle were critical towards both the sociology of knowledge, as defined by their German colleagues, and the sociology of science, which was often identified with the former. The report on the first Circle meeting summarizes Bogdan Suchodolski's discussion of a work by Max Scheler. See Bogdan Suchodolski: "Review: Max Scheler, *Die Wissenschaftsformen und die Gesellschaft: Probleme einer Soziologie des Wissens* (Leipzig, 1926)", *Nauka Polska*, vol. 11 (1929), 379–381, as well as Sprawozdanie z działalności Koła Naukoznawczego, *Nauka Polska* vol. 11 (1929), 353–355: 353 (not translated in this vol.). Despite many members of the Science Studies Circle shared a critical attitude towards theoretical assumptions by the sociology of science, a considerable number of articles in Nauka Polska and elsewhere, most importantly Stanisław Ossowski: "Funkcja dziejowa nauki," *Nauka Polska*, vol. 4 (1923), 8–35 and id., "Nauki humanistyczne a ideologia społeczna," *Nauka Polska*, vol. 20 (1937), 1–24, as well as ► Paweł Rybicki: "Science Science and the Forms of Social Life" (1929), i.e. "Nauka a formy życia społecznego. Kilka zagadnień z pogranicza socjologii i teorii nauki," *Nauka Polska*, vol. 11 (1929), 24–64 and id., "Nauka a element irracjonalny," *Nauka Polska*, vol. 24 (1939), 1–26. See also Florian Znaniecki: "Uczeni polscy a życie polskie (I)," in: *Droga*, vol. 15/2–3 (1936), 101–116 and id., "Uczeni polscy a życie polskie (II)," in: *Droga*, vol. 15/4 (1936), 255–271, as well as id., "Społeczne role uczonych a historyczne cechy wiedzy," *Przegląd Socjologiczny*, vol. 5 (1937), 2–56.

tionnaire on organizing scientific conventions by 66 international and Polish scholars.[23] Drawing on their international networks, Michalski and his colleagues secured the cooperation of eminent researchers such as the French mathematician Emile Borel, British science historian George Sarton, Danish physiologist and Nobel laureate August Krogh, and ethnologist Bronisław Malinowski, who then resided in London. Michalski also had arranged for contributions by the creator of quantum theory, Niels Bohr, and the well-known astrophysicist Arthur S. Eddington, which, however, were not published before the Second World War.[24]

Polish Interwar Science Studies Beyond Warsaw

Although the Mianowski Fund formed the central science studies hub in Interwar Poland, there were also other initiatives for inquiries into science and the practical forms of its implementation. Several scientific or scholarly conventions and periodicals took similar, though often less influential approaches, especially concerning links to politics. An especially active centre was the philosophical milieu of Lviv, which gathered the most influential figures of interwar philosophy in Poland. However, although they organized separately from the Mianowski Fund and formed no formal links, many of the Lviv-Warsaw philosophers frequently contributed to Nauka Polska and meetings of the Circle, for example Jan Łukasiewicz and Tadeusz Kotarbiński.

Apart from that, questions of science were prominently discussed at Polish philosophical congresses. For example, Władysław Witwicki opened the First Convention of Polish Philosophers, which was organized by the Polish Philosophical Society in Lviv in May 1923, with a lecture on Aspects of the Philosophy of Science (*Z filozofii nauki*). In this talk, he focused on the psychology of creative scientific activity, and especially on differences between scientific and common thinking. Topics related to science studies were also raised at the second and third conventions in Warsaw (1927) and Cracow (1936). At the latter, Jerzy Szydłowski proposed to form an institute for scientific organization.[25] The Polish Philosophical Society also had a Science Theory

---

[23] See section "Organisation des Congrès Scientifiques," *Organon*, vol. 2 (1938), 133–236.

[24] Piskurewicz: *W służbie nauki i oświaty* (footnote 15), Warszawa 1993, 102. For two actually published texts see ▸ Emile Borel: "Contribution (Documents sur la Psychologie de l'Invention Dans Le Domaine De La Science)" (1936), i. e. Borel: "Contribution (Documents sur la Psychologie de l'Invention Dans Le Domaine De La Science)," *Organon*, vol. 1 (1936), 33–42, ▸ August Krogh: "Visual Thinking. An Autobiographical Note" (1936), i. e. "Visual Thinking," *Organon*, vol. 2 (1938), 86–94..

[25] Ryszard Jadczak: *Polskie zjazdy filozoficzne*, Toruń 1995. For example, Jan Łapszyn spoke about "Intuition in Scientific, Technological and Philosophical Creativity" ("Intuicja w twórczości naukowej, technicznej i filozoficznej", *Przegląd Filozoficzny* vol. 31/1 (1928), 112–115), Bogumił Jasinowski on "Science and Philosophy" ("Nauka a filozofia", *Przegląd Filozoficzny*, vol. 39/4 (1936), 368–370), and Jan-Franciszek Drewnowski, a Thomist philosopher, later doctoral student of Kotarbiński about a separate "Technique of Knowledge" ("Technika wiedzy", *Przegląd Filozoficzny*,

Section (*Sekcja Teorii Nauki*), which was run by philosopher, mathematician and painter Leon Chwistek.

Apart from Nauka Polska and Przegląd Filozoficzny, the periodical *Przegląd Współczesny* (Modern Review) was most open for discussions on science studies. Beside articles published in a permanent section on the "Organization of Science," Przegląd Współczesny published works dealing with the philosophical and sociological aspects of practicing science. The contributors were mostly professors affiliated with Cracow's Jagiellonian University: philosophers Joachim Metallmann and Bolesław Gawecki as well as sociologist Paweł Rybicki, who also published in Nauka Polska. In the early 1930s, when the educational reform by minister Janusz Jędrzejewicz stirred discussions about academic freedom, this periodical initiated a series on the meaning and function of universities – a topic later revisited by the Science Studies Circle.[26]

Around these periodicals a whole discourse of science studies formed, as cross references suggest. For instance, writing for Nauka Polska on creative efforts in science, psychologist Stefan Błachowski from the University of Poznań referred to Jan Łukasiewicz's article On Science (O nauce)[27], and also made use of scholars' autobiographies published in the preceding volumes of Nauka Polska.[28] Pedagogue Mirosław Sekreta proceeded similarly in his treatise on an "irrational coefficient of scientific creativity," which he published in the journal *Chowanna*.[29] Paweł Rybicki's article Science and Forms of Social Life (*Nauka a formy życia społecznego*)[30] referred to Znaniecki's study on the science of knowledge.[31] Przegląd Współczesny witnessed heated polemics between Tadeusz Bilikiewicz and Ludwik Fleck, two exponents of the medical sciences, over the influence of social environment on scientific creativity.[32] This clash of views encouraged the development of specific questions in science studies and shaped contexts of naukoznawstwo – a broad term for research on science – as a separate research area.

For the greater part of the Interwar period, journal articles were the preferred form of publication in the context of science studies in independent Poland. To many au-

---

vol. 39/4 (1936), 414–415), which would, based on an analysis of results from the exact sciences, help to organize other areas of knowledge. For Szydłowski's call see *Przegląd Filozoficzny*, vol. 39/4 (1936), 541.

[26] See ► List of Sessions.

[27] Jan Łukasiewicz: "O Nauce," in: *Poradnik dla Samouków* (wyd. nowe, vol. 1), Warsaw, 1915, XV–XXXIX.

[28] See ► Stefan Błachowski, "The Problem of Scientific Creativity" (1928), 213–216, i. e. "Zagadnienie twórczości naukowej," *Nauka Polska*, vol. 9 (1928), 1–67.

[29] Mirosław Sekreta: "Irracjonalny współczynnik naukowej twórczości," *Chowanna*, vol. 8/9 (1937), 396–404, and id: "Irracjonalny współczynnik naukowej twórczości (Dokończenie)," *Chowanna*, vol. 8/10 (1937), 433–441.

[30] ► Rybicki: "Science and the Forms of Social Life" (footnote 22), 240–245.

[31] ► Znaniecki: "The Subject Matter and Tasks of the Science of Knowledge" (footnote 12).

[32] For an English edition of the debate see Ilana Löwy (ed): *The Polish School of Philosophy of Medicine. From Tytus Chałubiński (1820-1889) to Ludwik Fleck (1896-1961)*, Dordrecht 1990, 229–275 and Löwy's comments on 215–227.

thors it seemed to be simply too early for a comprehensive take. As Bogdan Suchodolski observed in a discussion of Max Scheler's *Die Wissensformen und die Gesellschaft: Probleme einer Soziologie des Wissens* in the first ever session of the Science Studies Circle, research into science was still a very young field:

> In this field, which is revealed by the recognition that science is not a matter that takes place solely on the plane of the object examined and the investigating reason, but that the social plane also somehow slips into it, there is of course still much, or rather everything, to be done. The study of science is only taking its first steps.[33]

After a general overview by the speaker, in which he pointed out the book's "significance as a synthetic work on results of works and opinions of various authors", Suchodolski criticized the very personal metaphysical background, which he found in Scheler's writing. The discussion following Suchodolski's talk seemingly seconded his diagnosis on the pitfalls of all too early generalization:

> The discussion further emphasized the lack of originality in Scheler's work and the riskiness of some of his claims, which are more aphorisms of a journalistic nature than the results of serious scientific research. Scheler's main merit is his boldness in writing about such difficult issues and organizing a joint effort to work on them.[34]

Despite all doubts about the right time for systematization in Poland at the time, one of the later classic monographs in science studies was written by Lviv biologist and epistemologist Ludwik Fleck, who – perhaps underlining said doubts about the early stages of the field – decided to publish his seminal work in German.[35] Fleck deals with the methodology and sociology of science in the context of the discovery of the Wassermann reagent. He argues that facts are proposed, in the first place, through the researcher's convictions, cognitive habits and preferences, which are determined by a 'thought style' and the culture to which the researcher belongs (his identification and reference group). Thus, rather than stemming from logical procedures, notions and beliefs appear to follow from psychosocial traditions and processes. In Fleck's view, this entanglement of cognition in the sphere of values and emotions reinforces its effects (when it comes to theories, for instance), whilst also causing their resistance to change, and paradigmatic quality. The potential for change, for the development of new theories, would primarily depend on transformations in a society's or community's culture of thought and the thought style of a given research collective. Having

---

[33] Bogdan Suchodolski: "Review: Max Scheler, *Die Wissenschaftsformen und die Gesellschaft: Probleme einer Soziologie des Wissens* (Leipzig, 1926)", *Nauka Polska*, vol. 11 (1929), 379–381: 381.

[34] Ibid.

[35] Ludwik Fleck: *Entstehung und Entwicklung einer wissenschaftlichen Tatsache: Einführung in die Lehre vom Denkstil und Denkkollektiv*, Basel 1935. For the first Polish edition see *Powstanie i rozwój faktu naukowego. Wprowadzenie do nauki o stylu myślowym i kolektywie myślowym*, transl. by Maria Tuszkiewicz, Lublin 1986. On the historical context of Fleck's study see Bernhard Kleeberg, Sylwia Werner (eds): *Gestalt, Ritus, Kollektiv: Ludwik Fleck im Kontext der Ethnologie, Gestaltpsychologie und Soziologie seiner Zeit*, special issue of *NTM Zeitschrift* für Geschichte der Wissenschaften, Technik und Medizin, vol. 22/1–2 (2014).

passed almost unnoticed before the Second World War, Fleck's study gradually gained interest from the 1960s onwards, which mainly went to the credit of Thomas Kuhn, who mentioned it as a source of inspiration for his own theory of scientific revolutions.[36]

Outlook: Science Studies in Poland since 1945

Summarizing the major achievements of Polish science studies from the two decades between the World Wars, organizational efforts, especially the formation of societies and journal springs to the eye: not only did they provide a forum for discussions on the role of science, but they decisively influenced the development of science itself. Under the auspices of the Mianowski Fund the probably most active forum emerged, which, however, mainly consisted of philosophers and sociologists, most of whom represented the Lviv-Warsaw School with its broad vision of theoretical reflexion on science.

In the long run, the program for a "science of science" as formulated by Maria Ossowska and Stanisław Ossowski might have had the strongest influence on further developments. Its longevity is attested by its republication in 1964 by the British *Minerva* and, in the following year, in Norman Kaplan's edited volume *Science and Society*.[37] Some commentators deem this article to have initiated international science studies as a discipline. In the opinion of science historian Derek John de Solla Price, the Ossowski couple's study was far ahead of its time.[38]

Thus, the interwar period's achievements, as far as the emerging 'science of science' is concerned, were not lost after the war, although the division of Europe severely restricted contacts between Eastern and Western scholars, especially during Stalinist times. It was only in the 1960s, in the time of political thaw in Poland, that major texts by Polish authors were made available again to West-European readers. At the same time, first unrestricted visits of Polish scholars to the West were possible.

Discussions on the science of science continued in post-war Poland, in spite of the forced closure of the Mianowski Fund in the early 1950s and Stanisław Michalski's death in 1949. Many former members of the Science Studies Circle took important positions in national science institutions, which were then undergoing reformation. The major political transitions after the end of Stalinism were crucial. Starting with

---

[36] Thomas Kuhn: *The Structure of Scientific Revolutions*, Chicago 1962, IX.

[37] See Maria Ossowska and Stanisław Ossowski: "The Science of Science," *Minerva: A Review of Science, Learning and Policy*, vol. 3/1 (1964), 72–82, as well as the same title in: *Science and Society*, ed. by Norman Kaplan, Chicago 1965, 19–29.

[38] Derek J. de Solla Price: "The History of Science as Training and Research for Administration and Political Decision-Making," *Organon*, vol. 1 (1964), 21–24: 22. See also John D. Bernal and Alan L. Mackay: "Na drodze do naukoznawstwa," *Zagadnienia Naukoznawstwa* vol. 2/1–2 (1966), 9–17: 9 resp. id.: "Toward a Science of Science," *Organon* 3 (1966), 9–17: 9. ▶ Friedrich Cain, Bernhard Kleeberg: "Introduction", 21–28.

the Thaw in October 1956, many researchers who had been banned from teaching before could resume their activities to the full. In 1957, Tadeusz Kotarbiński became president of the Polish Academy of Sciences in Warsaw. He introduced a Laboratory for General Questions of Work Organization (*Pracownia Ogólnych Problemów Organizacji Pracy*), which was later transformed into a Department of Praxeology (*Zakład Prakseologii*).

Since 1945, the tradition of Science Studies in Warsaw was in partly held up by members of the Warsaw School of Ideas (Leszek Kołakowski, Andrzej Walicki, Bronisław Baczko and others), particularly in the 1960s[39], but also with the Institute for the History of Science, Education and Technology (*Instytut Historii Nauki, Oświaty i Techniki*) of the Polish Academy of Sciences (*Polska Akademia Nauk*, PAN), today the Ludwik and Aleksander Birkenmajer Institute for the History of Science (for example Stefan Amsterdamski, Stefan Zamecki, Janusz Skarbek and many others). Most of these scholars shared a "secular", non-ideological approach[40] to science and worldview problems, which facilitated their assimilation with so-called 'real socialism' and omnipresent Marxism. Yet, most of them resisted the communist system – clandestinely or even openly as for example the Ossowskis did.

The Mianowski Fund was integrated in the newly founded Polish Academy of Sciences in early 1953 together with the Warsaw Scientific Society and the Academy in Cracow (*Polska Akademia Umiejętności*, PAU). While the journal Nauka Polska had to give way to the History of Science and Technique Quarterly (*Kwartalnik Historii Nauki i Techniki*) of the Academy Institute, the Organon was restarted in 1964 by Bogdan Suchodolski and others, again as an international review. After the transformation, the Mianowski Fund was reactivated on 20 May 1991. Only shortly later, the Fund started publishing Nauka Polska again, which in 2021 counted 30 post-1991 and an overall number of 55 volumes.

---

[39] Ryszard Sitek: *Warszawska szkoła historii idei. Między historią a teraźniejszością*, Warszawa 2000.

[40] This wording refers back to the so-called "laic intelligentsia" of the Interwar time, which developed a specific "humanist" agenda. After 1945, this humanism was often used to set a contrast to communist propaganda (to which the term itself was not unknown).

A look back on *Koło naukoznawcze* after the Science Wars

Andreas Langenohl

The question informing the present reflections on the publication of this volume is the following: Why should we look at the Science Studies Circle (*Koło naukoznawcze*) as an epistemological project concerned with the preanalytical conditions of scientific analysis *today*, that is, after having experienced an explosion of concerns regarding the viability of modern science since the 1960s? Is the Science Studies Circle's approach not hopelessly outdated? What meaning can it have for today's knowledge-theoretical episteme for which the critique of science's claims to knowledge is quite common parlance, at least in the social sciences and the humanities?

Arguably, the Science Studies Circle was more nuanced and differentiated academic and intellectual movement than these somehow blatant questions indicate. Yet, instead of paying tribute to this heterogeneity, I want to focus on some aspects of the Science Studies Circle, in particular in its formative period, which might bear important and rather general lessons for today's study of science. Due to the purpose of the present paper, I see those lessons not so much in certain facets of the Science Studies Circle's research agenda which undoubtedly have enormous potential to inform today's discussions, like the question of the social emergence and epistemological significance of 'creativity' in science's epistemic procedures. Rather, this short paper depicts them in a certain, from today's perspective highly unlikely, combination of a constructivist approach to science with a tendency to endorse the modernist project of science.

I. Critiques of scientific knowledge practices since the 1960s

Since the 1960s, the knowledge claims of science and its authority to describe and explain the ways of the world have been challenged in various ways, and coming from rather different directions. I will mention only a few examples. Thomas Kuhn's immensely influential book on scientific revolutions drove home the point that changes in scientific approaches need not necessarily be embraced as scientific progress, but should rather be understood as epistemic turning points that follow their own cataclysmic logic which in large parts is due to the way modern science is organized.[1]

---

[1] Thomas S. Kuhn: *The Structure of Scientific Revolutions*, Chicago 1962.

While Kuhn's approach did not necessarily encompass a radical deconstructivist critique of scientific analyses, it opened a way to think about science and epistemic procedures as being impacted by dynamics that had nothing to do with either the objects under investigation or with the accumulation of knowledge. From the 1970s onwards, the Sociology of Scientific Knowledge, later known as Science and Technology Studies (STS), attacked more directly the correspondence-theoretical assumptions especially of the natural sciences. Its protagonists, among them David Bloor, Bruno Latour, Steve Woolgar, Andrew Pickering, Karin Knorr Cetina, and others, held the view that scientific knowledge gets created together with the objects under investigation.[2] The notion of the "epistemic object" coined by Hans-Jörg Rheinberger[3] served as a shorthand version for the argument that the objects of scientific analysis – like, for instance, substances in a laboratory – become constituted and transformed in one and the same process through which they are analyzed. From this point of view, it became difficult to uphold any great esteem for a traditional, correspondence-theoretical understanding of scientific objectivity. Instead, objectivity appeared as an effect of the illusion that the isolated epistemic object is pre-given.[4]

For Bruno Latour, the notion of 'knowledge' seemed to disappear from the agenda of Science and Technology Studies altogether. The more he broadened out his understanding of the social as being intimately bound up with non-social and non-human entities, an analysis he first had presented in his research on laboratories,[5] the more he became interested in the ontology of the social, and indeed of the world as such for which the process of scientific knowledge production served merely as an astute example.[6] The deconstruction of scientific knowledge as a correspondence-theoretical achievement thus gave way to the rise of a "flat ontology"[7] where knowledge merely designates a 'practice' that, as any practice, interconnects with other practices, instead of cognitively penetrating the deep structure of nature, society, or the human condition.

---

[2] David Bloor: *Knowledge and social imagery*, Chicago 1991 (1976), Bruno Latour: "The Politics of Explanation: An Alternative," in: *Knowledge and Reflexivity: New Frontiers in the Sociology of Knowledge*, ed. by Steve Woolgar, London et al. 1988, 155–176, id.: *Pandora's Hope: Essays on the Reality of Science Studies*, Cambridge, Mass. 1999, Steve Woolgar: "Reflexivity is the Ethnographer of the Text," in: *Knowledge and Reflexivity: New Frontiers in the Sociology of Knowledge*, ed. by id., London et al. 1988, 14–34, Andrew Pickering: *The Mangle of Practice: Time, Agency, and Science*, Chicago, London 1995, id.: "The Politics of Theory," *Journal of Cultural Economy*, vol. 2 (2009), 197–212, Karin Knorr Cetina: *Epistemic Cultures: How the Sciences Make Knowledge*. Cambridge, Mass. 1999, id.: "Objectual Practice," in: *The Practice Turn in Contemporary Theory*, ed. by Theodore Schatzki, London, New York 2001, 175–188.

[3] Hans-Jörg Rheinberger: *Toward a History of Epistemic Things: Synthesizing Proteins in the Test Tube*. Stanford, Cal. 1997.

[4] Lorraine Daston and Peter Galison: *Objectivity*, New York 2007.

[5] Bruno Latour and Steve Woolgar: *Laboratory Life: The Construction of Scientific Facts*, Beverly Hills, Cal. 1979.

[6] Latour: *Pandora's Hope* (footnote 2), id.: "Why Has Critique Run out of Steam? From Matters of Fact to Matters of Concern," *Critical Inquiry*, vol. 30 (2004), 225–248.

[7] Manuel DeLanda: *Intensive Science & Virtual Philosophy*, New York 2002, 47.

While the natural sciences and their knowledge claims came under deconstructive fire mostly from the perspective of sociology and historiography, the humanities were attacked in particular by cultural anthropology, postcolonial literary studies, and feminist political theory. In the course of the "Writing culture" debate that preoccupied cultural anthropology in the 1980s and 1990s, it was argued that the ways cultures are represented in anthropology is more indicative of anthropology as an order of knowledge organized along the lines of rhetorical and stylistic conventions[8] – while any claim to evidence with respect to the ways these cultures actually are appeared as dubious and shot through with power differentials. Postcolonial criticism, especially Colonial Discourse Analysis, insisted on the strictly discursive order of western representations of the colonized.[9] Feminist interventions – most prominently, those of Judith Butler[10] – targeted hierarchical gender orders that were thought of as being deeply imbricated in gendered representations claiming the status of 'knowledge' for themselves.[11]

Not least, Pierre Bourdieu issued a critique of any social science that does not engage in a reflection on the researcher's positionality in the professional field to begin with. In contrast to the anthropological debate, he however insisted that descriptive objectivity, and ensuing analytical precision, might be rescued for sociology, under the condition that the researcher starts out with a rigorous analysis of their own positionality in the field of academia.[12]

While Bourdieu deviated from the general line of the critique of science that gained ground since the 1960s in that he did not question the possibility of objectivism in science per se, the general tendency of those critical interventions into the correspondence-theoretical epistemology of the traditional sciences, including also

---

[8] James Clifford and George E. Marcus (eds): *Writing Culture: The Politics and Poetics of Ethnography*, Berkeley 1986, Johannes Fabian: *Time and the Other: How Anthropology Makes Its Object*. New York 2002 (1983). Cf. also D. Wade Hands: "The more things change, the more they stay the same: Social realism in contemporary science studies," in: *Fact and Fiction in Economics: Models, Realism and Social Construction*, ed. by Uskali Mäki, Cambridge et al. 2002, 341–355, Michael Carrithers: "The anthropologist as author: Geertz's 'Works and Lives'," *Anthropology Today*, vol. 4 (1988), 19–22 and Akbar S. Ahmed and Cris N. Shore: "Introduction: Is anthropology relevant to the contemporary world?," in: *The Future of Anthropology: Its Relevance to the Contemporary World*, ed. by id., London 1995, 12–45.

[9] Edward W. Said: *Orientalism: Western Conceptions of the Orient*, London et al. 1995 (1978), Homi K. Bhabha: "Of mimicry and man: The ambivalence of colonial discourse," in: *The Location of Culture*, ed. by id., London, New York 1994 (1987), 85–92, Gayatri Chakravorty Spivak: "Can the subaltern speak?," in: *Marxism and the Interpretation of Culture*, ed. by. Lawrence Grossberg and Cary Nelson, Basingstoke, London 1988, 271–313.

[10] Judith Butler: *Gender Trouble: Feminism and the Subversion of Identity*, London, New York 1990.

[11] Cf. Margery Wolf: *A Thrice Told Tale: Feminism, Postmodernism and Ethnographic Responsibility*. Stanford, Cal. 1992.

[12] Pierre Bourdieu and Loïc Wacquant: *An Invitation to Reflexive Sociology*, Chicago 1992, Pierre Bourdieu: *Pascalian Meditations*, Stanford, Cal. 2000, cf. Andreas Langenohl: "Die Reflexivität Pierre Bourdieus: Soziologische Objektivität wider die Kulturwissenschaften," in: *Pierre Bourdieu und die Kulturwissenschaften. Zur Aktualität eines undisziplinierten Denkens*, ed. by Daniel Šuber, Hilmar Schäfer and Sophia Prinz, Konstanz 2011, 319–338 for a critique.

the social sciences and the humanities, had a twofold effect. First, these interventions demanded a more or less tight circumscription of the legitimate reach of science's knowledge claims, calling for a calibration of scientific knowledge claims with other modes of representation. Second, these interventions cautioned against the hegemonic effects of science when embracing a traditional correspondence-theoretical epistemology.

II. The Science Studies Circle's epistemological project

These two features did not hold true for the Science Studies Circle – indeed, for the Science Studies Circle, especially in its formative phase, they applied the other way round. The Science Studies Circle was explicitly meant to empower science in society, not to restrain its claims and its effects; and it sought to strengthen the grounds of science's claims to knowledge, not to undermine them (here displaying a certain affinity to Bourdieu's understanding of sociology as a reflexive science). I will substantiate this argument through a quick run-down of early key epistemological and methodological approaches and suggestions of the Science Studies Circle. This serves not the purpose of inadequately rejecting the plurality and heterogeneity of approaches that were then labelled as belonging to the Science Studies Circle, but rather to flesh out the character of the (especially early) Science Studies Circle as an *epistemological project* that, in contrast to more recent epistemological discussions, believed in the perfectibility of science.

First, the Science Studies Circle proposed a reflection on the positive conditions of science, by which it mostly meant 'creativity' (*twórczość*). Creativity appeared as that element in scientific endeavors that had so far escaped the attention of historians and sociologists of science, and at the same time as a regulator that could be used in order to enhance science's normal conduct. Particular methodologies, much unlike since the 1960s, did hardly attract the attention of Science Studies Circle researchers because creativity was conceived of as residing in a pre-analytical and pre-methodical stage of the epistemic process. Furthermore, creativity was modeled after the logic of discovery, while the aspect of the construction or production of scientific knowledge figured only in elaborations on methodology.

Second, the relation between science and society, due to the Science Studies Circle's premises, ought to be worked toward science gaining greater importance for society, its institutions, and its conduct. Society would be rationalized by dint of a science that had clarified, understood, and optimized its own origin and condition of possibility in creativity. Again, methodological issues were not so much critically reflected but rather envisaged as instruments to achieve this grand aim (see Paweł Kawalec's contribution in this volume[13]). The viability of 'science of science' for soci-

---

[13] See ▶ **IN THIS VOL.** Paweł Kawalec: "Philosophical Perspectives".

ety thus crucially depended on a *functioning* epistemology, not on the radical *questioning* of epistemology. An apt example is the way some authors from the Science Studies Circle relate to anthropology.[14] While there is a general appreciation of a new quality that this methodology might introduce to the 'science of science', also with respect to the organization of research, there is no epistemological reflection, but rather a vague hint at the German tradition of *Wissenssoziologie*.[15] For instance, it is suggested to observe school children in order to gain insights into how creativity emerges and develops[16] – but the peculiar epistemological characteristics of observation (including its limits) that has characterized anthropological discourse so profoundly remain unaddressed. In other words, observation is meant to enhance the process of knowledge production for the 'science of sciences', but it does not merit much consideration of the ways that certain methodologies bring about certain epistemological problems. Instead, epistemological problems are, as a tendency, discussed as possibly erring trajectories of creativity, that is, as "errors and faults"[17].

Third, the Science Studies Circle did not engage in a questioning of (politically, societally and culturally declared) boundaries. While especially since the 1980s, one major concern for most critiques of scientific knowledge production was an epistemological bias resulting from constructions of self and other (the imperial metropolis vs. the colonies, or the juxtaposition of nation-states with one another) in which science was embedded, the Science Studies Circle, at least some of its protagonists, accepted such boundary constructions.[18] Indeed, it is not inaccurate to say that, insofar the Science Studies Circle was an epistemological project, it embraced them, especially the one that has become the *bête noire* of postcolonial and feminist critiques: the nation-state. For the Science Studies Circle deliberately dedicated itself to supporting the nation- and state-building process in Poland.

Fourth, this had to do with another of the Science Studies Circle's preoccupations that has meanwhile become highly questionable, namely, its modernization-theoretical ethos. According to proponents of the Science Studies Circle, the 'science of science' corresponded to, or reflected, a certain stage in societal development and consciousness, and claimed avant-gardism in this stage for itself:

Modern culture enables man in an ever growing degree to transform his environment according to his own aims instead of adjusting himself to it. In relation to former cultures this is a rather quantitative difference, but in such a powerful degree that it is impossible not to see in it a qualitative difference as well. Not limiting itself to ruling over an extra-human environ-

---

[14] ▶ María Ossowska and Stanisław Ossowski: "The Science of Science" (1935), 173, i.e. "Nauka o nauce," *Nauka Polska*, vol. 20 (1935), 1–12.

[15] ▶ Ibid., 175.

[16] ▶ Antoni B. Dobrowolski: "The Urgent Need for Mental Education in Poland" (1918), 296, i.e. "O pilnej potrzebie wychowania umysłowego w Polsce: o konieczności zasadniczej reformy nauczania w szkołach średnich oraz stworzenia w związku z ową reformą nowych placówek pracy naukowej," *Nauka Polska*, vol. 1 (1918), 489–502.

[17] Ibid., 301.

[18] Ibid.

ment, science endeavours to assume control over human instincts, over social and economic forces. In the last few decades scientific plans have been forming for the organisation of human life on a world scale. This was unknown to any pre-scientific culture.[19]

The authors take this characterization of modern society's proneness to scientific manageability as a core argument for the autonomy, and necessity, of the 'science of science'. This argument is fairly close to the foundations of sociology as presented by Auguste Comte and Émile Durkheim. According to Comte, the advent of the science of sociology signaled a new era in the cognitive development of humankind, because sociology endeavored nothing less ambitious than an exact science even of those phenomena which were most complex, namely, social ones.[20] Durkheim, in turn, distinguished between those social phenomena which are synchronized with the overall development of the modern social order and those that, like relics, extend from the past into the present while having lost their functionality; and he self-consciously aligned sociology with modernity – which founded his insistence on the pedagogical and educational significance of sociology.[21] Ludwik Fleck has been shown to adhere to an optimism regarding scientific progress that was based on the claim of a coming together of an increasing "differentiation of society, division of labor and specialization"[22]. Consequently, the Science Studies Circle targeted those factors of scientific knowledge production that had so far escaped scientific scrutiny because they counted as non-rationalizable, like 'creativity'. In Dobrowolski's terms: "This is why the time has come to finally *find out what we are actually doing with the brains of our youth in school*. We will gain that knowledge once the education of thinking becomes connected to its indispensable foundations, i.e. our understanding, at least in broad terms, of the mind's ideal activities and of the grammar of actual thinking."[23] This modernization-theoretical ethos was meant to render Poland as a genuinely, if not exemplarily, modern society, a part of 'modern European culture'.

Fifth and finally, the 'science of science' staged science's own possible merit for a modern society, crucially including that of the Science Studies Circle itself. This applied, for instance, to the understanding of the scientific process itself, which was represented as proceeding in strictly defined stages and along the trajectories of dif-

---

[19] ▶ Ossowska and Ossowski: "The Science of Science" (footnote 14), 181.

[20] Auguste Comte: *Rede über den Geist des Positivismus*, Hamburg 1974 (1844).

[21] Émile Durkheim: *The Rules of Sociological Method, And Selected Texts on Sociology and its Method*, London, Basingstoke 1982 (1895), for a critique cf. Andreas Langenohl: "Divided Time: Notes on Cosmopolitanism and the Theory of Second Modernity," in: *European Cosmopolitanism in Question*, ed. by Roland Robertson and Anne Sophie Krossa, Basingstoke 2012, 64–85.

[22] Julian Bauer: "'Gerichtetes Wahrnehmen', 'Stimmung', 'soziale Verstärkung'. Zur historischen Semantik einiger Grundbegriffe der Lehre vom Denkstil und Denkkollektiv", *N.T.M. Zeitschrift für Geschichte der Wissenschaften, Technik und Medizin*, vol. 22 (2014), 87–109: 99, translation by A. L. Cf. Ludwik Fleck: „In der Angelegenheit des Artikels von Frau Izydora Dąmbska in 'Przegląd Filozoficzny' (Jg. 40, Heft III) (1937)," in: *Ludwik Fleck. Denkstile und Tatsachen. Gesammelte Schriften und Zeugnisse*, ed. by Sylwia Werner and Claus Zittel, Berlin 2011, 320–326.

[23] ▶ Dobrowolski: "The Urgent Need for Mental Education in Poland" (footnote 16), 292, emphasis in the original.

ferent types of 'creativity' within a normative framework set by scientific progress.[24] The valorization of modernity and modernization reveals itself in the conceptualization of the relation between science and society, too, which was conceptualized as a process of an individualistic popularization of science. According to Dobrowolski, the impact of science on society was envisaged as an influence of science on individuals' trajectories, which "will contribute significantly to the fundaments of the *theory* of thinking [...]: an ever more detailed understanding of the norms of intellectual activity in different real-life situations; an augmented catalogue of concrete faculties of the mind; an ever improving state of research on the mind's natural tendencies, commonly referred to in terms of 'ability,' as well as its errors and faults."[25] In other words, science was thought of as an educational agent. In a certain sense, this strategy corresponded to the Science Studies Circle's epistemological baseline, which was the individual and his or her creative intuitions.

To summarize the Science Studies Circle's project of epistemology, one may say that its attention was focused on the individual as the seat of creativity, though not in isolation, but as part of a social milieu. Creativity was seen as the nucleus of scientific production and at the same time as so far unaccounted for in research on science. It was precisely this concentration on the 'source' of scientific endeavors that effectively neutralized any possible qualms about biasing effects resulting from scientific practice; rather, and the other way round, scientific practice, namely the 'science of science', promised to rationalize this source in a bold claim to modernity. It was therefore not so much scientific *practice* that the Science Studies Circle was interested in, but scientific *inspiration*.

Yet, in principle this opened up an avenue to conceptualize scientific inspiration as a social practice. It is true that the Science Studies Circle never came up with a suggestion that would be comparable to today's practice-theoretical approaches, which relate social practices to the "routine grounds of everyday activities"[26] but also to activities connected to subjectivity and the body that Foucault might have called "technologies of the self".[27] Yet, Florian Znaniecki noted that the correspondence-theoretical "attitude" of the scientist is itself the product of socialization; and Ludwik Fleck's fame results not least from his notion of 'Denkstil' which, as Julian Bauer has reconstructed, assembled various theorems concerning the societal, historical and affectual conditionality of knowledge strategies.[28] So, the Science Studies Circle did pose the

---

[24] ▶ Stefan Błachowski: "The Problem of Scientific Creativity" (1928), i. e. "Zagadnienie twórczości naukowej," *Nauka Polska*, vol. 9 (1928), 1–67.

[25] ▶ Dobrowolski: "The Urgent Need for Mental Education in Poland" (footnote 16), 301, emphasis in the original.

[26] Harold Garfinkel: "Studies of the Routine Ground of Everyday Activities," in: *Studies in Ethnomethodology*, Englewood Cliffs 1967: 35–75.

[27] Michel Foucault: „Technologies of the self," in: *Technologies of the Self: A Seminar with Michel Foucault*, ed. by Luther H. Martin, Huck Gutman and Patrick H. Hutton, Amherst, M. 1988, 16–49.

[28] Bauer: "'Gerichtetes Wahrnehmen'" (footnote 22). „The attitude of the positive scientist, the specialist in any field, is uniformly realistic. He always means to learn as exactly as possible about

question of how inspiration and intuition can be recast in the light of their social conditionalities – a question that has only comparatively recently attracted attention again in current Science and Technology Studies.[29]

So, even given all qualifications regarding the differences between the Science Studies Circle and current concerns in STS and other critical approaches to scientific epistemologies, the Science Studies Circle shared with those approaches an interest in those elements in scientific knowledge production which otherwise would mostly count as pre-given. The main difference, thus, is that the Science Studies Circle did never approach science from a *critical* perspective; it did not question the viability of methodology per se; and it did not address the so far unreflected elements in scientific knowledge (like creativity) with the intention to question the claims or even the possibility of scientific objectivity.

III. The question of the Science Studies Circle: What does critical epistemology mean?

This leads to a central question: What does it mean for current debates to depict in the Science Studies Circle a project that engaged in epistemological questions without assuming a politically or socially critical positionality? The "Writing Culture" debate, Bourdieu's interventions, postcolonial criticism, feminist epistemology all insist that epistemology have a critical imperative focused not merely on epistemological questions, but rather on epistemology as a site and mode of production for more encompassing power differentials in society. Connecting the critique of science and social and political criticism, they argue that those interventions that question science must necessarily question dominant social orders, as science has been part and parcel of the modernist project as patriarchal, imperialist, orientalist, racist, or classist domination. Even Latour, insisting that much of the constructivism informing critical social and cultural studies is misdirected in its generalizing skepticism regarding the potentially hierarchical, marginalizing, and exclusivist ways that 'facts' are represented, did discover the register of critique for himself.[30]

---

reality such as it is, independently of him. Whether his object-matter be nature or culture, he eliminates himself entirely, tries to behave not as a human being who wishes reality to accord with his particular prejudices, but as an impersonal "knower." If he notices that his experience and activity do affect his data, he treats this as a source of error to be avoided. Thus the astronomer corrects his "personal equation" in observations, and the psychologist or sociologist tries not to influence personally the people whose behavior he studies. In short, all science tends alike to approach complete objectivity. The difference in question does not lie in the attitude of the scientist but exclusively in the character of reality itself as given to the scientist when it is made the object-matter of impersonal investigation." (Florian Znaniecki: *The Method of Sociology*. New York 1934: 35)

[29] Cf. for instance Knorr Cetina: "Objectual Practice" (footnote 2).
[30] Latour: "Why has Critique Run out of Steam" (footnote 6).

Such considerations mattered little for the Science Studies Circle's epistemological project. According to it, epistemology could and indeed should be affirmative of modern societal structures. It was supposed to optimize the chances to give society a genuinely modern shape. In this respect, if we try to relate the Science Studies Circle to that which came after ('Writing culture', Bourdieu, postcolonial criticism, feminist critiques of science), we should not forget to relate it to what came before – for instance, Comte and Durkheim with their outspoken optimism regarding the transformative impact of science on society[31] – while also accounting for the "relations between science, society and culture that, for the period from 1880 to 1930, are being discussed under the keyword of a general modernism"[32].

If we thus want to extract a critical concern from the Science Studies Circle, we may cite Bruno Latour in his urge to make the concern of critique the determination "to detect *how many participants* are gathered in a *thing* to make it exist and to maintain its existence"[33], because from this perspective the Science Studies Circle's project of *epistemology* reveals itself as a project of *facilitation* and at the same time one that *reflects* such facilitation. This can be seen more clearly when thinking of the Science Studies Circle as a project that emerged not in the center of the western metropolises, but rather at its margins. It was a movement that materialized in a nation-state which was viewed by the metropolises as being not yet fully modern, that is, on par with western nation-states. For the Science Studies Circle, science was meant to play a big role in the transformation and the catching up: it ought to be optimized, it ought to be used in education, and it ought to be rigorously planned. In other words: the Science Studies Circle had not the slightest doubt about the constructedness of the nation and the state, because it fully realized that nations and states were not simply there but indeed had to be designed, planned, and constructed. And it was fully aware of – and, from today's perspective, overconfident in – the power of science to contribute to this project. It even went so far as to not let the rise of a national science be left to the chance that a great mind happened to be born – instead, it wanted to understand, and construct, the very origins from which great minds emerge. In this sense, it was a truly constructivist project, and a critical one in a fundamentally epistemological sense – just not a socially, politically, or culturally critical one.

IV. The legacy of the Science Studies Circle: Constructivism, modernism, and the politics of applied science

The Science Studies Circle's epistemological legacy for contemporary Science Studies, which after the Science Wars have learned the lessons of constructivism, is thus a

---

[31] See ▸ Florian Znaniecki: "The Subject Matter and Tasks of the Science of Knowledge" (1925), 115–116, i.e. "Przedmiot i zadania nauki o wiedzy," *Nauka Polska*, vol. 5 (1925), 1–78.
[32] Bauer: "'Gerichtetes Wahrnehmen'" (footnote 24), 101, translation by A. L.
[33] Latour: "Critique Run out of Steam" (footnote 6), 246.

complicated one, prone to irritate some of today's commonplaces about critical epistemologies in the social sciences and the humanities. The Science Studies Circle teaches us the following lessons about constructivism: that constructivism and social criticism are not natural allies; that constructivism may maintain a deep affinity with modernism, that is, the idea and utopia that things human, natural and social are not pre-given and can thus be changed through conscious, planned and sustained intervention; and that, as a consequence and reminder, constructivist critiques ought to reflect on its potential, if unaccounted for, modernism.[34]

In terms of the politics of science, the Science Studies Circle confronts Science Studies, in most of its contemporary incarnations, with the embarrassing question of application. For the Science Studies Circle, the application of the results of the 'science of science' was a matter of course – and, on top of that, a question of national importance.[35] This orientation sits extremely uneasily with the critical-constructivist branch in today's Science Studies, according to which the deconstruction of scientific orders of knowledge is an end in itself. But the Science Studies Circle's appraisal of the contribution of the 'science of science' to projects of societal modernization is also at odds with the ethnographic and micro-analytical orientation in many of today's investigations especially in STS, which often restrain themselves to a re-narration and second-order problematization in a sociological or anthropological idiom of meandering epistemic processes which, for scientists, are meandering as well but *still* a matter of course. The 'science of science' was meant to be an *applied* science – and neither critical-constructivist Science Studies nor STS dare to make a comparable bid.

This chasm can, in turn, be explained historically. Today's reservations against the application of Science Studies – and against the exposure especially of humanist and social-scientific 'theory' to the demand to be applied – might be made sense of as a reaction to demands for applied scientific results that are formulated not in scientific institutions but in politics (or, for many protagonists more eerily, in the economy). In utter contrast to that, the Science Studies Circle was not only self-conscious, but *anticipatory*, in its demand to have its results recognized and appreciated as applicable ones by political and educational institutions. The critique of the application of scientific results, we thus need to learn, requires to be historicized, too. For the Science Studies Circle, application was a matter not only of gaining societal impact and political recognition, but of *taking the initiative* in the question of what science is for. For today's (critical) social sciences and humanities, application often appears as a specter forcing the reluctant insight that initiative has shifted away from the scientific system – which is why their critique is so often reactive and ineffective. In this im-

---

[34] Which it often does, but not always. Cf. Andreas Langenohl: "Zweimal Reflexivität in der gegenwärtigen Sozialwissenschaft: Anmerkungen zu einer nicht geführten Debatte," *Forum Qualitative Sozialforschung/Forum: Qualitative Social Research*, vol. 10 (2009). https://doi.org/10.17169/fqs-10.2.1207; accessed: 03 Aug. 2017.

[35] ▸ Ossowska and Ossowski: "The Science of Science" (footnote 14), 181–182.

passe, the Science Studies Circle might become a guiding light. We don't need to embrace its spectacular modernism, which has become untenable today, in order to receive from it the inspiration and encouragement to claim the initiative in questions of what science is for.

The Original Conception of the Science of Science and Innovation Studies[1]

Paweł Kawalec

I. The scope of the science of science and its economization

In this chapter science studies are understood broadly as interdisciplinary research on science. This includes theoretical studies on the foundations of science, studies on different forms and stages of scientific research, as well as practically oriented investigations into the most efficient forms of science governance and organization – forms that take into account the social and economic consequences of scientific research, new knowledge and innovations, which originate with scientific research. This comprehensive conception of the science of science comprises the following key areas: foundational (ontological, epistemological, methodological and moral), social and cultural, economic and political, legal, organizational and administrative. The study of those areas has recently matured into several firmly established fields of research: history and philosophy of science (HPS), general methodology of science, research ethics, science and technology studies (STS), science policy and innovation studies, responsible innovations, intellectual property (IP) and innovation management, science journalism and public communication of science, scientometrics and creativity studies.[2]

The comprehensive conception of the science of science reflects how the notion of science has evolved since the Second World War. With the advent of knowledge society and innovation economy, the notion of science has broadened to capture not only the stage of basic, or even applied, research – it now extends to cover all stages of 'the innovation process' (see Fig. 1). But we must distinguish between invention and innovation (as schematically represented by the dashed line on Fig. 1). Invention is new knowledge (basic, applied or implemented in a small number of objects) while innovation is new knowledge that has demonstrated its market value (for both the innovator and the end users). Thus the concept of 'innovation' entails that innovative

---

[1] Acknowledgment: The author gratefully acknowledges the support of the Polish National Science Center (NCN) under the grant no 2014/15/B/HS1/03750.

[2] For a more detailed presentation see Paweł Kawalec (ed): *Podstawy naukoznawstwa* (2 vol.s), Lublin 2011 and id. "The Science of Science – Some Recent Advances," *Ruch Filozoficzny*, vol. 75/2 (2019): 33–57; Sarah Trousset: "Current Trends in Science and Technology Policy Research: An Examination of Published Works from 2010–2012," in: *Policy Studies Journal*, vol. 42/S1 (2014), 87–117.

products that embody embedding new knowledge have reached critical mass for diffusion on the market.³

Twentieth-century philosophy of science was dominated by a focus on inventions at the stage of basic research. Logical empiricists emphasized the 'logic of science,' limiting themselves to the stages at which law-like regularities are discovered and hypotheses or theories are formed. Their paradigm privileged explanations and predictions that promised to lead to a 'logical reconstruction' of key scientific concepts and types of inference. As I shall discuss below, it was on account of this perspective that important advancements in emerging areas of the science of science were overlooked.⁴

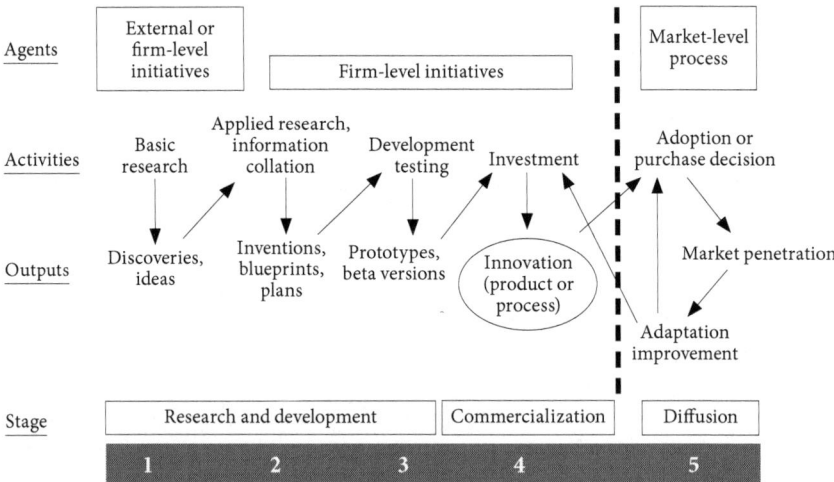

Figure 1. A multidimensional view of the stages of the innovation process. Source: Christine Greenhalgh and Mark Rogers: *Innovation, Intellectual Property, Economic Growth*, Princeton 2010, 7.

As the concept of science became broader, a major shift occurred in research interests and priorities within the science of science. The Second World War is recognized as instrumental in establishing governments' willingness to finance science, which in turn brought science policy and the economics of science to the fore within science studies the science of science studies.⁵ The following statement illustrates the upshot of this process:

---

³ See Robert D. Atkinson and Stephen J. Ezell: *Innovation Economics: The Race for Global Advantage*, New Haven 2012, 8–9.

⁴ K. Szaniawski succinctly characterizes the turning points of the entire process. See Klemens Szaniawski: "Preface," in: *Polish Contributions to the Science of Science*, ed. by Bohdan Walentynowicz, Warsaw, Dordrecht 1982, VII–X.

⁵ Homer A. Neal, Tobin L. Smith and Jennifer B. McCormick: *Beyond Sputnik: U.S. Science Policy in the Twenty-First Century*, Ann Arbor 2008, 12–13.

Policy makers responsible for channelling governmental funds into research have recently rebranded science and technology policy into innovation policy. This should remind academic scientists and technologists that their inventions require transformations into something with value and impact.[6]

In the remainder of this chapter I refer to *the economization of the science of science* as the process by which scientific knowledge comes to be seen exclusively as a commodity enhancing social welfare and economic growth (economic value) – a process that limits the scope of research on science to economic investigation. The prevalence of such economization within the science of science is epitomized in the dominant role that innovation studies (in particular the economics of innovation) play in interdisciplinary research on science.

While the historical processes leading to such economization are complex,[7] two arguments seem particularly influential in the economization of the science of science. One is represented in V. Bush's 1946 report *The Endless Frontier*, which paved the way for the US government's massive engagement in science investment and policy.[8] It legitimized public funding of basic research, and subsequently led to the institutionalization of US science policies. In effect, the National Science Foundation (NSF) was formed, which was to become a blueprint for public policies concerning basic research across the world.

The key publications that recognized the economic value of scientific inventions were authored by prominent economists at the National Bureau of Economic Research, such as Zvi Griliches and Robert Solow.[9] Their government's preoccupation with efficacy and outcomes appears to be at the root of today's innovation studies and their predominant leaning towards economic value.[10]

---

[6] Dimitris Assimakopoloulos, Ilan Oshri and Krsto Pandza: "Management of Emerging Technologies for Economic and Social Impact: An Introduction," in: *Managing Emerging Technologies for Socio-Economic Impact*, ed. by id., Cheltenham 2015, 1–21: 2. For a systematic historical survey see e.g. Wesley M. Cohen: "Fifty Years of Empirical Studies of Innovative Activity," in: *Handbook of the Economics of Innovation* (vol. 1), ed. by Bronwyn H. Hall and Nathan Rosenberg, Amsterdam 2010, 129–213.

[7] For a first attempt at a systematic overview see Benoît Godin: *Innovation Contested: The Idea of Innovation over the Centuries*, New York 2014, see also Esther-Mirjam Sent: "Economics of Science: Survey and Suggestions," in: *Journal of Economic Methodology*, vol. 1/6 (1999), 95–124. The main reasons for the economization of science are discussed in Paula E. Stephan: "The Economics of Science," in: *Handbook of the Economics of Innovation* (vol. 1), ed. by Bronwyn H. Hall and Nathan Rosenberg, Amsterdam 2010, 217–273: 219.

[8] Bush, as the head of Office of Scientific Research and Development during the Second World War, was the chief presidential science advisor, overseeing all R&D military expenditures, and initiated the Manhattan Project.

[9] See e.g. Zvi Griliches: "Hybrid Corn: An Exploration in the Economics of Technological Change," in: *Econometrica* 4/25 (1957), 501–522, id.: "Hybrid Corn and the Economics of Innovation," in: *Science*, vol. 132/3422 (1960), 275–280; Robert M. Solow: "A Contribution to the Theory of Economic Growth," in: *The Quarterly Journal of Economics*, vol. 1/70, 65–94.

[10] Ben R. Martin: "The Evolution of Science Policy and Innovation Studies," in: *Research Policy* vol. 7/41 (2012), 1219–1239.

A parallel advancement – conceptual as well as practical – by the US government in the 1940s was the successful performance of large-scale social experiments. These had a sound theoretical underpinning in the experimental and behavioural economics developed by prominent social researchers such as James Coleman, George Katona, Paul Lazarsfeld, Margaret Mead and Kurt Lewin. Examples of these experiments included the modification of Americans' eating habits,[11] the engineering of human-machine interactions,[12] tax-policy interventions studied by the members of the Cowles Commission[13] and successful interventions in the process of innovation diffusion.[14] These experiments led to A. Tversky's and D. Kahneman's groundbreaking work on behavioural economics.[15] As I argue elsewhere, these experiments, beginning with the Cowles Commission's publications, provided a rationale for an interventionist approach to technology shocks as causal dependencies determined by macroeconomic models.[16] The Solow-type macroeconomic models, which represented technology shocks as outcomes of government policy and interventions, established the economic value of knowledge as the proper subject-matter of innovation studies as we know them today.[17]

These conceptual foundations had two effects: first, they enabled a new understanding of the aim of research on science (namely to determine the most effective science policy instruments); second, the belief took root that science and government intervention could be successfully used to achieve social transformation. These two effects combined to stimulate the 'economics of innovation' as a new field of research – research that focused on the last stage of the innovation process (stage 5 on Fig. 1) as crucial for bringing added value to society and contributing to social welfare.

This economization of research on science diverged from earlier approaches to science that had mostly been concerned with theoretical and foundational issues. The latter kind of approach is not only reflected in historical writings such as Aristotle's, but also in the programmatic papers of the Polish founders of the science of science

---

[11] The change from meats to 'variety' meats (esp. hearts, kidneys, brains, stomachs, intestines, and even the feet, ears, and heads of cows, hogs, sheep, and chickens) was indeed dramatic – with 33 % increase during the war, and 50 % by 1955; see Brian Wansink: "Changing Eating Habits on the Home Front: Lost Lessons from World War II Research," in: *Journal of Public Policy & Marketing*, vol. 1/21 (2002), 90–99.

[12] Floris Heukelom: *Behavioral Economics: A History*, New York 2014, 73.

[13] On behalf of the Cowles Commission the research was carried out by G. Katona, who thereby initiated behavioral economics; see George Katona: *War without Inflation: The Psychological Approach to Problems of War Economy*, New York 1942.

[14] Bryce Ryan and Neal C. Gross: "The Diffusion of Hybrid Seed Corn in Two Iowa Communities," in: *Rural Sociology*, vol. 1/8 (1943), 15–24; Griliches: "Economics of Technological Change" (footnote 9), id.: "Economics of Innovation" (footnote 9).

[15] See chapter 4 in Heukelom: *Behavioral Economics* (footnote 12).

[16] Paweł Kawalec: "On the origin and meta-principles of causal inference. The case of T. Haavelmo," in: *Zagadnienia Naukoznawstwa*, vol. 53/4 (2017), 453–466.

[17] Cohen: "Empirical Studies" (footnote 6).

in the twentieth century.[18] Of course, the two major developments discussed in the preceding paragraph, which had a significant impact on the economics of innovation that was later promoted as core research within science studies, chronologically followed the programmatic papers by Znaniecki and the Ossowskis (in this volume), who, understandably, could not have anticipated those developments some ten or twenty years ahead of time. The question thus arises whether Znaniecki and the Ossowskis were misguided in their approach to the science of science. Did they fail to recognize the directions of future developments in the new interdisciplinary research that they advocated? Before I turn to this question in Section II below, I will briefly characterize the original conception of the science of science as formulated by Znaniecki and the Ossowskis in their programmatic papers.

II. Research priorities of the science of science at its inception

Although systematic philosophical inquiry on science has a long tradition, in this chapter I focus on two seminal papers that are commonly recognized as programmatic for the inception of the research project we call 'science of science'.[19]

Znaniecki's motivation for a science of science is clearly theoretical and foundational:

> To be sure, we are still dealing with an accumulation of miscellaneous observations rather than with a systematically and consciously developed scientific whole, but gradually an order is emerging from this chaos and there is beginning to take shape a concept of a single, general theory of knowledge as a separate branch of human culture, endowed with special empirical properties and permitting of empirical study.[20]

For Znaniecki it is theoretical and foundational issues that offer solutions to the applied problems of science governance and organization. He characteristically understands the science of science as an empirical field that determines what factors stimulate scientific inventions. While he recognizes that the traditional discussion on "the logical ideals" of scientific knowledge (referring to e.g. John Dewey, George Herbert Mead, Ernst Mach and Henri Poincaré) is relevant, he also emphasizes that this discussion must always be approached in the context of the actual historical development of knowledge, rather than from an essentialist standpoint. Taking both a synchronic and diachronic perspective, he elaborates on the role of creativity studies and the conditions in which knowledge is generated. Among these theoretical and foun-

---

[18] Marta Skalska-Zlat: "Nalimov and the Polish Way towards Science of Science," in: *Scientometrics*, vol. 2/52, 211–223.
[19] See Bohdan Walentynowicz: "Editor's Note," in: *Polish Contributions to the Science of Science*, ed. by id., Warsaw, Dordrecht 1982, xi–xii.
[20] ▶ IN THIS VOL. Florian Znaniecki: "The Subject Matter and Tasks of the Science of Knowledge" (1925), 113–114, i.e. "Przedmiot i zadania nauki o wiedzy," *Nauka Polska*, vol. 5 (1925), 1–78.

dational issues Znaniecki also identifies the role of methodology for establishing the emerging science of science as an autonomous area of research. It is only these theoretical and foundational developments that reveal how decision on practical organizational and managerial issues can constitute a causal set-up for the optimal stimulation of creative acts of invention. Practical issues are nonetheless inherently relevant to the science of science, as they provide guidance on decisions regarding science and technology governance (which in turn condition whether and how cognitive values are produced) and on generating intellectual capital ("cognitive dispositions") and its diffusion (in the sense of Gabriel Tarde).

Theoretical and foundational issues – including philosophical, historical, psychological, sociological and methodological perspectives – similarly prevail in 'The Science of Science' by Maria Ossowska and Stanisław Ossowski.[21] In contrast to Znaniecki, the Ossowskis' paper explicitly discusses science policy and the role of the state in regulating the place of science within the economic system. Clearly though, also on the Ossowskis' account the resolution of practical problems related to science governance and organization follows their antecedent theoretical conceptualization.

These two papers are complimentary in the way they anticipate subsequent developments in science studies broadly understood. For instance, Znaniecki's call for a causal study of scientific knowledge is based on his criticism of the rationality of scientific progress that prefigured Kuhn's later approach; he also anticipates the dynamic development of creativity theory and indicates credibility (and the repeatability of results) as a key criterion for policy considerations; he explicitly considers the influence of technological advancement upon research, and finally – although with some hesitation – he points towards diffusion studies, citing groundbreaking works by Tarde. The Ossowskis, meanwhile, explicitly recognize the mutual influence between economic systems and science, and anticipate applied research on science management and organization. Both papers also represent the science of science as an empirical endeavour that needs to account for the multidimensional aspects of science.

The two papers build on a shared notion of interdisciplinary research on science – research that prioritizes theoretical and foundational studies. But they also focus on a narrow understanding of science as invention – science as basic research. This original conception of the science of science differs from today's economized view in that it sets research priorities and narrows down the scope of science. Viewed from the currently dominant perspective of economization, both papers may seem to miss the target, as they address inventions but fail to take into account the full scope of innovations. Needless to say, Znaniecki and the Ossowskis were very much aware of the political and economic impact of the science of science. This impact also preoccupied Stanisław Michalski, the 'mastermind' of the Polish science of science movement, who first developed the plan to consider political and economic aspects. Michalski

---

[21] ▶ Maria Ossowska and Stanisław Ossowski: "The Science of Science" (1935), i.e. "Nauka o nauce," *Nauka Polska*, vol. 20 (1935), 1–12.

proceeded gradually, identifying the first investigators in the science of science, then founding the journal *Organon*, and finally, in the 1930s, planning to set up the world's first professional science of science research institute in Poland.[22] In the final Section III I argue that given the distortions of scientific practice, such as plagiarism or disintegrity of pharmaceutical research, are negative consequences of the economized view, it is indeed the original value-based approach to the science of science that yields more promising prospects for the future development of this interdisciplinary field of research. At the same time, I take it for granted that the original notion of science needs to be expanded to account for both scientific inventions and innovations.[23]

III. Towards a value-based approach in the science of science

Returning to the question at the end of Section 1, I should stress that the Polish founders of the science of science in the 1920s and 1930s were in no position to foresee the key factors that would lead to the economization of the science of science during and after the Second World War.[24] Their concept of the science of science is in principle opposed to economization. By reconstructing the argument according to which the science of science was initially understood in Poland as a value-based approach to the study of science I hope to offer a viable and perhaps more productive alternative to the widespread economized approach.

The Ossowskis identify two basic pathways to explore the science of science.[25] The "epistemological point of view" examines science as a way of generating knowledge about the world, while the "anthropological point of view" asks how science fosters the development of human culture "in the fullness of historical circumstances".[26] They then attribute the "anthropological" point of view to Znaniecki's paper.[27]

Here I will only highlight the key points of Znaniecki's argument, referring the interested reader to the original paper. In his search to establish autonomy of the science of science as a separate empirical field of research he focuses on a detailed examination of its subject matter. He rejects essentialism, i.e. the traditional expectation that it is possible to formulate a speculative and context-free definition of the

---

[22] Jan Piskurewicz: „Stanisław Michalski w dziejach nauki polskiej pierwszej połowy XX wieku," in: *Kwartalnik Historii Nauki i Techniki*, vol. 1/35, 55–92.

[23] This tendency is commonly recognized by science scholars and its origin in the second half of the twentieth century is often attributed to T. Kuhn.

[24] Wanda Osińska: "Les Débuts de Recherches Systématiques Sur La Scienciologie Dans Le Milieu Varsovien Au Tournant Des XIXe et XXe Siècles," in: *Organon*, vol. 6 (1969), 279–295.

[25] ► Ossowska and Ossowski: "Science of Science" (footnote 22), 172–173.

[26] Ibid., 173.

[27] Ibid., 175.

essence of scientific knowledge. Nonetheless, Znaniecki seeks to identify an inherent and unique feature of the subject matter of the science of science:

[...] we must find a purely formal criterion that, without presupposing anything in advance about the essence of the phenomena comprising 'knowledge' in a historico-cultural humanistic construction, would still allow for distinguishing these phenomena from among any other cultural (economic, legal, social, linguistic, esthetic, religious, etc.) phenomena, that would allow a judgement in every particular case concerning whether the given phenomenon constitutes a proper subject of study for the theory of knowledge or whether it belongs in the domain of a different humanistic discipline.[28]

Znaniecki further characterizes this distinguishing feature of the subject matter of the science of science as "trueness". When phenomena possess the characteristic of "trueness" for certain human subjects, this trueness "manifests itself empirically in the intensional utterances of these human subjects [for whom the phenomena possess the characteristic of 'trueness'] or in the use which they make of the phenomena in question".[29] A phenomenon possessing this feature is defined as "cognitive value". For Znaniecki this is only a formal characterization of the subject matter of the science of science; the task of determining its content falls to individual scientists and depends on the context. A researcher representing the science of science has to take this context-relative determination for granted as part of the empirical data she seeks to accounted for.

"Human operations," which generate, transform, combine and use cognitive values to develop culture, "constitute a special category of cultural activities" – namely "cognitive operations" – distinct from social, economic, legal or artistic activities.[30] According to Znaniecki, cognitive values and cognitive operations form the proper subject matter of the science of science. Even though, they manifest themselves empirically, they have an inherent mental and intensionalintentional component. Hence, the empirical science of science study must take into account 'the humanistic coefficient':

In studying cognitive operations, just as with cognitive values, we must maintain the humanistic viewpoint, i.e., regard as cognitive those operations which are entitled to that characterization in the conviction of the acting or 'thinking' subjects themselves. [...] The theoretician of knowledge [...] is studying not absolute cognition in itself but historically given cognition as it presents itself to historically living people.[31]

This concept of the subject matter of the science of science and its 'humanistic' (or 'anthropological') methodology suggests that an economized approach to the interdisciplinary study of scientific research is deeply misguided. Economization misidentifies the crucial feature of knowledge across all stages of the innovation process

---

[28] ▶ Znaniecki: "The Subject Matter and Tasks of the Science of Knowledge" (footnote 21), 122.
[29] Ibid., 122.
[30] Ibid., 123.
[31] Ibid., 125.

and applies an oversimplified methodology to its study, ignoring the 'humanistic coefficient' and thus stakeholders' perception of possible negative consequences of the research projects in question.

We are now in a position to indicate some negative consequences of the economization of the science of science – problems inherent to the very idea of economization as well as external problems. External problems are most clearly visible in "innovation market failures" that occur when the market fails to provide sufficient incentives to stimulate innovation with widespread benefit.[32] Internal problems of economization are summarized in Philip Mirowski's pithy statement that "markets [...] can equally well be deployed to produce ignorance".[33] This appears to be a straightforward consequence of the misconception according to which the subject matter of the science of science would consist in the economic value of scientific knowledge as a commodity. Such misconception ignores completely the "trueness" of scientific knowledge, which for Znaniecki constitutes cognitive values as the proper subject matter of the science of science. By the same token, Mirowski's statement points at the rationale of the major forms of fraud in science, which – along with the increasing pressure on researchers to generate economically viable outcomes – have become so prevalent that they may even threaten the very nature of the scientific enterprise.[34]

Given today's broad definition of science and its social and economic consequences, we should consider whether the arguments that Znaniecki and the Ossowskis proposed against the economization of the science of science in the narrow sense of 'science' focused on the early stage of invention, would also hold for the remaining stages of the innovation process. It seems that such an extension of the initial argument is natural on Znaniecki's standpoint. Elsewhere I argue that the difference between cognitive and economic values can be consistently identified for all stages of the process of innovation.[35] I demonstrate why cognitive value remains constant throughout the innovation process, while the economic value undergoes several changes between the stages of invention, reaching its climax in diffusion of innovation, and then its subsequently rapidly declining with the advent of new innovation on the market.

Znaniecki's characterization of the subject matter of the science of science – that is to say his emphasis on the humanistic approach – has much in common with the recent methodological approach in interdisciplinary studies known as 'mixed methods research'. It is intended to capture the complexity of social phenomena by integrating results of qualitative and quantitative investigations within a single research

---

[32] Christine Greenhalgh and Mark Rogers: *Innovation, Intellectual Property, Economic Growth*, Princeton 2010, 17–18.
[33] Philip Mirowski: *Science-Mart*, Cambridge, Mass. 2011, 318.
[34] See Raphael Sassower: *Compromising the Ideals of Science*, Basingstoke 2015.
[35] Paweł Kawalec: "Ambivalued Innovation and Interactive Research Design," in: *Social responsibility and science in innovation economy*, ed. by id. and Rafał P. Wierzchosławski, Lublin 2015, 335–352.

project. Although the mixed methods approach has not yet been systematically explored within the science of science, there are successful examples of its application.[36] One particular area where the 'humanistic' value-based approach, analogous to the one advocated by the Polish founders of the science of science, clearly dominates over the economization is research on responsible innovations.[37] In order to counterbalance negative consequences, the stakeholders' perspective and values are taken into account throughout the process of design and development of innovations. A version of the mixed methods approach accommodated to match the initial objectives of the science of science promises to open new prospects for fruitful and responsible research on science.

---

[36] For a survey see Paweł Kawalec: "Metody mieszane w kontekście procesu badawczego w naukoznawstwie," in: *Zagadnienia Naukoznawstwa*, vol. 1/50, 3–22.

[37] See e.g. Richard Owen, Phil Macnaghten and Jack Stilgoe: "Responsible Research and Innovation: From Science in Society to Science for Society, with Society," in: *Science and Public Policy*, vol. 6/39, 751–760.

Editorial Remarks and Sources

The original texts do not follow a fully standardized bibliographical practice. If necessary, the bibliographical information has been completed and moved from the text into footnotes without indication. The edition should adapt the individual character of the original texts and give room to the idiosyncratic styles of writing. While square brackets mark linguistic clarifications, minor typographic changes were not indicated.

Original footnotes are numbered with Arabic numerals. Editorial notes are added in original footnotes after three vertical bars, or else are given in their own footnotes with Latin letters. Translator's remarks by T'ulsi (Tuesday) Bhambry [T.B.] and Christopher Kasparek [C.K.] follow the same scheme.

Friedrich Cain contributed the editorial introductions to the sources.

The list below gives the original bibliographical information of the sources edited in this volume. If only a Polish reference is given, the text was translated for this volume by Tul'si (Tuesday) Bhambry. Znaniecki's text was translated by Christopher Kasparek. Two references to a text indicate contemporary translations from (or to) the Polish language, which were included in the volume.

Bracketed page numbers in the original references indicate partial translations in this edition.

The original material from both *Nauka Polska* and the *Organon* are available online at the *Śląska Biblioteka Cyfrowa* (Silesian Digital Library):
  Nauka Polska: https://www.sbc.org.pl/publication/20689
  Organon: https://www.sbc.org.pl/dlibra/publication/68769

w/o author: *Editorial Introduction* (1923)
   "Wstęp redakcyjny," *Nauka Polska*, vol. 4 (1923), VII–IX.

w/o author: *Report on the activities of the* Science Studies Circle (1929)
   "Sprawozdanie z działalności Koła Naukoznawczego," *Nauka Polska*, vol. 11 (1929), 353–355 [353, 354–355].

w/o author: *Second Report on the activities of the* Science Studies Circle (1930)
   "Sprawozdanie drugie z działalności Koła Naukoznawczego," *Nauka Polska*, vol. 13 (1930), 166–169 [166–167].

w/o author: *Preface* (1936)
"Preface," *Organon*, vol. 1 (1936), V–VI.

Białobrzeski, Czesław ["C.B."]: *An autobiographical sketch and remarks on scientific creativity* (1927)
"Szkic autobiograficzny i uwagi o twórczości naukowej," *Nauka Polska*, vol. 6 (1927), 49–76 [49–53, 56–59, 61–64, 68–76].

Błachowski, Stefan: *The Problem of Scientific Creativity* (1928)
"Zagadnienie twórczości naukowej," *Nauka Polska*, vol. 9 (1928), 1–67 [1–23, 32–67].

Borel, Emile: *Contribution (Documents sur la Psychologie de l'Invention Dans Le Domaine De La Science)* (1936)
"Contribution (Documents sur la Psychologie de l'Invention Dans Le Domaine De La Science)," *Organon*, vol. 1 (1936), 33–42.

Bujak, Franciszek: *The Man of Action and the Student* (1929/1936)
"Działacz i badacz," *Nauka Polska*, vol. 11 (1929), 11–23.
"The Man of Action and the Student," *Organon*, vol. 1 (1936), 20–32.

Dobrowolski, Antoni B.: *The Urgent Need for Mental Education in Poland. The Necessity for a Fundamental Teaching Reform in Middle Schools and New Institutions for Scientific Research in Relation to this Reform* (1918)
"O pilnej potrzebie wychowania umysłowego w Polsce: o konieczności zasadniczej reformy nauczania w szkołach średnich oraz stworzenia w związku z ową reformą nowych placówek pracy naukowej," *Nauka Polska*, vol. 1 (1918), 489–502.

Dobrowolski, Antoni B.: *An Archive of Materials for Research on Creativity* (1927)
"Archiwum materjałów do badania twórczości," *Nauka Polska*, vol. 6 (1927), 140.

Dobrowolski, Antoni B. ["A.B.D."]: *My "Scientific Biography"* (1928)
"Mój 'życiorys naukowy'," *Nauka Polska*, vol. 9 (1928), 68–216 [68–70, 154–158].

Kotarbiński, Tadeusz: *On the Skills of a Researcher* (1929)
"O zdolnościach cechujących badacza," *Nauka Polska*, vol. 11 (1929), 1–10.

Krogh, August: *Visual Thinking: An Autobiographical Note* (1938/1939)
"Visual Thinking," *Organon*, vol. 2 (1938), 86–94.
"Myślenie wzrokowe," *Nauka Polska*, vol. 14 (1939), 35–42.

Ossowska, Maria and Stanisław Ossowski: *The Science of Science* (1935/1936)
"Nauka o nauce," *Nauka Polska*, vol. 20 (1935), 1–12.
"The Science of Science," *Organon*, vol. 1 (1936), 1–12.

Rybicki, Paweł: *Science and the Forms of Social Life: Issues at the Intersection of Sociology and Theory of Science* (1929)
"Nauka a formy życia społecznego. Kilka zagadnień z pogranicza socjologii i teorii nauki," *Nauka Polska*, vol. 11 (1929), 24–64 [24–32, 60–64].

Suchodolski, Bogdan: *Investigation and Teaching* (1936/1938)
  "Badanie a Nauczanie," *Nauka Polska*, vol. 21 (1936), 1-44.
  "Investigation and Teaching," *Organon*, vol. 2 (1938), 43-78.

Znaniecki, Florian: *The Subject Matter and Tasks of the Science of Knowledge* (1925/1982)
  "Przedmiot i zadania nauki o wiedzy," *Nauka Polska*, vol. 5 (1925), 1-78.
  "The Subject Matter and Tasks of the Science of Knowledge," transl. by Christopher Kasparek, in: *Polish Contributions to the Science of Science*, ed. by Bohdan Walentynowicz, Dordrecht et al. (1982), 1-81.

Biograms – A New Organon

Tul'si (Tuesday) Bhambry, raised in Poland, Germany and India, studied English and French in the UK and France. After an MPhil in European Literature in Cambridge she got a PhD in Polish Literature at UCL in 2013. In 2015 she won the Harvill Secker Young Translators' Prize and moved on to translate poetry and song lyrics, short stories and comics. She also translates non-literary writing, such as books in the humanities and social sciences, art museum catalogues, historical source texts as well as writings related to anti-discrimination and empowerment. She also works as an interpreter and moderator, and she is in the process of becoming a state-certified translator in Berlin.

Marta Bucholc is professor of sociology at the Faculty of Sociology, University of Warsaw, and associate researcher at the Centre de recherche en science politique, Université Saint-Louis Bruxelles. From 2015 through 2020 she was research professor at Käte Hamburger Centre for Advanced Studies „Law as Culture", University of Bonn. She was visiting scholar at the universities of Jena, Cambridge, Graz and Saint-Louis Bruxelles, as well as Bronisław Geremek Fellow of the Institute of Human Sciences in Vienna. Her research focus is sociology of law and historical sociology. She is the PI in a research project "National habitus formation and the process of civilization in Poland after 1989: a figurational approach" funded by Polish National Science Centre, and the Polish PI in the Volkswagen Foundation project "Towards Illiberal Constitutionalism in East Central Europe: Historical Analysis in Comparative and Transnational Perspectives". In 2022, she started a five-year ERC Consolidator project "Using Human Rights to Change Abortion Law: Involvement Patterns and Argumentative Architectures in the Global Figuration of Human Rights (Abortion Figurations)".

Friedrich Cain is a historian of science and the humanities. His research interests cover the history of science studies during the Cold War and political epistemologies of Central and Eastern Europe since the late 19th century. He was awarded a doctorate from Konstanz University for a thesis on clandestine research of Polish scholars under German occupation during World War II (Wissen im Untergrund. Praxis und Politik klandestiner Forschung im besetzten Polen (1939–1945), Mohr Siebeck, 2021). He is currently a PostDoc Assistant at the University of Vienna.

Paweł Kawalec – professor, Department of Epistemology, Faculty of Philosophy, John Paul II Catholic University of Lublin. Member of the Science of Science Committee of the Polish Academy of Sciences and the editorial board of the journal *Science of Science Quarterly*. Publishes on the theory and methodology of science, Bayesian epistemology, argumentation theory and early analytic philosophy (esp. R. Carnap, N. Goodman). Books include: *Integral Methodology* (2018, in Polish); *Causality and Explanation* (2006, in Polish) and *Structural Reliabilism* (2003, Springer). Editor and translator of the Polish editions of Carnap's *Der logische Aufbau der Welt* (2011) and *Logical foundations of probability* (2024).

Bernhard Kleeberg is professor for the History of Science at the University of Erfurt. He held a junior professorship for the history of science and humanities at the University of Konstanz, has taught and done research at the Max Planck Institute for the History of Science, the London School of Economics, the Justus Liebig University Giessen, and the Universities of Basel and Zurich. His current research focuses on praxeologies of truth, political epistemologies of Central and Eastern Europe, and the history of social psychology. He is editor-in-chief of the book series *Historical Knowledge Research* and of *NTM – Journal for the History of Science, Technology, and Medicine*.

Andreas Langenohl is professor of Sociology at Faculty 03 Social Sciences and Cultural Studies and head of graduate studies at the International Graduate Centre for the Study of Culture (GCSC), Justus Liebig University Giessen. He also has the position of professor extra-ordinary of Political Studies at North-West University, South Africa. Andreas Langenohl held research fellowships at Cornell, Konstanz, and Freiburg universities. His fields of research and teaching include social and cultural theory, research on political public communication, economic sociology and the sociology of finance, transnationalism, and the epistemology of the social sciences.

Jan Piskurewicz – full professor in the Institute for the History of Science of the Polish Academy of Sciences, member of the Warsaw Society for Sciences, member of the Society for the History of Education. Main interests: history of science and history of education.

Jan Surman is a historian of science and scholarship, focusing on Central and Eastern Europe in the nineteenth and early twentieth centuries. Surman holds a PhD in history from the University of Vienna and has most recently been working at the Masaryk Institute and Archives of the Czech Academy of Sciences, Herder-Institute, Marburg; IFK, Vienna; and the National Research University Higher School of Economics, Moscow. His research focuses on scientific transfer, academic mobility and socialist scientific internationalism.

Leszek Zasztowt – historian, ordinary professor, and a former director of the Ludwik & Aleksander Birkenmajer Institute for the History of Science (2007-2015), Polish Academy of Sciences, and professor at the University of Warsaw of the Center for East European Studies. President of the Mianowski Fund – Foundation for the Promotion of Science, general secretary and vicepresident of the Warsaw Scientific Society. The President of the Academic Council of the Polish Academy of Sciences' Archives. His research concerns the history of East Central Europe and Russia since the 18th to the 21st c., especially embracing political, social, confessional, educational, academic, and scholarly life.

Index

Abramowski, Edward   201, 202
d'Alembert, Jean-Baptiste le Rond   196, 197
Ampère, André-Marie   177
Amsterdamski, Stefan   360
Amundsen, Roald   77
Arctowski, Henryk   78
Aristarchus of Samos   130
Aristotle   1, 6, 140, 376
Astrow, Wladimir   82
Avenarius, Richard   77, 81, 82, 212

B., C.   see Białobrzeski, Czesław
B., F.   see Bujak, Franciszek
Babbitt, Irving   287
Bachelard, Gaston   81
Bacon, Francis   1, 6, 7, 24, 27, 65, 342
Bauer, Julian   367
Baczko, Bronisław   360
Baert, Patrick   54, 57, 58, 59, 60
Baden-Powell, Robert   80
Baldwin, James Mark   117, 118, 133
Balej, Stepan Maksym Volodymyrovyč: see Baley, Stefan
Baley, Stefan   36, 47
Bastgenówna, Zofia   85
Belar, Albin   33
van Beneden, Édouard   323
Berthelot, Daniel   203
Bernal, John D.   1, 16, 21, 23, 24, 25, 26
Bernard, Claude   203
Bernheim, Hippolyte   210
Berr, Henri   98
Białobrzeski, Czesław   17, 61, 62, 65, 91, 96, 101, 111, 183, 213, 214, 215, 221, 303, 316, 317, 341, 342, 343, 344, 355
Bilikiewicz, Tadeusz   357
Binet, Alfred   183
Birkenmajer, Aleksander   14, 42, 355, 360
Błachowski, Stefan   17, 18, 63, 67, 87, 102, 183, 206, 211, 213, 215, 357
Bloor, David   362

Blumenfeld, Izydor   82
Bogdanov, Alexander   81
Bohr, Christian   334, 336
Bohr, Niels   356
Boll, Franz   192, 193
Bolzano, Bernard   176
Borel, Emile   18, 42, 86, 325, 326, 333, 334, 356
Bouguer, Pierre   312
Bourdieu, Pierre   22, 58, 363, 364, 368, 369
Brentano, Franz   3
Brubaker, Rogers   36
Buckle, Thomas   351
Bujak, Franciszek   17, 18, 107, 108, 118, 183, 213, 214, 220, 226, 303, 317, 341, 342, 354, 355
Bujkowski, Zb.   95
Bukharin, Nikolai I.   26
Bunsen, Robert W. E.   193
Burgess, Ernest W.   169
Bush, Vannevar   375
Butler, Judith   363

Camic, Charles   58
de Candolle, Alphonse   118
Cantor, Georg   206, 329, 332
Cattell, Raymond B.   197
Cavendish, Henry   262
Chadwick, Mary   189
Chałasiński, Józef   13, 95
Charcot, Jean-Martin   204
Charléty, Sébastien   42
Chwistek, Leon   357
Claparède, Edouard   80, 196
Coleman, James   376
Coleridge, Samuel Taylor   210
Collins, Randall   58
Colozza, Giovanni Antoni   201
Columbus, Christopher   198
Comte, Auguste   76, 81, 92, 116, 118, 127, 129, 351, 366, 369

de Condorcet, Marie Jean Antoine 1
Cook, Frederick A. 77
Cook, James 208
Cooper, Frederick 36
Copernicus, Nicolaus 130, 135
Correns, Carl E. F. J. 219
Crookes, William 193
Curie-Skłodowska, Marie 12, 28, 33, 39, 77
Curie, Pierre 224
Czekanowski, Jan 38

D., A.B.: *see* Dobrowolski, Antoni Bolesław
Dabrowski, Patrice M. 36
Darwin, Charles 256
Daston, Lorraine J. 61
Daszyńska-Golińska, Zofia 77
Dawson, Walter 94
Decroly, Ovide 80, 83
Dedijer, Stevan 25
Dembowski, Jan 3
Descartes, René 307
Dewey, John 8, 80, 85, 95, 114, 169, 275, 377
Diderot, Denis 216
Dobrov, Gennadij M. 16, 24, 25, 28
Dobrowolski, Antoni Bolesław 7–8, 15, 17–18, 41, 42, 53–55, 62, 70–98, 101, 183, 192, 297, 298, 300, 303, 318, 333, 345, 352, 366, 367
Dobrowolski, Kazimierz 342, 343, 345
Doroszewski, Witold 342
Dreckmann, Johann 334
Drewnowski, Jan-Franciszek 356
Dugas, Ludovic 201
Dunkman, Karl 175
Durkheim, Émile 56, 118, 119, 163, 169, 239, 242, 243, 351, 366, 369
Dwelshauwers-Dery, Victor 32

Eddington, Arthur S. 356
Elias, Norbert 54, 56, 57, 59, 60, 62, 69
Engelmeyer, Peter 298, 300, 324
Engels, Friedrich 24
Euler, Leonhard 203, 225

Fabricius, Johann Andreas 216
Fechner, Gustav Theodor 203, 225, 226
Fehr, Henri 206, 225

Fichte, Johann Gottlieb 176
Fleck, Ludwik 3, 20, 36, 48, 63, 70, 91, 357–359, 366, 367
Follin, Henri Léon 34
Fontaine, Henri La 28
Foucault, Michel 59, 367
Fraunhofer, Johann von 193
Fresnel, Augustin Jean 194, 203
Fuchs, Lazarus 85

Galilei, Galileo 129, 131
Galison, Peter 61
Galton, Francis 66, 118, 195–197, 204–205, 209
Garfield, Eugene 22
Garrison, Noble Lee 272
Gaston, Jerry 16
Gauss, Carl Friedrich 221–223, 314
Gawecki, Bolesław 357
Gawroński, Andrzej 43, 118
Gelblum, Szymon (Siméon) 87–88, 90, 300, 322–324
Gide, André 83
Gilbert, William 218
Głowacki, Aleksander 350
Godlewski, Emil (Jr.) 40, 41
Goffman, Erving 36
Gombrowicz, Witold 36
Gordin, Michael 45
Górski, Artur 96
Grabowski, Tadeusz 217
Griliches, Zvi 375
Grossman, Henryk 27
Grünwald, Ernst 175
Guyau, Jean-Marie 83

Halecki, Oskar 33, 45
Heflich, Aleksander 116
Hegel, Georg Wilhelm Friedrich 129, 135
Heitzman, Marian 342
Helmholtz, Hermann von 85, 199, 209, 221, 223, 226, 227, 307
Heraclitus 113
Hermann, Tomáš 48
Hermite, Charles 206
Hertz, Aleksander 98
Hessen, Boris 26, 27
Hilbert, David 197

Hirschler, Jan   46
Hobhouse, Leonard Trelawny   117, 133
Hock, Alfred   183
Høffding, Harald   43
Hoffmann, Christoph   91, 334
Holzapfel, Rudolf Maria   81–84, 88, 92
Honigsheim, Paul   239
Hosiasson-Lindenbaum, Janina   95
Hughes, Jason   56
Hume, David   129
Huxley, Thomas Henry   262
Huygens, Christiaan   129, 131, 194–195

Jaensch, Erich Rudolf   196, 218
Jakubanis, Henryk   344
Janet, Paul   116, 217
Janikowski, Zygmunt   40
Jaspers, Karl   169
Jędrzejewicz, Janusz   92, 357
Jerusalem, Wilhelm   239, 243–244
Jones, Ernest   190

Kahneman, Daniel   376
Kant, Immanuel   114, 129, 134, 197–198
Kaplan, Norman   359
Kapuściński, Władysław   95
Katona, George   376
Katz, David   208
Kepler, Johannes   129, 131
Kerschensteiner, Georg   80
Kieszkowski, Bohdan   343, 344
Kirchhoff, Gustav   193
Klein, Felix   183, 206
Knorr Cetina, Karin   362
Kochanowski, Jan K.   118
Kołakowski, Leszek   360
Konopczyński, Władysław   46
Korczak, Janusz   80
Korzeniowski, Bohdan   95
Koselleck, Reinhart   58
Kotarbiński, Tadeusz   13, 17, 24–25, 37, 91, 101, 107, 108, 172, 176, 251, 258, 341, 342, 351, 354, 356, 360
Koyré, Alexander   91
Kozłowski, Władysław-Mieczysław   350
Kretschmer, Ernst   183
Krogh, August   9, 18, 42, 66, 333–334, 356
Krokiewicz, Adam   38

Kruszewska, Felicja   10
Krzywicki, Ludwik   239, 350
Kudrycka, Barbara   45
Kuhn, Thomas S.   22, 48, 65, 359, 361, 362, 378, 379
Kühne, Wilhelm   193
Kuratowski, Kazimierz   40
Kutrzeba, Stanisław   44

Lambert, Johann Heinrich   312
Łapszyn, Jan   356
Latour, Bruno   362, 368, 369
Lazarsfeld, Paul   376
Leibniz, Gottfried Wilhelm   130, 208, 307
Lenin, Vladimir Ilyich Ulyanov   24, 43
Lévy-Bruhl, Lucien   118, 169, 239, 243
Lewin, Kurt   376
Liebig, Justus von   200
Lindenbaum, Adolf   95
Linné, Carl   91
Lipmann, Otto   197
Litt, Theodor   248
Lombroso, Cesare   118
Lord Kelvin   273, 307
Loria, Stanisław   98
Lotze, Hermann   82, 316
Lovejoy, Arthur O.   58
Lubicz-Zaleski, Zygmunt   7, 86, 184, 201, 203–204, 325
Łukasiewicz, Jan   3, 47, 97, 185–188, 189, 343, 344, 345, 354, 356, 357
Lutosławski, Wincenty   43

Mach, Ernst   81–82, 84, 115, 187–188, 214, 377
Mackay, Alan L.   1, 16, 21, 23–24
Madeyski-Poray, Stanisław   35
Mahrburg, Adam   351, 352
Makowiecki, Tadeusz   342, 344
Malinowski, Bronisław   356
Małkowska, Janina   10, 41
Malsch, Fritz   220
Mann, Thomas, 82, 85
Mannheim, Karl   175
Marey, Étienne Jules   203
Marx, Karl   24
Maudsley, Henry   209
Mauss, Marcel   118, 239, 242, 243

Maxwell, James Clerk 307
Mayer, Alfred G. 256
Mayer, Robert 216
McMurry, Frank Morton 271, 275
Mead, George Herbert 114, 377
Mead, Margaret 376,
Mendel, Gregor 219
Merton, Robert K. 16, 22, 26
Metallmann, Joachim 357
Meumann, Ernst 183, 200
Meyerson, Émile 49
Mianowski, Józef 4, 13
Michalski, Konstanty 42
Michalski, Stanisław 5, 8, 10, 11, 12, 19, 71, 72, 74, 75, 86, 87, 103, 105, 116, 351, 352, 353, 355–356, 359, 378
Mickiewicz, Adam 211, 281
Mill, John Stuart 76
Minkiewicz, Romuald 38, 118
Mirowski, Philip 381
Monge, Gaspard 204
Monroe, Walter S. 272
Montessori, Maria 80
Moore, A. W. 114,
Müller-Freienfels, Richard 189, 202, 212, 217
Müller, Georg Elias 219
Müller, Max 205
Münsterberg, Hugo 83
Mysłakowski, Zygmunt 49

Nałkowski, Wacław 350
Bonaparte, Napoleon 224
Natanson, Władysław 33
Naville, Ernest 116
Newton, Isaac 26, 129, 131, 135, 195, 307
Niementowski, Stefan 45
Nietzsche, Friedrich 82, 83, 191, 262
Noack, Hermann 38
Nusbaum-Hilarowicz, Jósef 183, 351

Odin, Alfred 197
Olbers, Wilhelm 222, 314
Ossowska, Maria 1, 2, 3, 10, 13, 16, 17, 18, 21, 24–25, 28, 36, 56, 111, 176, 354, 355, 359, 360, 377, 378–379, 381,
Ossowski, Stanisław 1, 2, 3, 9, 10, 13, 16, 17, 18, 21, 24–25, 28, 36, 55, 56, 57, 111,

118, 341, 344, 354, 355, 359, 360, 377, 378–379, 381
Ostwald, Wilhelm 51, 118, 183, 206, 252, 308
Otlet, Paul 28, 34

Palmer, Anthony R. 271
Park, Robert E. 169
Parkhurst, Helen 80
Pascal, Blaise 307
Pasteur, Louis 203
Petrażycki, Leon 257
Pickering, Andrew 362
Piéron, Henri 204, 224
Piotrowicz, Ludwik 45
Piskurewicz, Jan 10, 86
Piłsudski, Józef 4, 11, 36
Plato 1, 134
Poe, Edgar Allan 85, 297, 298
Poincaré, Henri 85, 115, 183, 189, 197, 206, 207, 215, 222–223, 225, 297–298, 377
Prus, Bolesław: see Głowacki, Aleksander
Przybyłowicz, Wojciech 107, 108, 341
Pulikowski, Julian 95

Rádl, Emanuel 48, 177
Ramsay, William 208
Rauh, Frédéric 115
Regnault, Victor 337, 338
Reiset, Jules 337, 338
Reuleaux, Franz 88, 300, 324
Rheinberger, Hans-Jörg 362
Ribot, Théodule 201, 204, 209, 210
Richet, Charles 203, 208
Rignano, Eugenio 183, 205, 209, 214
Robinson, James Harvey 280
Roederer, Pierre-Louis 224
Rolland, Romain 82
Röntgen, Wilhelm Conrad 309
Roszkowski, Wacław 46
Rousseau, Jean-Jacques 129, 216, 217, 220, 221, 224, 283,
Rozwadowski, Jan 343
Rubczyński, Witold 191, 192
Rutkowski, Jan 343, 344
Rybicki, Paweł 17, 43, 107, 108, 239, 240, 247, 341, 344, 354, 357
Rychliński, Stanisław 111, 342

Saint-Paul, Georges 205
Saint-Simon, Henri de 1
Sarton, George 280, 356
Scheler, Max 175, 239, 242, 244, 355, 358
Schelting, Alexander von 175
Schiller, Ferdinand Canning Scott 114, 115
Schingnitz, Werner 176, 177, 355
Schultze, Otto 85, 95
Séailles, Gabriel 116, 217
Seegen, Josef 337
Sekreta, Mirosław 357
Siedlecki, Michał 39, 40, 41, 46
Simmel, Georg 118
Skarbek, Janusz 360
Smoluchowski, Marian 44
Snow, Charles Percy 25
Socrates 211, 277
De Solla Price, Derek John 16, 21, 23, 25, 28, 359
Solow, Robert M. 375
Somsen, Geert 34, 35
Sørensen, William 336
Spencer, Herbert 1, 76, 169, 283, 351
Spinoza, Baruch de 324
Spranger, Eduard 199, 226
Spring, Walthère Victor 323
Stasiak, Stefan 43
Stein, Ludwig 82
Steiner, Rudolf 80
Stern, William 169, 183
Storer, Norman W. 16
Suchodolski, Bogdan 13, 17, 18, 107, 108, 341, 343, 355, 358, 360
Sullivan, John William Navin 262
Świętochowski, Aleksander 350
Świętosławski, Wojcjech 11, 344, 354
Szaniawski, Klemens 374
Szweykowski, Zygmunt 342
Szydłowski, Jerzy 356, 357

Tarde, Gabriel 162, 169, 378
Tarski, Alfred 3
Tartini, Guiseppe 210
Terman, Lewis M. 197
Thomas, William I. 113, 353

Thorndike, Edward L. 169
Tönnies, Ferdinand 169
Trotsky, Leon 43
Tschermak, Erich 219
Tur, Jan 183, 35
Tversky, Amos 376
Twardowski, Kazimierz 3, 44, 206, 207

Vaihinger, Hans 174
Vincenz, Stanisław 82
Vierkandt, Alfred 175
De Vries, Hugo 219

Walicki, Andrzej 360
Ward, L. F. 197
Watt, James 300, 324
Weber, Max 57, 239, 248
Weber, Ernst Heinrich. 203, 225
Wells, Herbert George 98
Wertenstein, Ludwik 12
Whewell, William 187
Wiese, Leopold von 239, 245
Windelband, Wilhelm 217
Wiszniewski, Michał 301
Wittvogel, Karl August 175
Witwicki, Władysław 183, 356
Wolff, Christian 129
Woolgar, Steve 362
Wundt, Wilhelm 201

Xenophanes 173

Young, Thomas 194

Zamecki, Stefan 360
Zamenhof, Ludwik 51
Zawidzki, Jan 45
Zawirski, Zygmunt 344
Zdziechowski, Marjan 343
Zieliński, Tadeusz 343
Znaniecki, Florian 3, 17, 18, 28, 29, 36, 39, 48, 55, 56, 62, 116, 118, 119, 169, 175, 239, 240, 242, 353, 355, 357, 367, 377, 378, 380, 381
Żeromski, Stefan 11

Hirschler, Jan   46
Hobhouse, Leonard Trelawny   117, 133
Hock, Alfred   183
Høffding, Harald   43
Hoffmann, Christoph   91, 334
Holzapfel, Rudolf Maria   81–84, 88, 92
Honigsheim, Paul   239
Hosiasson-Lindenbaum, Janina   95
Hughes, Jason   56
Hume, David   129
Huxley, Thomas Henry   262
Huygens, Christiaan   129, 131, 194–195

Jaensch, Erich Rudolf   196, 218
Jakubanis, Henryk   344
Janet, Paul   116, 217
Janikowski, Zygmunt   40
Jaspers, Karl   169
Jędrzejewicz, Janusz   92, 357
Jerusalem, Wilhelm   239, 243–244
Jones, Ernest   190

Kahneman, Daniel   376
Kant, Immanuel   114, 129, 134, 197–198
Kaplan, Norman   359
Kapuściński, Władysław   95
Katona, George   376
Katz, David   208
Kepler, Johannes   129, 131
Kerschensteiner, Georg   80
Kieszkowski, Bohdan   343, 344
Kirchhoff, Gustav   193
Klein, Felix   183, 206
Knorr Cetina, Karin   362
Kochanowski, Jan K.   118
Kołakowski, Leszek   360
Konopczyński, Władysław   46
Korczak, Janusz   80
Korzeniowski, Bohdan   95
Koselleck, Reinhart   58
Kotarbiński, Tadeusz   13, 17, 24–25, 37, 91,
    101, 107, 108, 172, 176, 251, 258, 341, 342,
    351, 354, 356, 360
Koyré, Alexander   91
Kozłowski, Władysław-Mieczysław   350
Kretschmer, Ernst   183
Krogh, August   9, 18, 42, 66, 333–334, 356
Krokiewicz, Adam   38

Kruszewska, Felicja   10
Krzywicki, Ludwik   239, 350
Kudrycka, Barbara   45
Kuhn, Thomas S.   22, 48, 65, 359, 361, 362,
    378, 379
Kühne, Wilhelm   193
Kuratowski, Kazimierz   40
Kutrzeba, Stanisław   44

Lambert, Johann Heinrich   312
Łapszyn, Jan   356
Latour, Bruno   362, 368, 369
Lazarsfeld, Paul   376
Leibniz, Gottfried Wilhelm   130, 208, 307
Lenin, Vladimir Ilyich Ulyanov   24, 43
Lévy-Bruhl, Lucien   118, 169, 239, 243
Lewin, Kurt   376
Liebig, Justus von   200
Lindenbaum, Adolf   95
Linné, Carl   91
Lipmann, Otto   197
Litt, Theodor   248
Lombroso, Cesare   118
Lord Kelvin   273, 307
Loria, Stanisław   98
Lotze, Hermann   82, 316
Lovejoy, Arthur O.   58
Lubicz-Zaleski, Zygmunt   7, 86, 184, 201,
    203–204, 325
Łukasiewicz, Jan   3, 47, 97, 185–188, 189,
    343, 344, 345, 354, 356, 357
Lutosławski, Wincenty   43

Mach, Ernst   81–82, 84, 115, 187–188, 214,
    377
Mackay, Alan L.   1, 16, 21, 23–24
Madeyski-Poray, Stanisław   35
Mahrburg, Adam   351, 352
Makowiecki, Tadeusz   342, 344
Malinowski, Bronisław   356
Małkowska, Janina   10, 41
Malsch, Fritz   220
Mann, Thomas, 82, 85
Mannheim, Karl   175
Marey, Étienne Jules   203
Marx, Karl   24
Maudsley, Henry   209
Mauss, Marcel   118, 239, 242, 243

Maxwell, James Clerk 307
Mayer, Alfred G. 256
Mayer, Robert 216
McMurry, Frank Morton 271, 275
Mead, George Herbert 114, 377
Mead, Margaret 376,
Mendel, Gregor 219
Merton, Robert K. 16, 22, 26
Metallmann, Joachim 357
Meumann, Ernst 183, 200
Meyerson, Émile 49
Mianowski, Józef 4, 13
Michalski, Konstanty 42
Michalski, Stanisław 5, 8, 10, 11, 12, 19, 71, 72, 74, 75, 86, 87, 103, 105, 116, 351, 352, 353, 355–356, 359, 378
Mickiewicz, Adam 211, 281
Mill, John Stuart 76
Minkiewicz, Romuald 38, 118
Mirowski, Philip 381
Monge, Gaspard 204
Monroe, Walter S. 272
Montessori, Maria 80
Moore, A. W. 114,
Müller-Freienfels, Richard 189, 202, 212, 217
Müller, Georg Elias 219
Müller, Max 205
Münsterberg, Hugo 83
Mysłakowski, Zygmunt 49

Nałkowski, Wacław 350
Bonaparte, Napoleon 224
Natanson, Władysław 33
Naville, Ernest 116
Newton, Isaac 26, 129, 131, 135, 195, 307
Niementowski, Stefan 45
Nietzsche, Friedrich 82, 83, 191, 262
Noack, Hermann 38
Nusbaum-Hilarowicz, Jósef 183, 351

Odin, Alfred 197
Olbers, Wilhelm 222, 314
Ossowska, Maria 1, 2, 3, 10, 13, 16, 17, 18, 21, 24–25, 28, 36, 56, 111, 176, 354, 355, 359, 360, 377, 378–379, 381,
Ossowski, Stanisław 1, 2, 3, 9, 10, 13, 16, 17, 18, 21, 24–25, 28, 36, 55, 56, 57, 111,

118, 341, 344, 354, 355, 359, 360, 377, 378–379, 381
Ostwald, Wilhelm 51, 118, 183, 206, 252, 308
Otlet, Paul 28, 34

Palmer, Anthony R. 271
Park, Robert E. 169
Parkhurst, Helen 80
Pascal, Blaise 307
Pasteur, Louis 203
Petrażycki, Leon 257
Pickering, Andrew 362
Piéron, Henri 204, 224
Piotrowicz, Ludwik 45
Piskurewicz, Jan 10, 86
Piłsudski, Józef 4, 11, 36
Plato 1, 134
Poe, Edgar Allan 85, 297, 298
Poincaré, Henri 85, 115, 183, 189, 197, 206, 207, 215, 222–223, 225, 297–298, 377
Prus, Bolesław: see Głowacki, Aleksander
Przybyłowicz, Wojciech 107, 108, 341
Pulikowski, Julian 95

Rádl, Emanuel 48, 177
Ramsay, William 208
Rauh, Frédéric 115
Regnault, Victor 337, 338
Reiset, Jules 337, 338
Reuleaux, Franz 88, 300, 324
Rheinberger, Hans-Jörg 362
Ribot, Théodule 201, 204, 209, 210
Richet, Charles 203, 208
Rignano, Eugenio 183, 205, 209, 214
Robinson, James Harvey 280
Roederer, Pierre-Louis 224
Rolland, Romain 82
Röntgen, Wilhelm Conrad 309
Roszkowski, Wacław 46
Rousseau, Jean-Jacques 129, 216, 217, 220, 221, 224, 283,
Rozwadowski, Jan 343
Rubczyński, Witold 191, 192
Rutkowski, Jan 343, 344
Rybicki, Paweł 17, 43, 107, 108, 239, 240, 247, 341, 344, 354, 357
Rychliński, Stanisław 111, 342

Saint-Paul, Georges 205
Saint-Simon, Henri de 1
Sarton, George 280, 356
Scheler, Max 175, 239, 242, 244, 355, 358
Schelting, Alexander von 175
Schiller, Ferdinand Canning Scott 114, 115
Schingnitz, Werner 176, 177, 355
Schultze, Otto 85, 95
Séailles, Gabriel 116, 217
Seegen, Josef 337
Sekreta, Mirosław 357
Siedlecki, Michał 39, 40, 41, 46
Simmel, Georg 118
Skarbek, Janusz 360
Smoluchowski, Marian 44
Snow, Charles Percy 25
Socrates 211, 277
De Solla Price, Derek John 16, 21, 23, 25, 28, 359
Solow, Robert M. 375
Somsen, Geert 34, 35
Sørensen, William 336
Spencer, Herbert 1, 76, 169, 283, 351
Spinoza, Baruch de 324
Spranger, Eduard 199, 226
Spring, Walthère Victor 323
Stasiak, Stefan 43
Stein, Ludwig 82
Steiner, Rudolf 80
Stern, William 169, 183
Storer, Norman W. 16
Suchodolski, Bogdan 13, 17, 18, 107, 108, 341, 343, 355, 358, 360
Sullivan, John William Navin 262
Świętochowski, Aleksander 350
Świętosławski, Wojcjech 11, 344, 354
Szaniawski, Klemens 374
Szweykowski, Zygmunt 342
Szydłowski, Jerzy 356, 357

Tarde, Gabriel 162, 169, 378
Tarski, Alfred 3
Tartini, Guiseppe 210
Terman, Lewis M. 197
Thomas, William I. 113, 353

Thorndike, Edward L. 169
Tönnies, Ferdinand 169
Trotsky, Leon 43
Tschermak, Erich 219
Tur, Jan 183, 35
Tversky, Amos 376
Twardowski, Kazimierz 3, 44, 206, 207

Vaihinger, Hans 174
Vincenz, Stanisław 82
Vierkandt, Alfred 175
De Vries, Hugo 219

Walicki, Andrzej 360
Ward, L. F. 197
Watt, James 300, 324
Weber, Max 57, 239, 248
Weber, Ernst Heinrich. 203, 225
Wells, Herbert George 98
Wertenstein, Ludwik 12
Whewell, William 187
Wiese, Leopold von 239, 245
Windelband, Wilhelm 217
Wiszniewski, Michał 301
Wittvogel, Karl August 175
Witwicki, Władysław 183, 356
Wolff, Christian 129
Woolgar, Steve 362
Wundt, Wilhelm 201

Xenophanes 173

Young, Thomas 194

Zamecki, Stefan 360
Zamenhof, Ludwik 51
Zawidzki, Jan 45
Zawirski, Zygmunt 344
Zdziechowski, Marjan 343
Zieliński, Tadeusz 343
Znaniecki, Florian 3, 17, 18, 28, 29, 36, 39, 48, 55, 56, 62, 116, 118, 119, 169, 175, 239, 240, 242, 353, 355, 357, 367, 377, 378, 380, 381
Żeromski, Stefan 11